Petroleum Science and Engineering

Petroleum Science and Engineering

Editor: Andy Margo

www.callistoreference.com

Callisto Reference,
118-35 Queens Blvd., Suite 400,
Forest Hills, NY 11375, USA

Visit us on the World Wide Web at:
www.callistoreference.com

ISBN: 978-1-63239-887-1 (Hardback)

The publisher's policy is to use permanent paper from mills that operate a sustainable forestry policy. Furthermore, the publisher ensures that the text paper and cover boards used have met acceptable environmental accreditation standards.

Trademark Notice: Registered trademark of products or corporate names are used only for explanation and identification without intent to infringe.

Printed in the United States of America.

Cataloging-in-Publication Data

Petroleum science and engineering / edited by Andy Margo.
   p. cm.
Includes bibliographical references and index.
ISBN 978-1-63239-887-1
1. Petroleum engineering. 2. Petroleum. 3. Petroleum--Prospecting. 4. Petroleum--Geology. I. Margo, Andy.
TN870 .P48 2017
665.5--dc23

Table of Contents

Preface

Petroleum science refers to the exploration, extraction and refining of petroleum and natural gas. This book on petroleum science and engineering focuses on the methods, theories, concepts and applications that are used in the various processes related to petroleum manufacturing. All aspects of petroleum engineering, from reservoirs to drilling to the production of formation fluids are highlighted in this book. The aim of this text is to present researchers that have transformed this discipline and aided its advancement. It is a vital tool for all researching and studying this field. For all those who are interested in petroleum science and engineering, it can prove to be an essential guide.

The information contained in this book is the result of intensive hard work done by researchers in this field. All due efforts have been made to make this book serve as a complete guiding source for students and researchers. The topics in this book have been comprehensively explained to help readers understand the growing trends in the field.

I would like to thank the entire group of writers who made sincere efforts in this book and my family who supported me in my efforts of working on this book. I take this opportunity to thank all those who have been a guiding force throughout my life.

Editor

Investigation of shale gas microflow with the Lattice Boltzmann method

Xiao-Ling Zhang · Li-Zhi Xiao · Long Guo · Qing-Ming Xie

Abstract In contrast to conventional gas-bearing rocks, gas shale has extremely low permeability due to its nano-scale pore networks. Organic matter which is dispersed in the shale matrix makes gas flow characteristics more complex. The traditional Darcy's law is unable to estimate matrix permeability due to the particular flow mechanisms of shale gas. Transport mechanisms and influence factors are studied to describe gas transport in extremely tight shale. Then Lattice Boltzmann simulation is used to establish a way to estimate the matrix permeability numerically. The results show that net desorption, diffusion, and slip flow are very sensitive to the pore scale. Pore pressure also plays an important role in mass fluxes of gas. Temperature variations only cause small changes in mass fluxes. The Lattice Boltzmann method can be used to study the flow field in the micropore spaces and then provides numerical solutions even in complex pore structure models. Understanding the transport characteristics and establishing a way to estimate potential gas flow is very important to guide shale gas reserve estimation and recovery schemes.

Keywords Shale gas · Permeability · Adsorption · Desorption · Diffusion · Slip effect

X.-L. Zhang · L.-Z. Xiao (✉) · L. Guo
State Key Laboratory of Petroleum Resources and Prospecting, China University of Petroleum, Beijing 102249, China
e-mail: xiaolizhi@cup.edu.cn

Q.-M. Xie
Key Laboratory of Shale Gas Exploration, Ministry of Land and Resources, Chongqing Institute of Geology and Mineral Resources, Chongqing 400042, China

Edited by Jie Hao

1 Introduction

Shale gas, which is hard to exploit, will play an increasingly critical role in China's natural gas production in the future (Jia et al. 2012; Zou et al. 2010; Chen et al. 2012; Sun et al. 2013). The matrix of gas shale consists of fine clay minerals and organic matter, and pore space reduces largely due to the compaction and cementation of these very fine grains (Wang et al. 2013; Li et al. 2007). Organic-rich shale is treated as source rock which has potential to produce natural gas. The pore scale of gas shale is some two orders of magnitude smaller than that of conventional sandstone (Bustin et al. 2008; Loucks et al. 2009). The pore throat diameter of shale is also very small, only dozens of times the size of methane molecules (Curtis et al. 2010). Therefore, permeability of gas shale is extremely low, normally only about tens to hundreds of nanodarcy. The micropore structure and flow characteristics are very complex and not well understood (Huang et al. 2012; Li et al. 2013).

The shale gas production mechanism after fracturing is as follows. First, free gas in the formation fracture networks is transported to the wellbore driven by the pressure difference between the borehole and formation pressures. Then free gas in the pore network flows to the fracture network due to the gas concentration difference between the pore network and fracture network, resulting in reduction of pore pressure. Finally, gas adsorbed on the surface of organic matter will desorb to pore spaces to increase the pore pressure. Overall, natural gas moves from the pore network to the fracture network and finally flows to the wellbore to be collected (Zou et al. 2014).

To estimate how much gas can be produced, it is necessary to evaluate the gas transport ability of the formation rock. It is very difficult to estimate matrix permeability with traditional experimental methods and to predict flow ability after fracturing due to the complex flow mechanisms of shale gas. In recent years, a lot of studies have been carried out to study the micro-scale flow characteristics of shale gas, which focus on two aspects as, (1) Laboratory measuring methods consist of the pulse pressure decay method for cores and the gas expansion method for crushed samples. They were developed to overcome long measuring times through applying pulse pressure and reducing the volume of the samples. (2) Numerical simulation methods consist of the molecular dynamics method, finite difference method (Shabro et al. 2009, 2011), finite element method (Roy et al. 2003; Yao et al. 2013), level set method (Prodanovic and Bryant 2007; Prodanovic et al. 2008), and the Lattice Boltzmann method (Tolke et al. 2010; Fathi and Akkutlu 2011; Maier and Bernard 2010).

These methodologies resulted in some research progress which established the theoretical foundation of shale gas flow mechanisms, but quantitative research into adsorption mechanisms and influencing factors for micro-scale gas transport characteristics is still needed. Based on micro-scale gas flow mechanisms, this paper focuses on various possible factors that affect shale gas flow characteristics, and quantitatively analyzes the influence of each factor. Finally, Lattice Boltzmann simulation is used with complex models to estimate the matrix permeability.

2 Micro flow mechanisms

Darcy's law assumes that fluid properties are stable without any physical or chemical reaction with the rock. Only the viscous interaction forces between molecules control the flow without change in the fluid properties (viscosity) and pore geometry (cross-sectional area, length) in the regime of laminar flow. But physical adsorption of methane occurs and gas changes states on the surface of kerogen in organic-rich shale. Shale gas flow is also controlled by molecular diffusion. Due to these inherent characteristics of gas shale such as, (1) widely dispersed organic matter and (2) nano-scale pore system (Zou et al. 2013), shale gas transport mechanisms do not fully follow Darcy's law. We cannot simply use Darcy's law to calculate matrix permeability of gas shale because of these special phenomena and problems (Civan 2010; Civan et al. 2011).

Shale gas flow is controlled by flow mechanisms at different scales such as, (1) adsorption and desorption, (2) diffusion, (3) slip flow, and (4) nonlinear Darcy flow. All these flow mechanisms make the prediction of shale gas transport more and more complicated. Three flow mechanisms in nano-scale pores will be discussed in the following sections. First a parameter which distinguishes different flow regimes will be introduced.

2.1 Knudsen number

In 1934, Knudsen defined a dimensionless parameter (Knudsen parameter, K_n) which reflects the degree of collision between gas molecules when rarefied gas flows in narrow channels. Therefore, gas flow regimes at different scales can be determined by the value of K_n expressed as (Ziarani and Aguilera 2012)

$$K_n = \frac{\lambda}{l_{char}}, \tag{1}$$

where l_{char} is the characteristic value of the geometry through which fluids flow (such as channel height, tube radius, etc.), m; λ is the mean free path of gas molecules [namely, the average distance traveled by a moving particle (such as an atom or a molecule) between successive impacts (collisions)], m.

Different flow regimes classified by the value of K_n are shown in Table 1. The rarefied effect of gas molecules becomes more and more significant with an increase of K_n. For normal shale gas reservoir temperatures (350–450 K) and pressures (25–28 MPa), the mean free path of methane molecules would be between 0.9 and 4.4 nm. If the pore size is 50 nm diameter, K_n will be 10^{-3}–10^{-2} then the flow regime is slip flow. So in nano-scale pore spaces the continuum assumption of fluid flow breaks down and the Navier–Stokes (for short, N–S) equation with no-slip boundaries is no longer valid.

2.2 Adsorption and desorption

As shown in Fig. 1a, before well drilling there is dynamic equilibrium in pores of organic matter which means the adsorption and desorption rates are equal. But when the pore pressure reduces, this equilibrium breaks down and net desorption item occurs which will play a complementary role in the mass flux of the pores as shown in Fig. 1b (Parker et al. 2009).

Adsorption mechanisms are very important for shale gas flow. Physical adsorption of supercritical gases at high

Table 1 Flow regimes of Knudsen number

Flow regimes	Continuum flow	Slip flow	Transition	Free molecule flow
Knudsen number	0–10^{-3}	10^{-3}–10^{-1}	10^{-1}–10^1	10^1–∞
Equations	Navier–Stokes	Correctional Navier–Stokes	Burnett equation	Molecular dynamic
	Boltzmann equation			

Fig. 1 **a** Dynamic equilibrium, **b** Net desorption

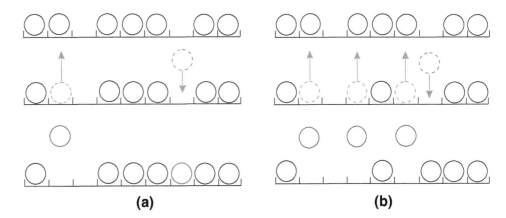

(a)　　　　　　　　　　　(b)

pressures is known to have some unusual features compared to subcritical fluid adsorption. This is important in the understanding of many adsorption processes and presents a challenge for fundamental theoretical research. All methods used for obtaining isotherms (gravimetric, volumetric, piezometric, and total desorption methods, column break-through methods, closed-loop recycle methods, and isotope exchange methods) measure the Gibbsian excess rather than the absolute amount adsorbed. In order to evaluate the amount of adsorbed gas in shale, it is necessary to know the absolute adsorption isotherm. A recovered equation of state (EOS) is needed to translate experimental data on excess adsorbed amounts to absolute adsorption data points. According to the Gibbs definition, the excess adsorption amount is given by

$$n_{ex} = \int \left(\rho(z) - \rho_g \right) dV, \tag{2}$$

where n_{ex} is the excess adsorbed amount; $\rho(z)$ is the density at coordinates z.

The main idea in this study is the application of the non-ideal EOS for the corresponding bulk phase to an adsorbed phase, but the parameters of the EOS are modified using the local density and the mean field approximations. The behavior of the adsorption phase obeys the EOS

$$RT \ln f^{(a)} + u = RT \ln f, \tag{3}$$

where $f^{(a)}$ and f are the fugacities of gas in a pore and in the bulk phase, respectively; u is the overall potential energy of the gas–solid interaction; T is the temperature, K; R is the universal gas constant. This equation is valid for subcritical fluids as well.

The force model consists of surface force and long-range Van der Waals forces, which are simulated by a Shan–Chen model (Shan and Chen 1993). On the basis of this operation, 0.5 nm (approximately the CH_4 molecule diameter) lattice unit length is selected. When more adsorbed molecules remain on the surface, other molecules are less likely to be attracted. Therefore, surface force is proportional to θ which can be written as

$$\theta = \frac{P}{P + P_L} - P \cdot A, \tag{4}$$

where P is the pressure of gas, Pa; P_L is the Langmuir pressure constant, Pa. By fitting a Langmuir isotherm to experimental data, constants P_L and A can be obtained. The bond force between the molecules and the surface is confined to the first layer, and the long-range Van der Waals forces hold other layers. The interactions between molecules are governed by

$$\psi = \exp(-1/\rho), \tag{5}$$

where ψ is a function of apparent density; ρ is the density of each lattice unit. For high Knudsen numbers, a combination of bounce back and specular reflection was used to generate the slip effect at the wall. The temperature of the whole gas flow is assumed to be constant. Thus, the pressure can be expressed as

$$P_{ads} = \frac{\rho_{ads}}{3} + \frac{G}{6} \left(\Psi(\rho_{ads}) \right)^2. \tag{6}$$

The drop of pore pressure leads to a higher desorption velocity compared to adsorption velocity which causes an added flux in the pores as shown in Fig. 2.

2.3 Diffusion

From the perspective of molecular dynamics, gas diffusion is actually the result of irregular thermal motion of gas molecules. According to the value of K_n, diffusion can be divided into Fickian diffusion, Knudsen diffusion, and transitional diffusion. In nano-scale pores with high Knudsen number, Knudsen diffusion is the dominant flow mechanism.

Knudsen diffusion in nanometer pores can be written in the form of a pressure gradient (Roy et al. 2003). In nanometer pores ignoring the viscosity effect, mass flux due to diffusion caused by molecular concentration difference (Javadpour et al. 2007) can be expressed as

$$J_d = \frac{MD_k}{RT} \frac{\Delta P}{L}. \tag{7}$$

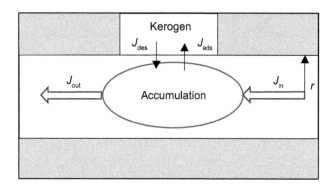

Fig. 2 Net desorption adds gas in the pore

The Knudsen diffusion coefficient D_k can be written as

$$D_k = \frac{2R_0}{3}\sqrt{\frac{8RT}{\pi M}}, \tag{8}$$

where R is the universal gas constant, J/kmol K; M is the molar mass of gas, kg/kmol; T is the temperature, K; ΔP is the pressure difference, Pa; R_0 is the pore radius, m; L is the length of the capillary, m; D_k is the diffusion coefficient, m²/s. It indicates that the Knudsen diffusion coefficient D_k is proportional to the pore radius R_0 and the square root of the temperature T.

2.4 Slip flow

Free gas in the formation fracture network flows to the borehole due to the difference between the borehole and formation pressures.

The N–S equation is the momentum conservation equation which is used to describe the flow of incompressible viscous fluid (continuous flow). For long and straight capillary tubes with a circular cross section, the N–S equation can be simplified as

$$\frac{d^2v}{dr^2} + \frac{1}{r}\frac{dv}{dr} = -\frac{\Delta p}{\mu L}, \tag{9}$$

where v is the flow velocity per unit area, m/s; Δp is pressure difference between the ends of the capillary, Pa; μ is the gas viscosity under atmosphere pressure, Pa s; L is the length of the pore. The solution of the above nonlinear difference equations is given by

$$v = C_0 - \frac{\Delta p}{4\mu L}r^2, \tag{10}$$

where C_0 is an unknown constant which can be calculated under different boundary conditions. Velocity equations are solved under two different boundary conditions; continuum flow (Fig. 3a) and slip flow (Fig. 3b) (Zhang et al. 2012; Javadpour 2009).

For continuum flow, fluid can be treated as a unit. The velocities on the pore wall are zero. Mass flux $J_{a\_no\_slip}$ caused by pressure difference can be expressed as

$$J_{a\_no\_slip} = -\frac{\rho_{avg}R_0^2}{8\mu} \times \frac{\Delta p}{L} \tag{11}$$

and absolute permeability is given by

$$K_{no\_slip} = \frac{R_0^2}{8}. \tag{12}$$

In the case of continuum flow, the mass flux of the fluid depends on the proportional constant K, the so-called absolute permeability of the rock, mD. When the parameters such as cross-sectional area, length of the capillary, fluid viscosity, and the pressure difference are fixed, K is an inherent characteristic of the porous medium independent of the fluid used in the measurement.

For slip flow in nano-scale pores, the velocities on the pore wall are no longer zero. So the mass flux J_{a_slip} caused by pressure difference can be expressed as

$$J_{a\_slip} = -F \times \frac{\rho_{avg}R_0^2}{8\mu} \times \frac{\Delta p}{L}, \tag{13}$$

$$F = 1 + \sqrt{\frac{8\pi RT}{M}} \times \frac{\mu}{R_0 p_{avg}} \times \left(\frac{2}{\alpha} - 1\right). \tag{14}$$

And absolute permeability is given by

$$K_{slip} = F \times \frac{R_0^2}{8}, \tag{15}$$

where ρ_{avg} is the average density of the fluid; R_0 is the radius of the pore; F is the correction factor; R is the universal gas constant (Boltzmann's constant) = 8.314, J/mol K; M is the molar mass of the gas, g/mol; T is the temperature, K; p_{avg} [$=(P_1 + P_2)/2$] is the average pressure in the capillary, Pa; α is the dimensionless adjustment coefficient of angular momentum (affected by the surface smoothness of the pore walls, gas type, temperature, and pressure, the range is from 0.65 to 0.85).

When the pore scale reduces to nanometers, the slip effect can no longer be ignored and the mass flux due to pressure difference needs to be corrected. The smaller the pore radius is, the larger the value of the correction factor F will be. The flow velocity per unit area considering the slip effect will increase compared to the situation of continuum flow alone.

3 Lattice Boltzmann method

The Lattice Boltzmann method is a meso-scale approach to simulating fluid flows. This method intrinsically incorporates micro-scale and meso-scale physical mechanisms. This algorithm is simple, easy to program, highly scalable,

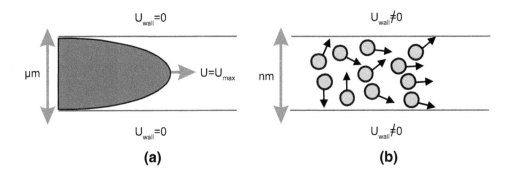

Fig. 3 Velocity profile.
a Continuum flow, **b** Slip flow

and convenient for dealing with the complex boundaries of porous media, making it very useful for shale gas simulation (Zhang et al. 2014).

The essential ingredients of the LBM are the Lattice Boltzmann equation, the space-filling lattice, and the local equilibrium distribution function. The general form of the Lattice Boltzmann equation is

$$f_a(x_b + \Delta t \overrightarrow{e_a}, t + \Delta t) = f_a(x_b, t) + \Omega(f_a) \quad a = 0, 1, \ldots, N,$$

(16)

where f_a is the distribution function of particles that travel with velocity $\overrightarrow{e_a}$; $\Omega(f_a)$ is the collision operator which describes changes in particle distribution due to particle collision.

To get correct flow equations it is necessary to use specific lattices that provide sufficient symmetry. The forms of such lattices depend on the dimension of the space. In two dimensions, examples are the square and hexagonal lattices. The nine-speed square lattice known as D2Q9 has been used extensively where each particle moves one lattice unit at its velocity defined by $\overrightarrow{e_a}$ [as Eq. (17)] and in one of the eight directions indicated with 1–8, particle at position 0 is called the rest particle that has a zero velocity.

$$e_a = \begin{cases} (0,0) & a = 9 \\ \left[\cos\dfrac{(a-1)\pi}{4}, \sin\dfrac{(a-1)\pi}{4}\right] & a = 1, 3, 5, 7 \\ \sqrt{2}\left[\cos\dfrac{(a-1)\pi}{4}, \sin\dfrac{(a-1)\pi}{4}\right] & a = 2, 4, 6, 8 \end{cases}.$$

(17)

The Bhatnagar–Gross–Krook (BGK) collision operator with a single relaxation time is often used. The BGK collision operator is derived by linearization of the collision operator around the equilibrium state, neglecting the higher-order terms, and assuming $\Omega_a(f_{eq})$ equal to zero. Therefore, the BGK collision operator can be written as Eq. (18)

$$\Omega_a = \frac{1}{\tau}\left(-f_a(x, t) + f_a^{eq}(x, t)\right),$$

(18)

where τ is the relaxation time; f_a^{eq} is the local equilibrium distribution function which defines what type of

flow equations are solved using the Lattice Boltzmann equation.

$$f_a^{eq} = t_a \rho(1 + \frac{3(\overrightarrow{e_a} \cdot \overrightarrow{u})}{e^2} + \frac{9(\overrightarrow{e_a} \cdot \overrightarrow{u})^2}{2e^4} - \frac{3(\overrightarrow{u} \cdot \overrightarrow{u})}{2e^2}),$$

(19)

where the weights t_a are equal to $t_9 = 4/9$, $t_1 = t_3 = {}_5 = t_7 = 1/9$, $t_2 = t_4 = tt_6 = t_8 = 1/36$, and $\overrightarrow{e_a}$ is the lattice velocity defined as the lattice size (Δx) over the lattice time step (Δt). Fluid density (ρ) and velocity ($\overrightarrow{u}$) are macroscale quantities that can be obtained by Eqs. (20) and (21)

$$\rho(x, t) = \sum_{a=1}^{9} f_a(x, t),$$

(20)

$$\overrightarrow{u}(x, t) = \frac{\sum_{a=1}^{9} f_a(x, t)\overrightarrow{e_a}}{\rho(x, t)}.$$

(21)

4 Results and discussion

In order to study gas flow characteristics in organic-rich shale, both analytical solution and numerical simulation methods are used to represent the transport phenomenon of shale gas. Analytical solutions can reflect the flux variation trend with different parameter values while numerical simulations can show the flow field distribution and estimate the matrix permeability of complicated pore models.

4.1 Analytical solutions and influencing factors

Influencing factors of mass flux in nano-scale pores were studied with analytical solutions based on the simplified capillary model.

Based on different pore scales, pressures, and temperatures, etc., mass flux generated by different forces will be analyzed respectively and quantitatively with analytical solutions. Nano-scale pores are very important for shale gas storage and production ability. Pore scale is treated as a sensitive factor to be changed to see how the mass flux changes.

As shown in Fig. 4, when the pore radius reduces to near 10 nm, net desorption and slip flow do not change too

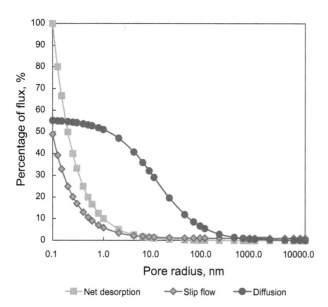

Fig. 4 Mass flux ratio of net desorption, diffusion and slip flow for different pore sizes

much but diffusion increases by 20 %. When the pore radius is less than 10 nm, net desorption, diffusion, and slip flow will increase significantly which means the smaller the pores are, the higher the mass fluxes are.

For gas shale under formation conditions, temperature and pressure increase and some shale gas plays are overpressured. Temperatures and pressures are among other factors that need to be discussed. The incremental steps of pressure and temperature are chosen as 0.2 kPa and 50 K.

From Fig. 5 it is easy to find the effect of pressure difference. Mass fluxes change substantially with the variation

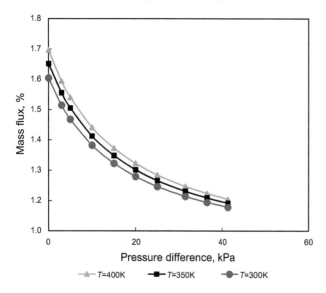

Fig. 5 Mass fluxes under different temperatures and pressures

of pressure difference, but change only slightly with temperature. In conclusion, flow flux of gas must be adjusted during the life of the shale gas plays, and permeability measurements in the laboratory must be taken in in situ pressure and temperature conditions, otherwise large errors will be introduced.

4.2 Numerical simulations

In the case of complex geometries including holes and cracks, analytical solutions cannot be used, while LBM can be used to simulate the microflow characteristics. The sphere accumulated model is used to simulate the flow in pore spaces. We carried out LBM simulation driven by a constant force on a 3-D model with $100 \times 100 \times 100$ lattices as shown in Fig. 6a. The radius of the sphere is 40 nm.

Figure 6a is the accumulation model. Figure 6b is the LBM simulation result of the velocity field, the deeper the color, the greater the velocity. Red lines represent the orientation direction of the velocity. For this regular model, porosity is 20 % and permeability calculated with the simulation is 0.003 mD.

Another model using spheres of random radius is simulated too as shown in Fig. 7. Pore spaces of the model reduce more than those of the model in Fig. 6 and the porosity is 8 %. After simulation the matrix permeability calculated is 23 nD which is much tighter than Fig. 6.

From the simulations above, LBM provides a feasible way to estimate the matrix permeability of gas in tight shale.

5 Conclusions

There are significant differences between gas-bearing shale and conventional sandstone. In particular, mobility of shale gas becomes a decisive factor in economic exploitation. In this paper, adsorption/desorption, diffusion, and slip flow mechanisms of shale gas are discussed and some factors like pore scale, pressure, and temperature are studied with analytical solutions. In nano-scale pore spaces of gas shale, gas transport is not only determined by the pore radius, but also significantly influenced by environmental conditions (temperature, pressure). Net desorption, diffusion, and slip play important roles in micro-scale flow which is higher than that calculated by Darcy's law. Lattice Boltzmann simulation on accumulated models is used to establish a technically feasible way to estimate the matrix permeability of complex pore structures.

Fig. 6 Regular accumulated model and flow field simulation

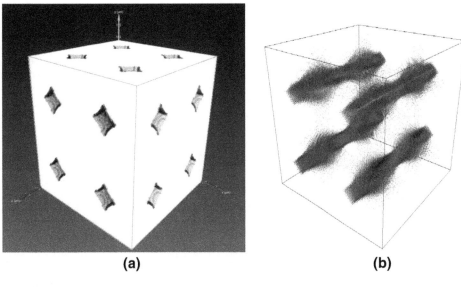

(a)　　　　　　　　　　　(b)

Fig. 7 Random accumulated model and flow field simulation

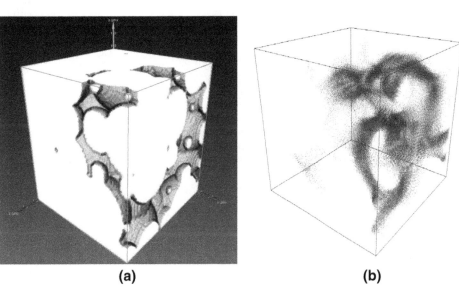

(a)　　　　　　　　　　　(b)

Acknowledgments This work was supported by the National Natural Science Foundation of China (Grant No. 41130417), "111 Program" (B13010) and Shell Ph.D. Scholarship. The authors thank the valuable comments from Professor Xiaowen Shan.

References

Bustin RM, Bustin AMM, Cui A, et al. Impact of shale properties on pore structure and storage characteristics. In: SPE Shale gas production conference. Fort Worth, Texas, USA (SPE 119892). 16–18 November 2008.

Chen XJ, Bao SJ, Hou DJ, et al. Methods and key parameters of shale gas resources evaluation. Pet Explor Dev. 2012;39(5):566–71 (in Chinese).

Civan F. Effective correlation of apparent gas permeability in tight porous media. Transp Porous Media. 2010;82(2):375–84.

Civan F, Rai CS, Sondergeld CH. Shale-gas permeability and diffusivity inferred by improved formulation of relevant retention and transport mechanisms. Transp Porous Media. 2011;86(3):925–44.

Curtis ME, Ambrose RJ, Sondergeld CH, et al. Structural characterization of gas shales on the micro- and nano-scales. In: Canadian Unconventional Resources & International Petroleum Conference. Alberta, Canada (SPE 137693). 19–21 October 2010.

Fathi E, Akkutlu IY. Lattice Boltzmann method for simulation of shale gas transport in kerogen. In: SPE annual technical conference and exhibition. Denver, Colorado, USA (SPE 146821). 30 October–2 November 2011.

Huang JL, Zou CN, Li JZ, et al. Shale gas generation and potential of the lower cambrian Qiongzhusi formation in Southern Sichuan Basin, China. Pet Explor Dev. 2012;39(1):69–75 (in Chinese).

Javadpour F. Nanopores and apparent permeability of gas flow in mudrocks (shales and siltstone). J Can Pet Technol. 2009;48(8):16–21.

Javadpour F, Fisher D, Unsworth M. Nanoscale gas flow in shale gas sediments. J Can Pet Technol. 2007;46(10):55–61.

Jia CZ, Zheng M, Zhang YF. Unconventional hydrocarbon resources in China and the prospect of exploration and development. Pet Explor Dev. 2012;39(2):129–36 (in Chinese).

Li XJ, Hu SY, Cheng KM. Suggestions from the development of fractured shale gas in North America. Pet Explor Dev. 2007;34(4):392–400 (in Chinese).

Li YJ, Feng YY, Liu H, et al. Geological characteristics and resource potential of lacustrine shale gas in the Sichuan Basin, SW China. Pet Explor Dev. 2013;40(4):423–8 (in Chinese).

Loucks RG, Reed RM, Ruppel SC, et al. Morphology, genesis, and distribution of nanometer-scale pores in siliceous mudstones of the Mississippian Barnett Shale. J Sediment Res. 2009;79(12):848–61.

Maier RS, Bernard RS. Lattice-Boltzmann accuracy in pore-scale flow simulation. J Comput Phys. 2010;229(2):233–55.

Parker MA, Buller D, Petre JE, et al. Haynesville shale-petrophysical evaluation. In: SPE annual technical conference and exhibition. Denver, Colorado (SPE 122937). 14–16 April 2009.

Prodanovic M, Bryant SL. Physics-driven interface modeling for drainage and imbibition in fractures. In: SPE annual technical conference and exhibition. Anaheim, California (SPE 110448). 11–14 November 2007.

Prodanovic M, Bryant SL, Karpyn ZT. Investigating matrix-fracture transfer via a level set method for drainage and imbibition. In: SPE annual technical conference and exhibition. Denver (SPE 116110). 21–24 September 2008.

Roy S, Raju R, Chuang HF, et al. Modeling gas flow through microchannels and nanopores. J Appl Phys. 2003;93(8):4870–9.

Shabro V, Javadpour F, Torres-Verdin C, et al. Diffusive-advective gas flow modeling in random nanoporous systems (RNPS) at different Knudsen regimes. In: Proceedings of the 17th International Conference of Composites or Nano-Engineering. 2009.

Shabro V, Torres-Verdin C, Javadpour F. Numerical simulation of shale-gas production: from pore-scale modeling of slip-flow, Knudsen diffusion, and Langmuir desorption to reservoir modeling of compressible fluid. In: SPE North American unconventional gas conference and exhibition. The Woodlands, Texas, USA (SPE 144355). 14–16 June 2011.

Shan XW, Chen HD. Lattice Boltzmann model for simulating flows with multiple phases and components. Phys Rev E. 1993;47(3):1815–9.

Sun LD, Zou CN, Zhu RK, et al. Formation, distribution and potential of deep hydrocarbon resources in China. Pet Explor Dev. 2013;40(6):641–9 (in Chinese).

Tolke J, Baldwin C, Mu YM, et al. Computer simulations of fluid flow in sediment: from images to permeability. Lead Edge. 2010;29(1):68–74.

Wang HY, Liu YZ, Dong DZ, et al. Scientific issues on effective development of marine shale gas in Southern China. Pet Explor Dev. 2013;40(5):574–9 (in Chinese).

Yao J, Sun H, Fan DY, et al. Numerical simulation of gas transport mechanisms in tight shale gas plays. Pet Sci. 2013;10(4):528–37.

Zhang WM, Meng G, Wei XY. A review on slip models for gas microflows. Microfluid Nanofluid. 2012;13(6):845–82.

Zhang XL, Xiao LZ, Shan XW, et al. Lattice Boltzmann simulation of shale gas transport in organic nano-pores. Sci Rep. 2014;4:4843.

Ziarani AS, Aguilera R. Knudsen's permeability correction for tight porous media. Transp Porous Media. 2012;91(1):239–60.

Zou CN, Dong DZ, Wang SJ, et al. Geological characteristics, formation mechanism and resource potential of shale gas in China. Pet Explor Dev. 2010;37(6):641–53 (in Chinese).

Zou CN, Yang Z, Zhang GS, et al. Conventional and unconventional petroleum "orderly accumulation": concept and practical significance. Pet Explor Dev. 2014;41(1):14–30 (in Chinese).

Zou CN, Zhang GS, Yang Z, et al. Geological concepts, characteristics, resource potential and key techniques of unconventional hydrocarbon: on unconventional petroleum geology. Pet Explor Dev. 2013;40(4):385–99 (in Chinese).

2

Symbol recognition and automatic conversion in GIS vector maps

Dun-Long Liu[1] · **Zi-Yong Zhou**[2] · **Qian Wu**[3] · **Dan Tang**[1]

Abstract Symbols are considered as the language of a map; hence, accurate understanding of the meaning of symbols is crucial when obtaining geographical information from a map: the symbolisation of spatial data is of key importance in cartography. A geographical information system (GIS) provides a convenient mapping platform and powerful functions for spatial data symbolisation, while the presence of various mapping standards impedes the understanding of maps and sharing of map information. On the other hand, the available GIS platforms find it difficult to deal with automatic conversion between maps and different mapping standards. To resolve this problem, an approach for symbol recognition and automatic conversion is proposed, and a conversion system based on the approach and the ArcGIS Engine platform is developed to realise automatic conversion between maps produced based on different mapping standards. To test these conversion effects of the proposed system, the petroleum sector is chosen as the research field and the mutual conversion of a map in practical work among the three mapping standards (i.e. the Chinese Petroleum, Shell and USGS standards) governing this field is taken as a case study. The results show that the conversion system has a high conversion accuracy and strong applicability.

Keywords GIS · Symbolisation · Map · Standard · Conversion

1 Introduction

A map is an abstract representation of the real world (Brewer et al. 2003; Goodchild 1999; Li et al. 2007; Chen et al. 2011; Petrovic 2003). Map symbols are usually considered as the language of a map (Yamada 1993; Tao et al. 2007; Robinson et al. 2011; Che et al. 2013) and used to represent real spatial phenomena (Qin et al. 2000; Stefanakis 2002; Tao et al. 2005; Dang et al. 2011; Schlichtmann 2004, 2009). Therefore, the symbolisation of spatial entities is a crucial procedure in cartography (Comentz 2002; Tsoulos et al. 2003). GIS provides powerful mapping capabilities and functions for spatial data management (Frehner and Brandli 2006; Gustavvson et al. 2006; Cheng and Zhang 2012; Zou et al. 2012): it is increasingly used in a wide range of applications. However, the same spatial object may be symbolised differently in different mapping standards. For instance, in the petroleum industry, various map standards are used, i.e. the Dutch Shell standard, the United States Geological Survey (USGS) standard and the Chinese Petroleum standard: as such, the "oilfield" object is symbolised differently (Fig. 1). The presence of different mapping standards puts obstacles in the path of spatial data exchange and information sharing and decreases user efficiency when map reading.

In practical work, adopted maps come from various sources with different mapping standards (Qin et al. 2000; Brewer et al. 2003; Gustavvson et al. 2006) (hereunder, the

✉ Zi-Yong Zhou
zhouziyong@263.net

[1] College of Software Engineering, Chengdu University of Information and Technology, Chengdu 610225, China

[2] State Key Laboratory of Petroleum Resources and Prospecting, China University of Petroleum, Beijing 102200, China

[3] Sichuan Institute of Geological Engineering Investigation, Chengdu 610072, China

Edited by Xiu-Qin Zhu

Fig. 1 Different expressions of an "oilfield" object under different mapping standards

map refers to the vector map made by a GIS platform, *.mxd format). To ensure consistency of the information contained in such maps, it is necessary to convert the maps according to a special mapping standard (e.g. convert USGS standard into Chinese Petroleum standard and vice versa). The conversion of a map between different standards is in fact the conversion of the symbols used therein. Theoretically speaking, each symbol in a map has its own specific code or name (Tao et al. 2005; Nass et al. 2011; Fan et al. 2011; Robinson et al. 2011), and it can be converted conveniently if the corresponding codes or names of the symbols are known. Where this is not the case, actually only the shape style and rendered information for each symbol are preserved in the map, while the symbol code and name are absent; thus, it is impossible to realise automatic batch conversion for a vector map according to the symbol encoding information. Therefore, the first question of symbols standard conversion is how to identify the symbols within the map for conversion. In order to solve this question, scholars have carried out a number of studies on symbol recognition and proposed a few technical solutions including the methods of statistics structure, template matching, neural network, line tracing and mathematical morphological (Yamada 1993; Zeng 2003; Llados et al. 2001a, b, 2002; Yang 2005; Liu et al. 2007; Wan and Liu 2007; Guo et al. 2012; Xie and Zhang 2014). However, the principle of these methods is so complex that it is difficult to be carried out. Additionally, a high requirement is need for the symbols recognition in these methods; hence, only a few symbols can be identified, resulting in a low recognition efficiency for real-time performance. Some scholars have proposed other ideas that are focused on symbol data structure, and established the description models of symbols based on various principles, such as XML, GML, SVG and TrueType, to design the universal map symbols (Yin et al. 2004; Tao et al. 2005; Antoniou and Tsoulos 2006; Mihalynuk 2006; Qin et al. 2008; Li et al. 2009; Chen et al. 2014). In theory, these approaches can provide favourable conditions and enlightenment for symbol recognition; however, only the description and design of symbols is discussed in this research, and no model or method is established yet. So far, research about the automatic recognition and conversion of symbols has not been reported. Currently, all the symbols within a map can only be manually converted one by one, which is not only laborious and time consuming but also burdensome, and one needs to understand the meaning of each symbol within the map in advance. If not, errors in symbol expression will arise, and the message from the map cannot be truly reflected. To accomplish symbol conversion rapidly and accurately, an approach using symbol recognition and automatic matching is proposed, and a conversion system integrated with the technical solution and an ArcGIS Engine platform is developed in this work: it aims to realise automatic batch conversion of the symbols within vector maps. In addition, as a case study, a practical map is converted among three petroleum standards (the Chinese Petroleum, Shell and USGS standards), which are frequently adopted in the petroleum industry, so as to illustrate the conversion accuracy and applicability of the conversion system.

2 Basic principles of symbol recognition and automatic conversion

2.1 Recognition of symbols in a map

As a result of symbols being graphical representations of spatial entities, the automatic conversion of symbols between different standards can be realised, as long as the corresponding symbol codes (names) of the same spatial entity in different mapping standards are clear (Stefanakis 2002; Schlichtmann 2004). Based on this idea, the corresponding relationship between symbol codes (names) in different mapping standards can be established, and then the mutual conversion of symbols between different standards can be realised by using the established relationship. So if the symbol for a given spatial entity is known, the corresponding symbol in the target mapping standard can be found through the relationship. Nevertheless, the symbol codes (names) are usually absent in maps made by common GIS platforms, only the symbol styles and rendering information are stored. However, the symbol libraries in GIS platforms store not only symbol codes (names) but also symbol styles and rendering information. Therefore, the critical problem to be solved is to identify the special symbols of spatial entities from a map and match them with the symbols stored in symbol libraries to get the codes (names) of the symbols in the map. To recognise a symbol, a pixel-by-pixel matching approach is proposed in which the symbols in a map, and those in the symbol library with the same mapping standard, are firstly transformed into BMP images, and then the symbol images within the map are matched with images from the library, to find out the correct symbol.

Table 1 shows an example of the recognition of the "oilfield" symbol in the USGS mapping standard.

Table 1 Symbol matching and recognition

Symbol on the map	Symbols in symbol library	Complete pixel matching	Recognised result
	⬤	×	
	⬤	×	
⬤	⬤	√	Symbol code (name)
	⬤	×	
	⬤	×	

2.2 Automatic conversion of symbols

Based on each mapping standard specification and symbol library in specific industries, the corresponding relationship to the same spatial entity is set up through the bidirectional mapping of the symbol codes (names) in different mapping standards, and the symbol codes (names) can be used as keywords to connect different mapping standards (Table 2). According to the identified symbol codes (names) and the relationship, the corresponding symbol codes (names) in other standards can be found, and then those symbols meeting the target standard can be obtained from the symbol library. The symbols in the converted map can be replaced by the target symbols to accomplish the conversion but accurate symbol recognition is crucial to the success of any conversion.

3 Approach of symbol recognition and automatic conversion

According to the aforementioned principle, the procedure of automatic symbol conversion can be deduced as follows: symbol association table construction, symbol matching, recognition and symbol standard conversion. Firstly, the corresponding relationship between symbols in different mapping standards should be constructed according to the symbol codes (names). Secondly, each symbol used in a map should be matched with the symbols stored in the symbol library that have the same standard and category as the map, so that the code (name) of each symbol can be acquired. Finally, based on the relationship of symbols across different mapping standards, the obtained codes (names) are used as keywords to deduce the corresponding target symbols, and then the target symbols found are used to symbolise the spatial entities expressed thereby. The "oilfield" symbol used in a USGS standard map is taken as an example to illustrate the implementation (Fig. 2): the other mapping symbols underwent the same type of conversion.

3.1 Construction of symbol association tables

Constructing the symbol association tables for different mapping standards is the basis of the symbol standard conversion. Actually, it is a bidirectional mapping relationship for symbols in different mapping standards, or say, the different symbolisations of the same spatial entity, and the codes (names) of symbols are the unique keywords with which to associate the tables. The tables are stored in

Table 2 Associated symbol mapping table

ID	Mapping standard A	Mapping standard B	Mapping standard	Meanings
1	Symbol code (name) a_1	Symbol code (name) b_1	Symbol code (name)	Spatial entity s_1
2	Symbol code (name) a_2	Symbol code (name) b_2	Symbol code (name)	Spatial entity s_2
3	Symbol code (name) a_3	Symbol code (name) b_3	Symbol code (name)	Spatial entity s_3

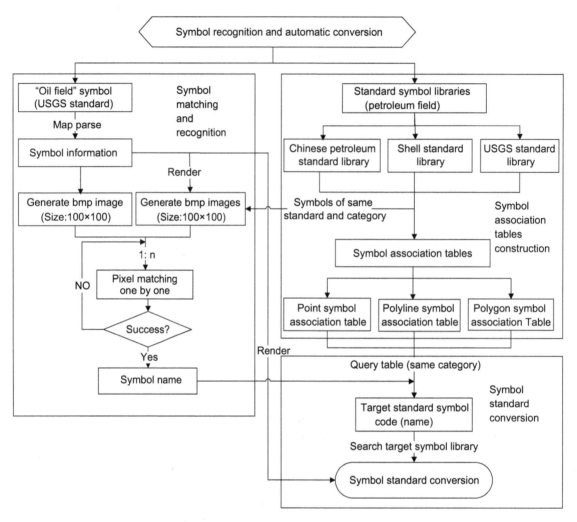

Fig. 2 Flowcharts for symbol recognition and automatic conversion

the database according to the structure outlined by Table 2. To improve query efficiency, three tables are established corresponding to the point, polyline and polygon spatial entities.

3.2 Automatic matching and recognition of symbols

The automatic symbol matching and recognition is the key to the scheme designed in this work. The purpose of this stage is to get the code (name) of the symbol from the USGS standard symbol library. To enhance the query and matching efficiency, the symbol can be matched with those sharing the same category. The automatic symbol matching and recognition procedure is as follows:

(1)   To analyse the symbol to get symbol information. The "oilfield" symbol can be analysed by a compiled computer program to obtain the symbol style (e.g. shape) and rendering information (e.g.

size, fill colour, texture, etc.), which varies with different symbol categories.

(2)   To generate a BMP image of the symbol. Based on the style and rendering information from Step (1), a BMP image of an appropriate size is drawn. The image size should be larger than the shape of the symbol itself, but not be so large as to influence the matching efficiency. In this work, the size is set to 100×100 pixels.

(3)   To generate BMP images of the symbols in the USGS standard symbol library. Firstly, search the symbol library to find symbols within the same category as the "oilfield" symbol. Namely, if it is a point symbol, then find all the point symbols from the standard symbol library; if it is a polyline symbol, then obtain all the polyline symbols from its library and so on. Secondly, generate BMP images in the same way as Step (2) according to the style and rendering information from Step (1).

(4) To match each pixel one by one and recognise the symbol. Match the picture generated in Step (2) with each picture obtained in Step (3) by pixel matching. After satisfactory matching, the code (name) of the "oilfield" symbol in its standard symbol library can be obtained according to the matched results.

3.3 Conversion of symbol standard

According to the symbol code (name) obtained in the "automatic matching and recognition of symbols" process, the corresponding target symbol code (name) can be found by querying the relationship table. To improve the query efficiency, the corresponding relationship table will be searched according to the category of the "oilfield" symbol: if the symbol is a point, the point relationship table will be searched; if it is a polyline symbol, then the polyline relationship table would be searched and so on. Based on the target symbol code (name), the target symbol can be found in its standard symbol library and can be rendered using the relevant information for an "oilfield" symbol. Subsequently, the rendered symbol is used to visualise the corresponding spatial entity to complete symbol conversion. Each symbol used in the map can thus be converted, and symbol standard conversion between different mapping standards can be realised.

4 Verification of the symbol recognition and automatic conversion approach

4.1 Basic data preparation

There are three mapping standards in the petroleum field: the Chinese Petroleum standard, the Dutch Shell standard and the United States Geological Survey (USGS) standard. Therefore, the conversion of a practical map from the petroleum field among these standards was selected as the study case to test the conversion accuracy and working efficiency of the developed system in this work. The basic data for the conversion of a map among three petroleum standards are mainly the mapping specifications and symbol library files of the three standards, symbol association tables, and a map based on a standard. The mapping specifications, symbol library files (*.style format) and a map based on the Chinese Petroleum standard are provided by the Research Institute of Petroleum Exploration and Development (China), and the symbol library files (*.style format) are converted into files (*.serverstyle format) that can be recognised by the ArcGIS Engine platform. The files with *.serverstyle format are stored in the blob field of an Oracle database. According to

the three symbol libraries and the different symbol codes (names) of the same spatial entity in the three standards, the symbol association tables are constructed in the Oracle database. Based on the library files (*.serverstyle) and the association tables, the map can be conveniently mutually inter-converted among the three standards.

4.2 Verification of results

A map based on the Chinese Petroleum standard was selected for conversion to an equivalent map based on Shell and USGS standards, respectively (Fig. 3). The converted results show that the conversion is rapid, highly efficient, and most of symbols on the map, e.g., point symbols such as oilfield, gasfield, provincial capital, capital city and ocean; polyline symbols such as national borders, gas and oil pipelines; and polygon symbols such as oil basins, can be converted accurately. Only the point symbol for a local city and the polygon symbols for an oil field and small basin are not converted into their corresponding symbols (Table 3). Compared with the previous methods mentioned before, this proposed technical solution improves greatly in both conversion accuracy and working efficiency, and consequently, it can satisfy real-time demands and improve the conversion performance. According to the analysis of the results, and the conversion principle used, it can be deduced that if a symbol with a special style, namely only one symbol has the style in its symbol library, the symbol can be automatically matched and recognised. Based on the analysis of all the symbols in the three standard symbol libraries assessed, most of the point and polyline symbols have different styles, while some polygon symbols have the same style but with different sizes or colours. Therefore, most of the point and polyline symbols, and some polygon symbols, can be automatically matched and recognised by the conversion system. After the map was converted, human intervention is needed to manually match the several symbols that are not automatically converted by the system and to completely accomplish the conversion. In consideration of the fact that the manual matching method has several problems, the symbols that completely matched the unrecognised symbols can be displayed in list format, so that the cartographer can choose the appropriate symbol to accomplish the matching and conversion.

5 Discussion and conclusions

5.1 Discussion

The conversion system makes it possible to automatically convert vector maps between different mapping standards,

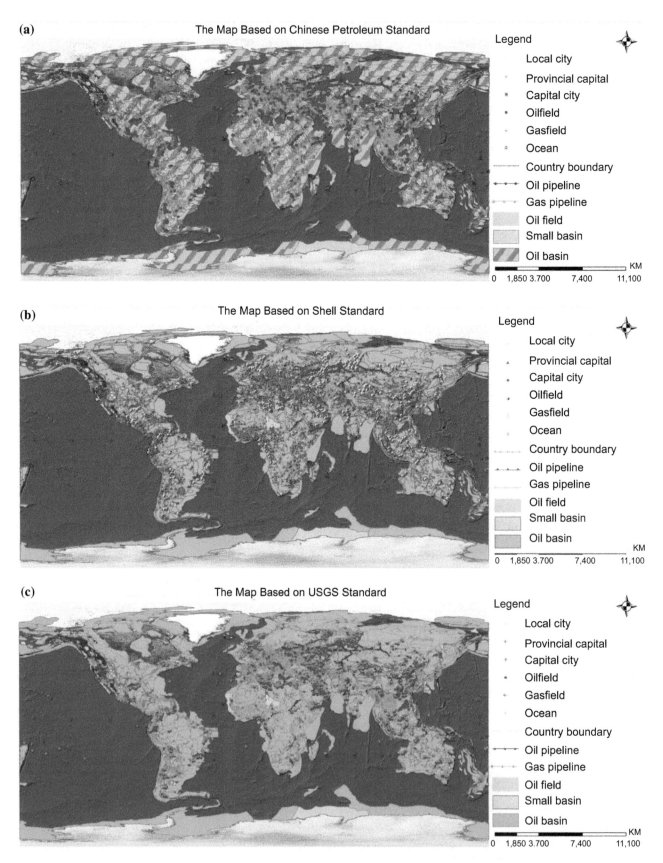

Fig. 3 Conversion results for the map based on the Chinese Petroleum standard. **a** Chinese Petroleum standard map. **b** Shell standard map. **c** USGS standard map

Table 3 Conversion results for map symbols based on the Chinese Petroleum standard

Layer names	Chinese Petroleum standard	Shell standard	USGS standard	Conversion success
Local city				No
Provincial capital				Yes
Capital city				Yes
Oilfield				Yes
Gasfield				Yes
Ocean				Yes
National boundary				Yes
Oil pipeline				Yes
Gas pipeline				Yes
Oil field				No
Secondary basin				No
Oil basin				Yes

so that vector maps can be conveniently converted from one standard to another by the system. So information contained in the converted map can be rapidly, accurately, and efficiently obtained. In addition, the feasibility of the proposed method and the reliability of the conversion system are proven by its practical operation. Nevertheless, in the process of symbol matching and recognition, some symbols (especially polygon symbols) with the same style and different sizes cannot be automatically identified, and manual intervention is needed. This suggests that the expressed meanings of the symbols should be clarified in advance, so as to choose the corresponding target symbols to allow complete conversion. However, in some mapping standards, the meanings of symbols may not be obtained

through their codes (names), and manual intervention is essential. To quickly and accurately choose the appropriate match for the converted symbols, a solution may be found as follows:

Based on each mapping standard in the field, a symbol table can be established to delineate the meanings of the symbols in the standard. When manual operation is needed in the matching process, the thumbnail images, names and meanings of the symbols that completely matched the converted symbol can be quickly and accurately obtained and displayed in a message box by querying the corresponding symbol table according to the appropriate standard (Fig. 4). To enhance the query efficiency, three types of symbol tables can be established to delineate the

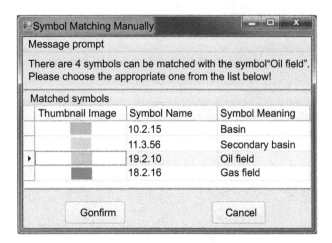

Fig. 4 Symbols information prompt window for manual operation

meanings of the symbols in each mapping standard according to the symbol categories (point, polyline and polygon).

From the aforementioned discussion, it may be deduced that the proposed technical solution could be classified as a semi-automatic conversion. In the next work, research into symbol analysis will be carried out to further improve and optimise the method of automatic symbol recognition, especially the recognition of polygon symbols, so that the conversion of maps between different mapping standards can be realised more rapidly and accurately.

5.2 Conclusions

A novel approach for symbol recognition and automatic conversion is proposed, and according to the approach, based on the ArcGIS Engine platform and the integrated development environment of Visual Studio 2010, a conversion system for a vector map standard is developed by using component object model (COM) technology. Taking the conversion of a practical map based on the Chinese Petroleum standard among the three standards frequently adopted in the sector as a test case, the applicability and conversion accuracy of the system have been analysed. Based on this analysis, the following conclusions are drawn:

(1)   The conversion system has high applicability and universality. Besides the petroleum field, this system also can be applied into fields that have different mapping standards, such as the field of electricity, architecture, traffic and engineering. As long as each mapping standard specification and each symbol library file are known, the symbol association tables can be constructed in the database. Based on the tables and the symbol libraries, the conversion system can be used to convert vector maps between

different mapping standards. Additionally, practical operation shows that the system has good encapsulation and is easily maintained and expanded.

(2)   Most of the symbols in the map could be automatically converted by the conversion system: it therefore has high conversion accuracy. The quality and efficiency of vector map conversion between different standards has been improved, and the previous challenge facing vector map standard conversion was overcome.

(3)   Some symbols in a symbol library may have the same style and different sizes or colours, especially polygon symbols. These symbols cannot be automatically matched and identified. In these cases, human intervention is needed to manually choose the appropriate symbol match. This is the main reason for the occasional incomplete vector map standard conversion when using this system.

Acknowledgments   This research was supported by the National Major Specific Project of Oil and Gas during the Twelfth Five-Year Plan (2011ZX05028-004) and the Major Science and Technology Program of PetroChina (2012D-4602-05) and the National Natural Science Foundation of China (61501064). The authors appreciate support for the data services provided by the staff of the Research Institute of Petroleum Exploration and Development.

References

Antoniou B, Tsoulos L. The potential of XML encoding in geomatics converting raster images to XML and SVG. Comput Geosci. 2006;32(2):184.

Brewer CA, Hatchard GW, Harrower MA. ColorBrewer in print: a catalog of color schemes for maps. Cartogr Geogr Inf Sci. 2003;30(1):5–32.

Comentz J. Cognitive geometry for cartography. Cartogr J. 2002;39(1):65–75.

Che S, Sun Q, Liu HY. Design of a parameter controlling symbol editing tool. Geomat Inf Sci Wuhan Univ. 2013;38(11):1326–9 (in Chinese).

Cheng W, Zhang YM. Research and implementation of oilfield basic platform based on integrated 2D with 3D of GIS. Procedia Eng. 2012;29:3651–8.

Chen TS, Lv GN, Wu MG. Research on GIS point symbol sharing. Geomat Inf Sci Wuhan Univ. 2011;36(2):239–43 (in Chinese).

Chen TS, Chen ML, Zhou J. Sharing ArcGIS point symbols based on PB symbols. J Geod Geodyn. 2014;34(2):69–73.

Dang LN, Dang GF, Wu F. The research on representation and realization of map symbol based on text. Procedia Environ Sci. 2011;10(C):2342–7.

Fan WF, Wang H, Ye FH. Research on style symbol library access and the implementation of the symbol selector. Bull Surv Mapp. 2011;11:25–31 (in Chinese).

Frehner M, Brandli M. Virtual database: spatial analysis in a Web-based data management system for distributed ecological data. Environ Model Softw. 2006;21:1544–54.

Gustavvson M, Kolstrup E, Seijmonsbergen AC. A new symbol-and-GIS based detailed geomorphological mapping system: renewal of a scientific discipline for understanding landscape development. Geomorphology. 2006;77(1–2):90–111.

Goodchild MF. Cartographic futures on a digital earth. In: Proceedings of 19th Int. Cartographic Conf. Section 2. Ottawa, 1999: 4–12.

Guo T, Zhang H, Wen Y. An improved example-driven symbol recognition approach in engineering drawings. Comput Graph. 2012;36(7):835–45.

Li L, Yin ZC, Zhu HH. Concept and schema of map-making markup language. Acta Geodaetica Cartogr Sin. 2007;36(1):108–11 (in Chinese).

Li QY, Su DG, Li HS, et al. Approach to general data model of GIS symbol library and symbol library data exchange XML schema. Geo-Spatial Inf Sci. 2009;12(4):235–42.

Liu WY, Wan Z, Luo Y. An interactive example-driven approach to graphics recognition in engineering drawings. Int J Doc Anal Recognit. 2007;9(1):13–29.

Llados J, Martí E, Villanueva JJ. Symbol recognition by error-tolerant subgraph matching between region adjacency graphs. IEEE Trans Pattern Anal Mach Intell. 2001a;23(10):1137–43.

Llados J, Valveny E, Sanchez G, et al. Symbol recognition: current advances and perspectives. In: Blostein D, Kwon Y-B. (eds.), GREC 2001. LNCS, Springer, Heidelberg, 2002; 2390: 104–128.·

Mihalynuk MG. Geological symbol set for manifolds geographic information system. Comput Geosci. 2006;32:1228–33.

Nass A, van Gasselt S, Jaumann R, et al. Implementation of cartographic symbols for planetary mapping in geographic information systems. Planet Space Sci. 2011;59:1255–64.

Petrovic D. Cartographic design in 3D maps. In: Proceedings of 21th International Cartographic Conference "Cartographic Renaissance". Durban: 2003.

Qin XJ, Wang QN, Wang KQ, et al. On cartographic visualization. Geogr Res. 2000;19(1):15–9 (in Chinese).

Qin RF, Xu HP, Wang JL, et al. Design and implementation of universal map symbol library based on extensible markup language. J Tongji Univ (Nat Sci). 2008;36(8):1138–42 (in Chinese).

Robinson AC, Roth RE, Blanford J, et al. A collaborative process for developing map symbol standards. Procedia Soc Behav Sci. 2011;21:93–102.

Schlichtmann H. On the semantic analysis of map symbolism: order by oppositions. In: Wolodtschenko A, Schlichtmann H, editors. Diskussions beitrage zur Kartosemiotik und zur Theorie der Kartographie, vol. 7. Dresden: Selbstverlag der Technischen Universitat Dresden; 2004. p. 20–34.

Schichtmann H. Overview of the semiotics of maps. In: Proceedings of 24th International Cartographic Conference. Santiago, Chile, 2009; 15–21.

Stefanakis E. Representation of map objects with semi-structured data models. In: Richardson D, van Oosterom P, editors. Advances in spatial data handling (10th International Symposium on Spatial Data Handling—SDH2002). Canada: Springer; 2002. p. 547–62.

Tao T, Lv GN, Li YN. Study on symbol sharing based on common integrated symbol editor. Geogr Geo-Inf Sci. 2005;21(4):28–31 (in Chinese).

Tao Tao, Guonian Lv, Shuliang Zhang, et al. Research Progress and prospect of map symbol sharing in GIS. J Image Graph. 2007;12(8):1326–32 (in Chinese).

Tsoulos L, Spanaki M, Skopeliti A. An XML-based approach for the composition of maps and charts. In: The 21st International Cartographic Conference (ICC), Durban; 2003.

Wan Z, Liu WY. A new vectorial signature for quick symbol indexing, filtering and recognition. In: Proceedings of 9th International Conference on Document Analysis and Recognition. Los Alamitos: IEEE Computer Society Press, 2007; 1: 536–540.

Xie YW, Zhang H. Research on symbol fuzzy recognition in vector drawings based on 2-neighborhood local structures. J Comput Aided Design Graph. 2014;26(10):1613–23.

Yamada H. Directional mathematical morphological and reformalized hough transformation for the analysis of topographic maps. IEEE Trans PAMI. 1993;15(4):380–7.

Yang S. Symbol recognition via statistical integration of pixel-level constraint histograms: a new descriptor. IEEE Trans Pattern Anal Mach Intell. 2005;27(2):278–81.

Yin ZC, Li L, Zhu HH. Description model of map symbols based on SVG. Geomat Inf Sci Wuhan Univ. 2004;29(6):544–7 (**in Chinese**).

Zeng YS. Research on the algorithm of extraction and recognition for map symbols. Thesis for Degree of Doctor of National University of Defense Technology; 2003 (in Chinese).

Zou Q, Wang Q, Wang CZ. Integrated cartography technique based on GIS. Energy Procedia. 2012;17(A):663–70.

3

Petroleum geology features and research developments of hydrocarbon accumulation in deep petroliferous basins

Xiong-Qi Pang · Cheng-Zao Jia · Wen-Yang Wang

Abstract As petroleum exploration advances and as most of the oil–gas reservoirs in shallow layers have been explored, petroleum exploration starts to move toward deep basins, which has become an inevitable choice. In this paper, the petroleum geology features and research progress on oil–gas reservoirs in deep petroliferous basins across the world are characterized by using the latest results of worldwide deep petroleum exploration. Research has demonstrated that the deep petroleum shows ten major geological features. (1) While oil–gas reservoirs have been discovered in many different types of deep petroliferous basins, most have been discovered in low heat flux deep basins. (2) Many types of petroliferous traps are developed in deep basins, and tight oil–gas reservoirs in deep basin traps are arousing increasing attention. (3) Deep petroleum normally has more natural gas than liquid oil, and the natural gas ratio increases with the burial depth. (4) The residual organic matter in deep source rocks reduces but the hydrocarbon expulsion rate and efficiency increase with the burial depth. (5) There are many types of rocks in deep hydrocarbon reservoirs, and most are clastic rocks and carbonates. (6) The age of deep hydrocarbon reservoirs is widely different, but those recently discovered are predominantly Paleogene and Upper Paleozoic. (7) The porosity and permeability of deep hydrocarbon reservoirs differ widely, but they vary in a regular way with lithology and burial depth. (8) The temperatures of deep oil–gas reservoirs are widely different, but they typically vary with the burial depth and basin geothermal gradient. (9) The pressures of deep oil–gas reservoirs differ significantly, but they typically vary with burial depth, genesis, and evolution period. (10) Deep oil–gas reservoirs may exist with or without a cap, and those without a cap are typically of unconventional genesis. Over the past decade, six major steps have been made in the understanding of deep hydrocarbon reservoir formation. (1) Deep petroleum in petroliferous basins has multiple sources and many different genetic mechanisms. (2) There are high-porosity, high-permeability reservoirs in deep basins, the formation of which is associated with tectonic events and subsurface fluid movement. (3) Capillary pressure differences inside and outside the target reservoir are the principal driving force of hydrocarbon enrichment in deep basins. (4) There are three dynamic boundaries for deep oil–gas reservoirs; a buoyancy-controlled threshold, hydrocarbon accumulation limits, and the upper limit of hydrocarbon generation. (5) The formation and distribution of deep hydrocarbon reservoirs are controlled by free, limited, and bound fluid dynamic fields. And (6) tight conventional, tight deep, tight superimposed, and related reconstructed hydrocarbon reservoirs formed in deep-limited fluid dynamic fields have great resource potential and vast scope for exploration. Compared with middle–shallow strata, the petroleum geology and accumulation in deep basins are more

X.-Q. Pang (✉) · W.-Y. Wang
State Key Laboratory of Petroleum Resources and Prospecting, Beijing 102249, China
e-mail: pangxq@cup.edu.cn

X.-Q. Pang · W.-Y. Wang
Basin and Reservoir Research Center, China University of Petroleum, Beijing 102249, China

C.-Z. Jia
PetroChina Company Limited, Beijing 100011, China

C.-Z. Jia
Research Institute of Petroleum Exploration and Development, Beijing 100083, China

Edited by Jie Hao

complex, which overlap the feature of basin evolution in different stages. We recommend that further study should pay more attention to four aspects: (1) identification of deep petroleum sources and evaluation of their relative contributions; (2) preservation conditions and genetic mechanisms of deep high-quality reservoirs with high permeability and high porosity; (3) facies feature and transformation of deep petroleum and their potential distribution; and (4) economic feasibility evaluation of deep tight petroleum exploration and development.

Keywords Petroliferous basin · Deep petroleum geology features · Hydrocarbon accumulation · Petroleum exploration · Petroleum resources

1 Introduction

As the world demands more petroleum and petroleum exploration continues, deep petroleum exploration has become an imperative trend. As it is nearly impractical to expect any major breakthrough in middle or shallow basins (Tuo 2002), petroleum exploration turning toward deep basins has become inevitable. After half a century's exploitation in major oilfields across the world, shallow petroleum discoveries tend to be falling sharply (Simmons 2002). Nor are things optimistic in China, where the rate of increase of mid-and-shallow petroleum reserves is increasingly slowing down (Wang et al. 2012). At the same time, the world's petroleum consumption continues to increase. According to BP Statistical Review of World Energy 2014, from 2002 to 2012, the world's petroleum consumption increase was virtually the same as its petroleum output increase, with annual average oil and natural gas consumption increases of 1.35 % and 3.14 %, respectively, compared to the annual average output increases of 1.47 % and 3.23 %. In China, however, the petroleum consumption increase is far greater than its output increase. According to National Bureau of Statistic of China statistics in 2013, from 2000 to 2013, China's average annual oil and natural gas consumption increases were 6.1 % and 14.6 %, respectively, compared to the annual average output increases of 1.93 % and 11.9 %. Deep petroleum, as one of the strategic "three-new" fields for the global oil industry (Zou 2011) as well as one of the most important development areas for China's oil industry, forms the most important strategically realistic area for China's oil industry to lead future petroleum exploration and development (Sun et al. 2013). All indicate that deep hydrocarbon exploration is an inevitable choice toward ensuring energy supply and meeting market demands.

After half a century's effort, gratifying achievements have been made in deep petroleum exploration throughout the world, despite being faced with challenges and problems today. The Former Soviet Union discovered four 6,000 m or deeper industrial oil–gas reservoirs out of its 24 petroliferous basins (Tuo 2002). The oil discovered in deep basins in Mexico, the USA, and Italy contributes more than 31 % of their present recoverable oil reserves (Kutcherov et al. 2008) and the natural gas discovered there makes up approximately 47 % of their total proved natural gas reserves (Burruss 1993). China, too, has appreciable achievements in deep petroleum exploration. Compared with 2000, the deep reserves discovered in West China in 2013 increased an average of 3.5 times. The ratio of deep petroleum reserves increased from 40 % in 2002 to 80 % in 2013. Of the 156 well intervals in the Tarim Basin that have been tested so far, 58 have gone deeper than 5,000 m. The deep drillhole success rate in the Jizhong Depression is as high as 21.4 %. A 5,190-m-deep "Qianmiqiao buried hill hydrocarbon reservoir" was discovered in the Huanghua Depression (Tuo 2002). Despite these achievements, however, a lot of problems have also emerged in deep petroleum exploration. These include (a) the difficulty in understanding the conditions of deep oil–gas reservoirs and evolution due to the multiple tectonic events having taken place in deep basins (Zhang et al. 2000; He et al. 2005), (b) the difficulty in evaluating the resource potential and relative contribution due to the complex sources and evolution processes of deep petroleum (Barker 1990; Mango 1991; Dominé et al. 1998; Zhao et al. 2001; Jin et al. 2002; Zhao et al. 2005; Darouich et al. 2006; Huang et al. 2012; Pang et al. 2014a), (c) the difficulty in predicting and evaluating favorable targets due to the complex genesis and distribution of deep, relatively high-porosity and high-permeability reservoirs (Surdam et al. 1984; Ezat 1997; Dolbier 2001; Rossi et al. 2001; Moretti et al. 2002; Lin et al. 2012), and (d) the difficulty in predicting and evaluating the petroleum possibility in deposition targets due to the complex deposition mechanism and development pattern of deep petroleum (Luo et al. 2003, 2007; Ma and Chu 2008; Ma et al. 2008; Pang et al. 2008). All these problems provide a tremendous challenge to deep petroleum exploration.

With abundant resource bases and low proved rates, deep petroliferous basins are important for further reserve and output increases (Tuo 2002; Zhao et al. 2005; Dai 2006, Pang et al. 2007a; Zhu and Zhang 2009; Sun et al. 2010; Pang et al. 2014a). According to Dai (2006), the proved rate of the exploration concessions of PetroChina is 17.6 % for deep oil and 9.6 % for deep natural gas, far lower than their mid-and-shallow counterparts of 39.6 % for oil and 14.6 % for natural gas. Pang et al. (2007a, b) suggest that West China contains around 45 % of the residual petroleum resources of China, and 80 % of these residual resources are buried in deep horizons more than 4,500 m below the surface, yet the present proved rate is less than 20 %. As such, implementing deep petroleum resource research, tapping deep petroleum and increasing petroleum backup reserves are urgently needed if we ever want to relieve the nation's petroleum shortage and mitigate energy risks. Many scholars have investigated deep petroleum geologies and exploration (Perry 1997; Dyman et al. 2002; Pang 2010; Ma et al. 2011; He et al. 2011; Wang et al. 2012; Wu et al. 2012; Bai and Cao 2014). Our study in connection with the national "973 Program" (2011CB201100) involves a summary and description of the development and orientation of research by scholars in China and elsewhere with respect to petroleum geology and hydrocarbon accumulation in deep petroliferous basins.

2 Concept and division criteria of deep basins

Deep basins are also called deep formations by some scholars. The definition and criteria of deep petroliferous basins differ from country to country, from institution to institution and from scholar to scholar.

2.1 Concept and division criteria of deep basins proposed by overseas scholars

So far, there are two sets of definition and criteria for deep petroliferous basins outside China. One is according to the formation depth, i.e., formations within a certain limit of depth are called deep formations. However, the criteria for classifying deep basins also differ from scholar to scholar. Representative criteria include 4,000 m (Rodrenvskaya 2001), 4,500 m (Barker and Takach 1992), 5,000 m (Samvelov 1997; Melienvski 2001), and 5,500 m (Manhadieph 2001; Bluokeny 2001). Another is according to the formation age, i.e., for a given basin, formations older in age and deeper are called deep formations (Sugisaki 1981). Table 1 summarizes the criteria used by different institutions and scholars for deep basins from which it is easy to see that 4,000 and 4,500 m are the criteria accepted by more institutions and scholars.

2.2 Concept and division criteria of deep basins proposed by Chinese scholars

Chinese scholars use roughly the same criteria for deep petroliferous basins as their overseas counterparts. Most of them use three indicators: (1) formation depth (Wang et al. 1994; Li and Li 1994; Tuo et al. 1994; Zhou et al. 1999; Hao et al. 2002; Shi et al. 2005; Dai et al. 2005); (2) formation age (Kang 2003; Ma et al. 2007; Ma and Chu 2008); and (3) formation characteristics (Tuo et al. 1999a; Wang et al. 2001; Wang 2002; Pang 2010). Table 2 summarizes the criteria used for deep basins. Obviously, the concept of deep basins does not only differ from scholar to scholar, it also varies with the basin position and formation characteristics.

2.3 Importance of using the same concept and criteria in deep basins

No uniform concept or criteria have been agreed upon by scholars either in or out of China with respect to deep petroliferous basins, hence preventing further development and mutual promotion on science research. For this reason, we suggest using 4,500 m as the criterion for deep basins on grounds of the following considerations:

First, this classification represents a succession to previous findings. The U.S. Geological Survey and some former Soviet Union scholars used 4,500 m as the criteria for deep basins (Barker and Takach 1992). Chinese scholars, represented by Shi, Dai, and Zhao et al., also used 4,500 m to demarcate deep formations (Shi et al. 2005; Dai et al. 2005; Zhao et al. 2005). Chinese administrations like Ministry of Land and Resources even issued public documents that define deep petroliferous basins in West China to 4,500 m. Second, 4,500 m represents the general depth at which the hydrocarbon entrapment mechanism of a petroliferous basin transits from buoyancy accumulation to non-buoyancy accumulation. Above this depth, the porosities of the sand reservoirs are generally above 12 %; the permeabilities are higher than 1 mD; and the pore throat radii are larger than 2 μm. "High-point accumulation, high-stand closure, high-porosity enrichment, high-pressure accumulation" (Pang et al. 2014a) normal oil–gas reservoirs generally formed under the action of buoyant forces. Below this depth, to the contrary, "low-depression accumulation, low-stand inversion, low-porosity enrichment, low-pressure stability" unconventional oil–gas reservoirs generally formed. To make things easier, we divide a petroliferous basin into four parts according to the buried depth, using the criteria accepted by previous scholars: shallow (<2,000 m), middle (2,000–4,500 m), deep (4,500–6,000 m), and ultra-deep (>6,000 m). According to the maximum depths of basins, we divide them into shallow basins (<2,000 m), middle basins (2,000–4,500 m), deep basins (4,500–6,000 m), and ultra-deep basins (>6,000 m). Third, deep basins should be classified according to the depth rather than incorporating the geological aspects that constrain the depth distribution of hydrocarbon entrapment. For example, the fact that oil–gas reservoirs in East China basins are commonly shallow while those in West China basins are commonly deep is attributable to their respective unique basin evolution geologies such as the geothermal gradient, reservoir rock type, formation age, and evolution history. These should not form the basis for diverging the criteria for basin depths.

Table 1 Criteria for deep basins proposed by scholars outside China

Basis for deep basins	Criteria for deep basins	Targeted area	Researcher and year
Formation depth	>4,000 m	In former Soviet Union	Rodrenvskaya (2001)
	>4,500 m	Caspian Basin	
	>4,500 m	In the USA	
		Gulf of Mexico, USA	Barker and Takach (1992)
	>5,000 m		Samvelov (1997)
		West Siberia Basin, East Siberia Basin	Melienvski (2001)
	>5,500 m	South Caspian Basin	Manhadieph (2001)
		Timan-Pechora Basin	Bluokeny (2001)
Formation age	Stratigraphically old formations with large buried depths	In the USA	Sugisaki (1981)

Table 2 Criteria of deep basins proposed by Chinese scholars

Basis for deep basins	Criteria for deep basins	Targeted area or parameter features	Researcher and year
Formation depth	>2,500 m	Bohai Bay Basin	Qiao et al. (2002)
	>2,800 m	Songliao Basin	Wang et al. (1994)
	>3,500 m	Liaohe Basin	Li et al. (1999)
		Bohai Bay Basin	Tuo (1994)
		Bohai Bay Basin	Zhou et al. (1999)
		Yinggehai Basin	Hao et al. (2002)
	>3,500 m	East China basins	Ministry of Land and Resources (2005)
	>4,500 m	Junggar Basin	Shi et al. (2005)
		Tarim, Junggar, Sichuan basins	Dai (2003)
		Sichuan Basin	Zhao et al. (2005)
	>4,500 m	West China basins	Ministry of Land and Resources (2005)
Formation age & depth	Stratigraphically old with large buried depths	Varies from basin to basin	Kang (2003), Ma et al. (2007), Ma and Chu (2008)
Formation characteristics	Formation thermal evolution level	$R_o \geq 1.35\ \%$	Tuo et al. (1999b), Tuo (2002)
	Formation thermal evolution level or formation pressure	$R_o \geq 1.35\ \%$ or formation depth overpressure	Wang et al. (2001), Tuo et al. (1999b), Wang et al. (2002)
	Formation thermal evolution level and tightness level	$R_o \geq 1.35\ \%$ or sandstone formation $\Phi \leq 12\ \%$, $K \leq 1$ mD, $Y \leq 2$ μm	Pang (2010)

3 Exploration for deep oil–gas reservoirs

3.1 Exploration for deep oil–gas reservoirs across the world

Following the discovery of the first deep hydrocarbon field below 4,500 m in the USA in 1952, deep petroleum exploration boomed in many countries. Seventy countries tried deep exploration (Wu and Xian 2006). Echoing breakthroughs in deep well drilling and completion techniques, a succession of major breakthroughs have been made in deep hydrocarbon reservoir exploration (Dyman et al. 2002). First, major breakthroughs in drilling operation led to the discovery of a number of oil–gas reservoirs including a gas reservoir in the Cambrian–Ordovician Arbuckle Group dolomites at 8,097 m depth in the Mills Ranch gas field in the Anadarko Basin in 1977 (Jemison 1979). From 1980, deep petroleum exploration started to extend from onshore to offshore. Examples include a gas field discovered in Permian Khuff Formation limestones at 4,500 m in the Fateh gas reservoir in the Arabian-Iranian Basin in 1980, and an oil reservoir at a depth of 6,400 m was discovered in the Triassic dolomites of the Villifortuna-Trecate oilfield in Italy in 1984. Recently, major breakthroughs in deep oil exploitation have been reported in the deep and ultra-deep waters of the Gulf of Mexico, East Brazil, and West Africa (Bai and Cao 2014). According to IHS data as of 2010, 171 deep basins and 29 ultra-deep basins had been discovered out of the 1,186 petroliferous basins in the world. These deep basins are predominantly situated in the former Soviet Union, Middle East, Africa, Asia-Pacific, North America, and Central and South America (Fig. 1). A total of 1,290 oil–gas reservoirs have been discovered in deep basins and 187 oil–gas reservoirs in ultra-deep basins across the world. Breakthroughs are continuously reported around the world in the course of deep exploration. First, the drilling depth continues to increase, the maximum being deeper than 10,000 m, as exemplified by the deepest well with 12,200m drilling depth, SG-3 exploratory well. The deepest oil reservoir discovered so far is the Tiber clastic rock oil reservoir (1,259 m underwater and 8,740 m underground). The depth of gas wells continues to increase, and the deepest gas reservoir discovered so far (8,309–8,322 m) is a Silurian basin gas reservoir in the Anadarko Basin. Second, the manageable formation temperature and formation pressure in drilling operations are also continuously increasing. So far, the highest temperature encountered is 370 °C and the highest pressure encountered is 172 MPa (Table 3).

According to USGS and World Petroleum Investment Environment Database, from 1945 to 2014, the world's normal petroleum resource has increased from 96 billion ton in 1945 to 630 billion ton in 2014, the annual average increase being as high as 8.06 % (Fig. 2a), and the natural gas resource has also increased from 260 trillion m^3 in 1986 to 460 trillion m^3 in 2013, the annual average increase being as high as 2.85 % (Fig. 2b). Over the past years, the world has shown robust momentum for deep petroleum exploration. The number of oil–gas reservoirs discovered keeps growing fast (Fig. 3). According to data provided by Kutcherov et al. (2008), more than 1,000 hydrocarbon fields have been developed at depths of 4,500–8,103 m, the original recoverable oil reserve of which contributes 7 % of the world's total amount and the natural gas reserve makes up 25 %. According to IHS data, as of 2010, for the 4,500–6,000 m deep hydrocarbon fields in the world, the proved recoverable residual oil reserve is 83.8 billion ton or 35.5 % of the total recoverable oil reserve, and the natural gas is 65.9 billion ton oil equivalent or 44.4 % of the total productive natural gas reserve; for the 6,000 m or deeper hydrocarbon fields in the world, the proved recoverable residual oil reserve is 10.5 billion ton or 4.45 % of the total productive oil reserve, and the natural gas is 7 billion ton oil equivalent or 4.7 % of the total productive natural gas reserve (Fig. 4).

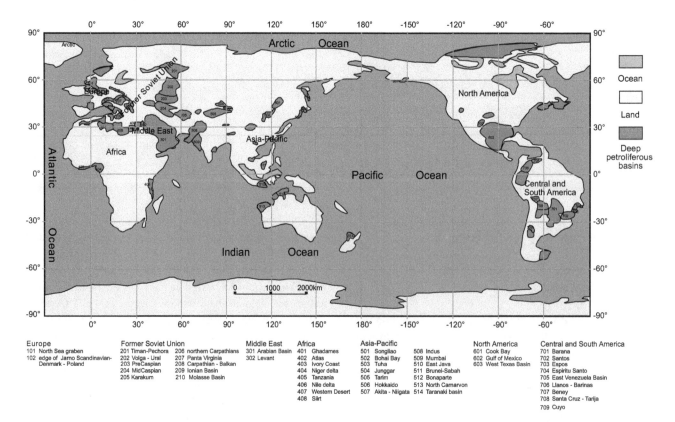

Europe	Former Soviet Union		Middle East	Africa		Asia-Pacific		North America	Central and South America
101 North Sea graben	201 Timan-Pechora	206 northern Carpathians	301 Arabian Basin	401 Ghadames		501 Songliao	508 Indus	601 Cook Bay	701 Barana
102 edge of Jarno Scandinavian-	202 Volga - Ural	207 Panta Virginia	302 Levant	402 Atlas		502 Bohai Bay	509 Mumbai	602 Gulf of Mexico	702 Santos
Denmark - Poland	203 PreCaspian	208 Carpathian - Balkan		403 Ivory Coast		503 Tuha	510 East Java	603 West Texas Basin	703 Espos
	204 MidCaspian	209 Ionian Basin		404 Niger delta		504 Junggar	511 Brunei-Sabah		704 Espiritu Santo
	205 Karakum	210 Molasse Basin		405 Tanzania		505 Tarim	512 Bonaparte		705 East Venezuela Basin
				406 Nile delta		506 Hokkaido	513 North Carnarvon		706 Llanos - Barinas
				407 Western Desert		507 Akita - Niigata	514 Taranaki basin		707 Beney
				408 Siirt					708 Santa Cruz - Tarija
									709 Cuyo

Fig. 1 Horizontal distribution of major deep petroliferous basins in the world

Table 3 Geological characteristics of world representative deep oil–gas reservoirs known so far

Feature	Name	Year	Parameters	Region
Deepest well	SG-3 exploratory well	1992	Completion depth 12,200 m	Kola Peninsula, Russia
Deepest oil reservoir	Tiber clastic rock oil reservoir	2009	Buried depth 8,740 m	Gulf of Mexico abyssal basin, USA
Deepest gas reservoir	Mills Ranch gas reservoir	1977	Buried depth 7,663–8,083 m	Western Interior Basin, USA
Deep hydrocarbon reservoir with highest porosity	Gaenserndorf Ubertief oilfield Hauptdolomit Formation gas reservoir	1977	Porosity 35 %–38 %	Vienna Basin, Austria
Deep hydrocarbon reservoir with lowest porosity	Mora hydrocarbon reservoir	1981	Porosity 2.6 %–4 %	Sureste Basin, Mexico
Deep hydrocarbon reservoir with highest permeability	Platanal oilfield 4830–Cretaceous hydrocarbon reservoir	1978	Permeability 7,800 mD	Sureste Basin, Mexico
Deep hydrocarbon reservoir with lowest permeability	Wolonghe Huanglong structural belt gas reservoir	1980	Permeability 0.01 mD	Sichuan Basin, China
Deep gas reservoir with highest temperature	Satis hydrocarbon reservoir, Tineh Formation gas reservoir	2008	Temperature 370 °C	Nile Delta Basin, Egypt
Deep oil reservoir with lowest temperature	Sarutayuskoye oilfield Starooskolskiy Group oil reservoir	2008	Temperature 47 °C	Pechola Basin, Russia
Deep gas reservoir with highest pressure	Zistersdorf Ubertief 1 oilfield Basal Breccia gas reservoir	1980	Pressure 172 MPa	Vienna Basin, Austria
Deep hydrocarbon reservoir with lowest pressure	Akzhar East oilfield Asselian VIII (PreCaspian) Unit hydrocarbon reservoir	1988	Pressure 8.4 MPa	Caspian Basin, Kazakhstan

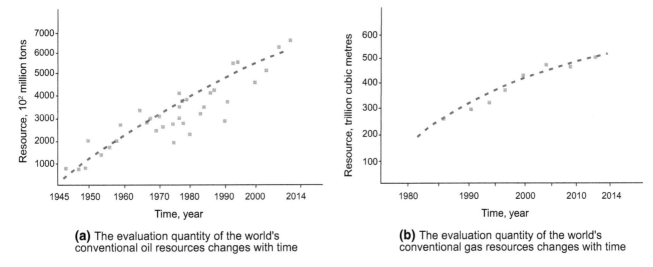

(a) The evaluation quantity of the world's conventional oil resources changes with time

(b) The evaluation quantity of the world's conventional gas resources changes with time

Fig. 2 World petroleum evaluation result and variation as a function of time

3.2 Exploration of deep oil–gas reservoirs in China

China started deep petroleum exploration from the late 1970s, having discovered a number of large deep oil–gas fields in the deep parts of some large sedimentary basins including Tarim, Erdos, and Sichuan basins, and made important progresses in the deep parts of the Daqing, Zhongyuan, Dagang, and Shengli fields in East China's petroleum region (Feng 2006; Song et al. 2008; Wu and Xian 2006). With the nation's breakthroughs in deep and ultra-deep well drilling techniques and equipment, onshore petroleum exploration has continued to extend toward deep and ultra-deep levels (Sun et al. 2010); petroleum exploration has also undergone a transition from shallow to deep and further to ultra-deep levels. On July 28, 1966, China's first deep well, Songji-6 of Daqing (4,719-m well depth), was completed, marking the transition of China's drilling operation from shallow wells to middle and deep wells, and signaling that China's petroleum exploration was turning from shallow toward deep

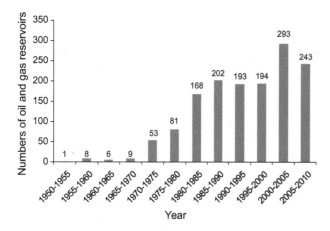

Fig. 3 Number of deep oil–gas reservoirs discovered in the world and its variation as a function of year

levels. From 1976, China's petroleum exploration marched toward ultra-deep levels. On April 30, 1976, China's first ultra-deep well, Nuji well in Sichuan (6,011 m well depth), was completed, marking the entry of China's petroleum exploration into ultra-deep levels (Wang et al. 1998). So far, China has drilled deep wells in 15 large basins with sedimentary thicknesses larger than 5 km (Pang 2010). Of the 176 deep exploratory wells drilled in the Jizhong Depression (the average well depth is 4,521 m), 37 have yielded industrial petroleum flows. The exploratory well success rate has reached 21.4 % (Tuo 2002). Of the 156 pay zone well intervals tested in the Tarim Basin, 58 have their bottom boundaries deeper than 5,000 m (Pang 2010). According to statistics, as of 2010, of the 47 petroliferous basins in China, seven deep basins have been discovered, out of which 210 deeper than 4,500 m oil–gas reservoirs have been identified. Shallow basins at a depth of 2,000 m or shallower are predominantly found in China's Inner Mongolia and Tibet; 2,000–4,500 m middle deep basins are typically located in the seas of East China; 4,500–6,000 m deep basins are

distributed in Central China and Southern North China. China also has a huge stock of ultra-deep petroleum resources, having discovered some ultra-deep basins with buried depths of more than 6,000 m, including the Tarim Basin and Songliao Basin. These are mostly located in Northwest and Northeast China (Fig. 5).

China is rich in deep petroleum resources with vast room for further exploration. According to a 2005 statistics of Shi et al., the deep oil resource within the mineral concession of CNPC is approximately 51.5×10^8 t or 12 % of the total; the deep natural gas reserve is 4.25×10^{12} m^3 or 19 % of the total. Zhu and Zhang (2009) suggest that China's deep petroleum resource reserves are extremely non-uniform and mostly found in Xinjiang. In the Junggar Basin, the middle–shallow and deep oil geological resources are 9.7×10^8 t or approximately 18 % of the basin's total amount; the deep natural gas resource is $2,081 \times 10^8$ m^3 or approximately 32 % of the basin's total amount. Pang (2010) discovered after studies that the Tarim Basin has the richest oil resources in the deep part at 33.7×10^8 t or 56 % of the basin's total oil resource, natural gas resources in the deep part at $29,244 \times 10^8$ m^3 or 37 % of the basin's total natural gas resource. Statistics indicate that China's deep petroleum resource is 30,408 million ton, which is 27.3 % of the nation's total oil resource (Fig. 6a); its deep natural gas resource is 29,120 billion m^3, which is 49.2 % of the nation's total natural gas resource (Fig. 6b). Since 2000, China's petroleum exploration has continued to extend toward deep and ultra-deep levels. In the Junggar Basin, the ratio of deep exploratory wells increased from 3 % in 2000 to 15 % in 2013 (Fig. 7a). In the Tarim Basin, this ratio increased from 65 % in 2000 to 92 % in 2013 (Fig. 7b). The ratio of newly increased petroleum reserves in deep formations has also continued to rise. In the Tarim Basin, the ratio of deep oil increased from 66 % in 2000 to 92 % in 2013 (Fig. 8a); the ratio of deep natural gas also increased from 66 % in 2004 to 92 % in 2013 (Fig. 8b).

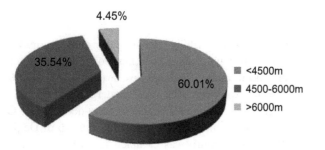

(a) The depth distribution of the world recoverable oil reserves

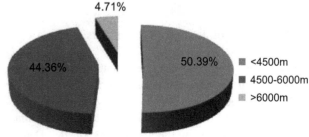

(b) The depth distribution of the world recoverable gas reserves

Fig. 4 Distribution of world recoverable oil and gas reserves at middle and deep depths of petroliferous basins

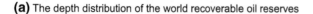

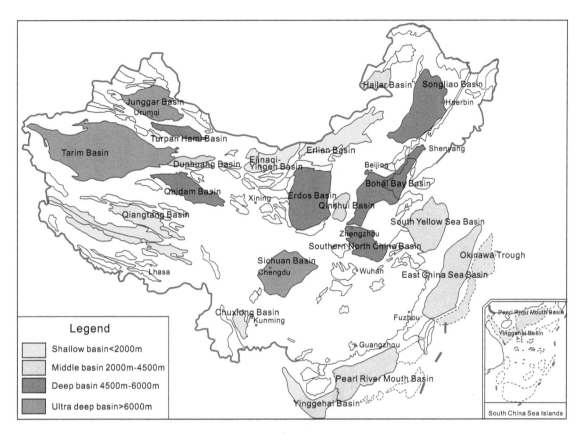

Fig. 5 Depth classification and horizontal distribution of petroliferous basins in China

4 Geological features of deep oil–gas reservoirs

Compared with middle or shallow petroliferous basins, deep basins have large buried depths and the features of high temperature, high pressure, low porosity, low permeability, complex structural styles, and highly variable sedimentary forms. These special properties have been responsible for the unique characteristics of deep basin oil–gas reservoirs compared with their middle or shallow counterparts. Many scholars (Zappaterra 1994; Dyman and Cook 2001; Liu et al. 2007a, b; Wang et al. 2012) have examined deep oil–gas reservoirs. Table 4 lists some typical hydrocarbon fields (reservoirs) discovered in deep petroliferous basins in the world, from which we can observe their differences and varieties in terms of formation age, lithology, buried depth, porosity, pressure, petroleum phase, trap type, and basin type.

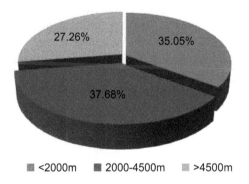

(a) The depth distribution of oil prospective resource

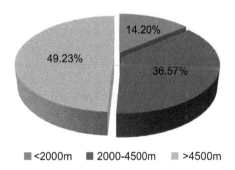

(b) The depth distribution of gas prospective resource

Fig. 6 Depth distribution of petroliferous basin resources in China

Fig. 7 Depths of exploratory
wells drilled in the Junggar and
Tarim basins. **a** Number and
depth of exploratory wells
drilled in the Junggar Basin over
time, **b** Number and depth of
exploratory wells drilled in the
Tarim Basin since 2000

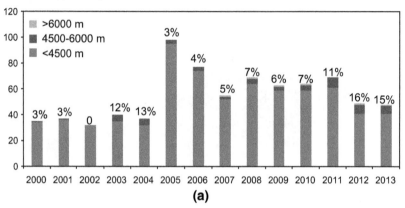

(a)

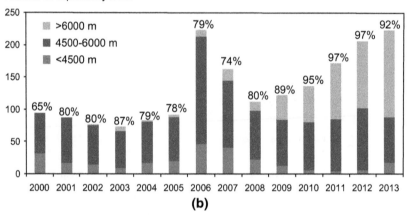

(b)

4.1 While oil–gas reservoirs have been discovered in many types of deep petroliferous basins, most have been discovered in low heat flux deep basins

Oil–gas reservoirs have been discovered in all types of deep petroliferous basins (Fig. 9). Based on the basin classification system of Ingersoll (1995), Bai and Cao (2014) classified 87 deep petroliferous basins into seven groups: continental rift, passive continental marginal, foreland, interior craton, fore-arc, back-arc, and strike-slip basins, of which the passive continental marginal basins (25) and foreland basins (41) are the richest in deep petroleum, followed by the rift basins (12). These three types contribute 47.7, 46.4, and 5.6 % of the world's deep proven and probable (2P) recoverable petroleum reserves. The deep 2P recoverable petroleum reserves in the back-arc basins (2), strike-slip basins (3), and interior craton basin (1) contribute merely 0.3 % of the world's total (Fig. 10).

As a matter of fact, the distribution divergences of deep oil–gas reservoirs in petroliferous basins are essentially decided by the geothermal gradients of the sedimentary basins. Compared with the higher geothermal gradient counterparts, lower geothermal gradient sedimentary

basins contain far more deep petroleum resources since when they reached the same buried depth, they had lower pyrolysis temperatures, and source rocks were richer in residual organic matter and hence had greater ability to generate and preserve hydrocarbon. Figure 11 compares the deep hydrocarbon potentials of source rocks in basins with different geothermal gradients in China as a function of depth, from which we can observed that "hot" basins expelled less hydrocarbon indicating they make up a smaller proportion of deep petroleum resources than their "cold" counterparts. As the geothermal gradient increases, the ratio of deep petroleum resources reduces. According to geothermal gradient records of 405 deep oil–gas reservoirs across the world, 318 % or 78.5 % were discovered in deep basins with geothermal gradients of 1–2 °C/100 m; 79 % or 19.5 % were discovered in deep basins with geothermal gradients of 2–3 °C/100 m; and 8 % or 2 % were discovered in deep basins with geothermal gradients larger than 3 °C/100 m (Fig. 12). In China, the geothermal gradient increases from the west toward the east. The number and reserves of deep oil–gas reservoirs discovered in West China basins is far larger than that discovered in East China basins (Fig. 13).

Fig. 8 Proved deep oil and gas reserves in the Tarim Basin discovered each year since 2000. **a** Depth distribution of proved oil reserves in the Tarim Basin discovered each year since 2000, **b** Depth distribution of proved natural gas reserves in the Tarim Basin

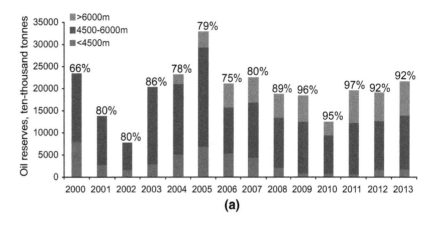

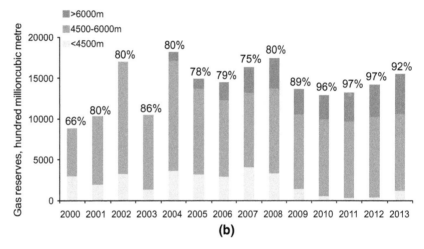

4.2 While many types of petroliferous traps are developed in deep basins, oil–gas-bearing features in deep basin traps are arousing increasing attention

Like their middle–shallow counterparts, deep basins also contain a variety of traps which, according to the conventional trap classification, include tectonic traps, stratigraphic traps, lithological traps, structural-lithological traps, structural-stratigraphic traps, and lithological-stratigraphic traps. Different types of traps differ significantly in terms of their reserves. After a statistical analysis on 837 deep oil–gas reservoirs in the USA, Dyman et al. (1997) discovered that structural traps and combination traps make up as much as 66.9 %. Only in Anadarko and California basins are there more lithological traps than structural ones (Fig. 14). Bai and Cao (2014), after summarizing the trap types and reserves of the world's deep oil–gas reservoirs, discovered that structural traps have 73.7 % of the world's deep recoverable 2P petroleum reserves, with structural-lithological traps and stratigraphic traps contributing 21.9 % and 4.4 %, respectively (Fig. 15).

Recently, with the discovery of tight, continuous oil–gas reservoirs in Canada's Alberta Basin, the USA's Red Desert Basin and Green River Basin, and China's Erdos, Sichuan, and Songliao basins, people have become more interested in these unconventional oil–gas reservoirs. First, these reservoirs have completely different genesis compared with conventional reservoirs, and their discovery has brought on a novel petroleum exploration field. Second, these reservoirs are widely and continuously distributed with vast resource potentials and great scope for petroleum exploration. Third, these reservoirs formed inside deep basin traps between the buoyancy accumulation threshold of a petroliferous basin and the basement of the basin. Their buried depths were quite large, but can be very shallow at present as a result of subsequent tectonic events in the basin. Deep basin traps are a special type of hydrocarbon trap in which the reservoir media have porosities smaller than 12 %, permeabilities smaller than 1×10^{-3} μm^2, and throat radii smaller than 2 μm. Hydrocarbon was not subject to buoyancy in its accumulation, thus making it possible to spread continuously. The more developed the sources rocks were in a deep basin trap, the more continuous the reservoirs were distributed close to the source rocks and the richer the petroleum resources they provide. Figure 16 gives a typical conceptual model and shows the difference about the development

Table 4 Geological features of representative deep hydrocarbon fields (reservoirs) in the world

Country/Area	Oil/gas field/reservoir name	Reservoir lithology	Target formation age	Buried depth, km	Porosity, %	Oil/gas phase	Fluid pressure	Trap type	Basin type
United States	Murphy Creek oilfield	Clastic rock, clastic limestone	J-K	>6.2	5–15	Gas/condensate gas/heavy oil	Constant pressure	Structural	Foreland basin
	Harrisville oilfield	Sandstone	J-E	>7.0	10–15	Gas/oil	Overpressure	Structural	Passive continental marginal basin
Former Soviet Union	Shebelinka oilfield	Clastic rock	C-P	>7.0	14–17	Gas/condensate gas	Overpressure	Structural	Rift basin
	Shebsh oil and gas field	Clastic rock	E	>6.5	12–18	Gas/condensate gas	Overpressure	Structural	Foreland basin
Central & South America	Ceuta oilfield	Limestone	K	>6.0	4–15	Oil	Overpressure	Structural	Foreland basin
	Luna oilfield	Dolomite	J-K	>6.5	8–12	Oil/gas	Overpressure	Structural	Passive continental marginal
West Europe	Malossa oilfield, Villafortuna-Trecate oilfield	Platform limestone	T	6.3	8–15	Gas/condensate gas/oil	Overpressure	Structural	Foreland basin
Mideast	Omani oilfield	Clastic limestone	O	6.0	5–10	Gas	Constant pressure	Structural	Passive continental marginal
Japan	Niigata Basin	Rhyolite, andesite and volcanoclastic rock	N_1	>4.7	4.3	Gas/condensate oil	Overpressure	Structural	Back-arc basin
Africa	Serir oilfield	Clastic rock	E	7.0	8–22	Gas	Overpressure	Structural	Rift basin
China	Puguang gas field	Dolomite	Z-P	7.2	7	Gas	Constant pressure	Structural/lithological	Craton-foreland basin
	Moxi Longwangmiao gas field	Dolomite	Z-∈	>4.8	4.8	Gas	Overpressure	Structural-lithological	Craton-foreland basin
	North Tarim gas field	Carbonate	O	7.1	4.8	Gas	Constant pressure	Structural	Craton-foreland basin
	Mosuowan oilfield	Volcanic rock	C	7.5	3.3	Gas	Overpressure	Structural	Foreland-rift basin
	Xinglongtai buried hill hydrocarbon reservoir	Metamorphic rock	γ	>4.5	3–8	Overpressure		Structural	Rift basin

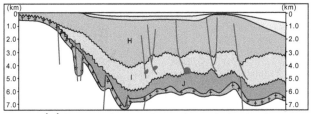

(a) The passive continental margin basin of the deep reservoirs in the Kamps Basin in Brazil

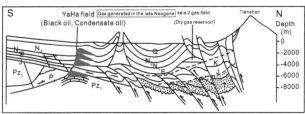

(b) The foreland basin of the deep reservoirs in the Kuqa Depression in China

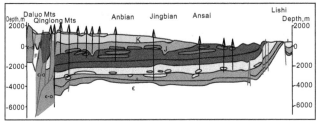

(c) The craton basin of the deep reservoirs in the Erdos Basin in China

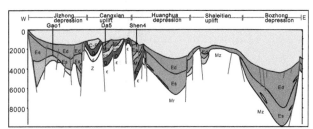

(d) The rift basin of the deep reservoirs in the Bohai Bay in China

Fig. 9 Deep oil–gas reservoirs developed in different types of basins (Li and Lü 2002; Li 2009)

Fig. 10 Deep hydrocarbon reservoir distribution in different types of basins in the world (Bai and Cao 2014)

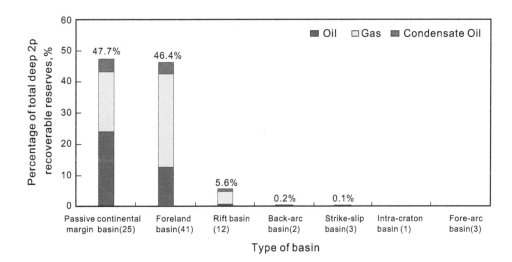

and distribution of a deep basin trap-controlled hydrocarbon reservoir in a petroliferous basin and conventional traps.

4.3 The composition of deep petroleum is widely different with more natural gas than liquid oil and the natural gas ratio increases with the buried depth

The composition of deep petroleum in petroliferous basins varies and includes gaseous hydrocarbon, condensate gas, condensate oil, liquid hydrocarbon, and oil–gas coexistence. Phase statistics of 1,477 deep oil–gas reservoirs in the world demonstrate that oil–gas miscible phases

contribute 54 % and gas phases contribute 40 %. The oil phase makes up a very small proportion of 6 % (Fig. 17). Generally, as the formation depth increases, natural gas makes up a larger proportion in deep petroleum and overtakes liquid hydrocarbon as the prevailing type of petroleum resources. Figure 18 shows how the oil and natural gas reserves discovered from different formations of East China's Bohai Bay Basin and West China's Tarim and Junggar basins vary as a function of depth. By and large, the ratio of older deep basin gas reservoirs increases due to pyrolysis of crude oil as a result of extended high temperature exposure of the oil reservoir in deep basins, or thermal cracking in the source rocks when they reached high maturity (Dyman et al. 2001). The Hugoton gas field

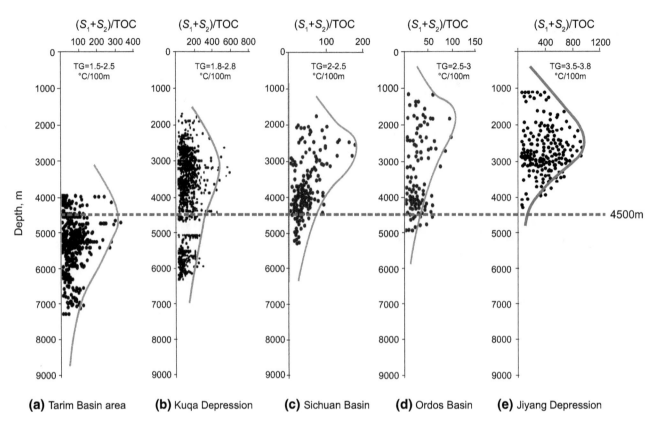

(a) Tarim Basin area   **(b)** Kuqa Depression   **(c)** Sichuan Basin   **(d)** Ordos Basin   **(e)** Jiyang Depression

Fig. 11 Comparison of hydrocarbon potentials of deep source rocks in basins in China with different geothermal gradients

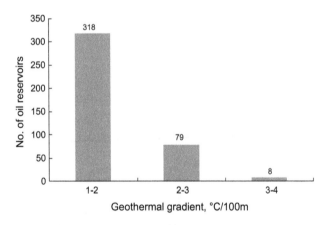

Fig. 12 Number of deep oil–gas reservoirs discovered in basins in the world with different geothermal gradients

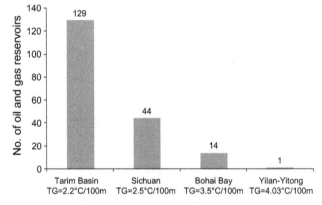

Fig. 13 Number of deep oil–gas reservoirs discovered in basins in China with different geothermal gradients

and Mills Ranch in the USA's Anadarko Basin, for example, are pure gas fields. Many deep gas reservoirs in China's petroliferous basins also originated from earlier oil or oil–gas reservoirs that were cracked into gas under high temperatures, as exemplified by the large Puguang carbonate gas field (Du et al. 2009), the large Kela-2 gas field (Jia et al. 2002), and the Hetian gas field (Wang et al. 2000). Huge liquid or condensate oil reservoirs have also been discovered in a number of deep basins such as the

USA's Rocky Mountain Basin, where gas wells make up only 34 % of the deep exploratory wells while most are oil wells and no pyrolysis has taken place deeper than 6,000 m. This is often because the low subsurface temperature or high pressure of the formation had prevented the crude oil from coming to its threshold pyrolysis temperature (Svetlakova 1987). A lot of factors can be related to deep basin petroleum phases. These include (1) the type of the original organic matter; (2) temperature and pressure; and (3) subsequent adjustment or reformation.

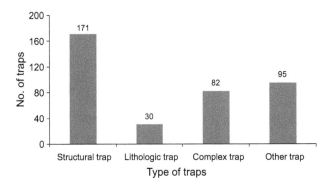

Fig. 14 Trap types of deep oil–gas reservoirs in the USA (Dyman et al. 1997)

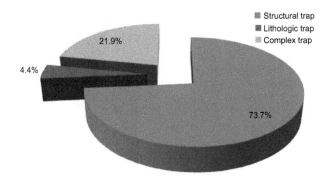

Fig. 15 Deep petroleum reserves in petroliferous basins in the world with different trap types

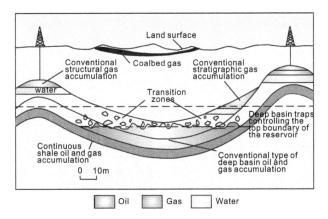

Fig. 16 Difference and comparison of conventional traps versus deep basin traps in petroliferous basins and their petroleum-control features

4.4 Residual organic matter in deep source rocks reduces but the hydrocarbon expulsion rate and efficiency continuously increase with the buried depth

As far as deep source rocks are concerned, they have much smaller measured residual hydrocarbon amount and hydrocarbon potential than their middle–shallow counterparts, and both reduce with the increase of the buried depth.

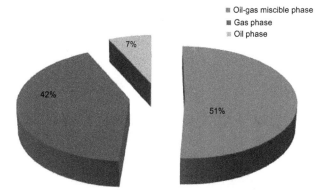

Fig. 17 Phase distribution of deep petroleum in petroliferous basins in China

They appear to have the following characteristics: (1) the residual hydrocarbon amount per unit parent material in the source rocks (total organic carbon, TOC) is represented by S_1/TOC or "A"/TOC (Dickey 1975; Hao et al. 1996; Durand 1988). The residual hydrocarbon amount in deep source rocks first increases then reduces with the increase of the depth or R_o and was already very small in deep basins (Fig. 19); (2) the hydrocarbon potential is represented by the H/C atomic ratio, O/C atomic ratio, and hydrogen index (Tissot et al. 1974; Tissot and Welte 1978; Jones and Edison 1978; Baskin 1997; Zhang et al. 1999). The H/C atomic ratio and O/C atomic ratio (Fig. 20) and hydrocarbon index HI (Fig. 21) reduce with the increase of the buried depth; and (3) the hydrocarbon potential index of source rocks is represented by $(S_1 + S_2)$/TOC (Zhou and Pang 2002; Pang et al. 2004). When the TOC is more than 0.1 %, the hydrocarbon potential index of deep source rocks shows a "big belly" profile of increasing followed by reducing with the increase of the depth or R_o (Fig. 22). In a word, as the buried depth increases and the hydrocarbon potential of source rocks gradually reduces, the accumulated oil–gas volume and hydrocarbon expulsion efficiency of source rocks appear to increase gradually with the increase of the buried depth (Fig. 23), reflecting the increase in the contribution made by source rocks to oil–gas reservoirs. This indicates that the quality and effectiveness of source rocks should be judged by investigating how much hydrocarbon was generated and expelled by source rocks rather than by relying on how much hydrocarbon or how much hydrocarbon potential is left of the source rocks.

4.5 While there are many types of rocks in deep hydrocarbon reservoirs, most of them are clastic rocks and carbonate

Deep target formations in petroliferous basins contain a variety of rocks, though clastic rocks and carbonate are the

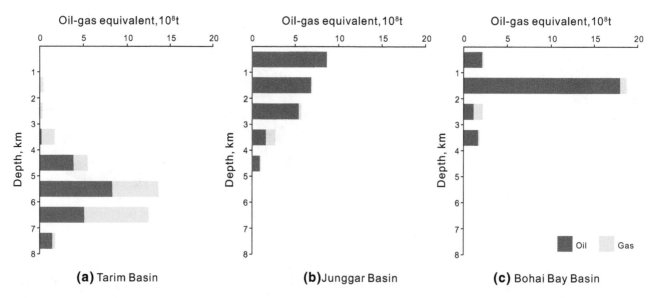

Fig. 18 Petroleum reserves of petroliferous basins in China as a function of buried depth

predominant types of deep hydrocarbon reservoirs discovered so far, with certain amounts of volcanic and metamorphic rocks too. Among clastic reservoirs, fractured sandstone reservoirs are the most favorable. Carbonate reservoirs include limestone and dolomite, with carbonate reservoirs extensively found in brittle fractures and karst caves taking the largest proportion. As of 2010, of the 1,477 deep oil–gas reservoirs discovered across the world, 1,035 % or 70.1 % were located in clastic reservoirs, 429 % or 29.0 % were in carbonate reservoirs, and 13 % or

0.88 % were in magmatic and metamorphic reservoirs (Fig. 24).

According to the latest nationwide petroleum resource evaluation made by the Ministry of Land and Resources (2005), deep hydrocarbon reservoirs discovered in China are predominantly carbonate and sandstone reservoirs (Fig. 25). Widespread marine carbonate reservoirs occur in Central West China basins and are responsible for a series of large marine carbonate hydrocarbon fields represented by Central and North Tarim hydrocarbon fields

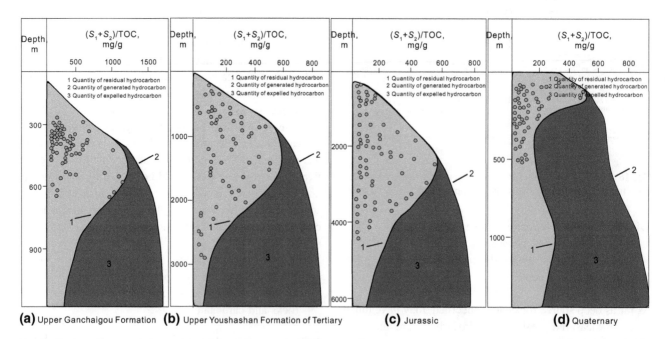

Fig. 19 Hydrocarbon generation, residual hydrocarbon, and hydrocarbon expulsion of source rocks in the Qaidam Basin as a function of buried depth

Fig. 20 Kerogen *H/C* atomic ratio and *O/C* atomic ratio of source rocks in the Tarim Basin platform area. **a** Change of *H/C* atomic ratio, **b** change of *O/C* atomic ratio

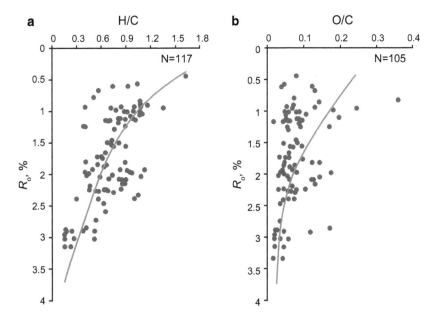

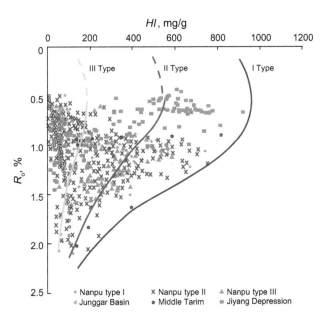

Fig. 21 Hydrogen indices of different types of source rocks as a function of thermal evolution level of the parent material

in the Tarim Basin, East and Central Sichuan reef-flat carbonate hydrocarbon fields. In the northwestern flank of the Junggar Basin, the Kuqa Depression of the Tarim Basin and the Xujiahe Formation of the Sichuan Basin, sandstone oil–gas reservoirs in clastic rocks are the predominant type. Besides, a lot of deep volcanic oil–gas reservoirs are also contained in the deep part of the Junggar and Songliao basins. Deep oil–gas reservoirs in bedrock metamorphic rocks have been discovered in places like the Liaohe Depression of the Bohai Bay Basin.

4.6 The age of deep hydrocarbon reservoirs is widely different, but those recently discovered are predominantly Paleogene and Upper Paleozoic

Oil–gas reservoirs in deep petroliferous basins are similar to their middle or shallow counterparts in terms of formation distribution. The ages of the reservoirs cover a wide range. Most of the deep oil–gas reservoirs discovered so far, however, are in five formation systems: Neogene, Paleogene, Cretaceous, Jurassic, and Upper Paleozoic, the deep 2P recoverable petroleum reserves of which account for 12.8 %, 22.3 %, 18.3 %, 12.8 %, and 22.2 % of the world's totals, respectively (Fig. 26). This suggests that the deep petroleum is mainly in Neogene and Upper Paleozoic formations. Also, as the reservoir ages become older, the ratio of deep natural gas in the total deep petroleum reserve tends to increase accordingly.

After summarizing the reservoir ages of oil–gas reservoirs in China's deep petroliferous basins (Table 5), we discovered that the reservoir ages of deep oil–gas reservoirs in Central and West China basins are predominantly Paleozoic, meaning the reservoirs are quite old; those in East China basins are predominantly Paleogene or Cretaceous, and reservoir ages of oil–gas reservoirs in the bedrock are predominantly Precambrian.

4.7 The porosity and permeability of deep hydrocarbon reservoirs are widely different, but they vary with the lithology and buried depth

The porosities and permeabilities of target formations for deep oil–gas reservoirs in petroliferous basins vary widely, ranging from high-porosity, high-permeability (with

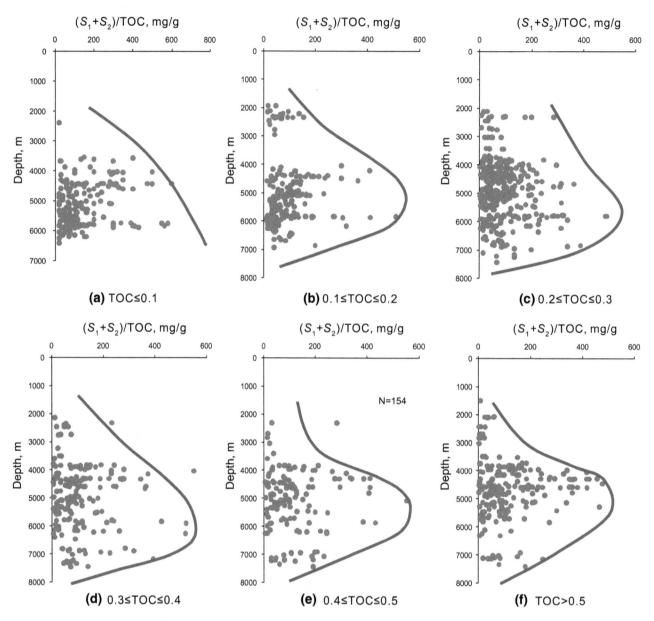

Fig. 22 Hydrocarbon potential of Cambrian–Ordovician carbonate source rocks in the Tarim Basin

porosity of 38 % and permeability of 7,800 mD) high-quality reservoirs to low-porosity, low-permeability (with porosity lower than 5 % and permeability less than 0.1 mD) tight reservoirs. High-porosity, low-permeability or low-porosity, high-permeability petroliferous reservoirs have also been identified. Deep drilling records across the world demonstrate that as the reservoir depth increases, the compaction effect and consequently the diagenesis intensify, so the porosity of deep rocks tends in general to decrease. The porosities of the world's deep petroliferous basins are mostly in the 10 %–12 % range (Wang et al. 2012). A summary of the porosities and permeabilities of 20,717 oil–gas reservoirs across the world as a function of depth revealed that the porosities and permeabilities of

these reservoirs tend to decrease with the increase of depth overall (Fig. 27), though this rule varies from one lithology to another in different areas. Clastic reservoirs, for example, show obvious porosity and permeability decreases in some reservoirs, but the porosity of other reservoirs varies little in deep or ultra-deep formations (Fig. 28), while carbonate reservoirs do not show obvious decreases in their reservoir properties as the depth increases (Fig. 29) due to their high rock brittleness, high compaction stability, and good solubility. The porosities of volcanic reservoirs do not vary much with depth (Fig. 30). The various petroliferous basins in China do not present the same characteristics. Figure 31 compares the reservoir porosity variations of different petroliferous basins in China as a function of

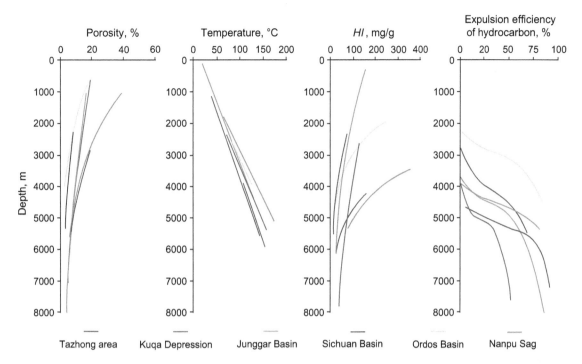

Fig. 23 Hydrocarbon expulsion efficiency of representative petroliferous basins in China as a function of depth

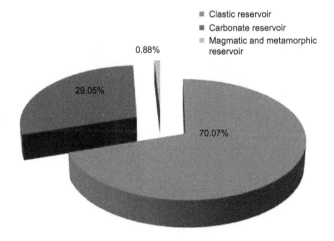

Fig. 24 Distribution of reservoir lithologies of deep oil–gas reservoirs in the world

depth. It shows that the porosity of sandstone reservoirs decrease with an increase in buried depth, while the porosity is in general retained in carbonate and volcanic rocks until below 6,500 m.

4.8 The temperatures of deep oil–gas reservoirs differ widely, but they typically vary with the buried depth and geothermal gradient

The temperature range of oil reservoirs in deep petroliferous basins has exceeded that of liquid hydrocarbon (windows) supposed by traditional kerogen theory

(60–120 °C, R_o = 0.6 %–1.35 %): the highest oil reservoir temperature discovered in the world so far is more than 200 °C. Compared with their middle–shallow counterparts, deep oil–gas reservoirs have even higher temperatures which vary even more widely. Statistics of the temperatures and pressures of 428 deeper-than-4,500 m oil–gas reservoirs in the world (Fig. 32) show that the temperatures of deep oil–gas reservoirs can be 200 °C maximum and those of a couple of gas reservoirs are more than 370 °C, compared with the lowest hydrocarbon reservoir temperature of 47 °C. Even at the same depth, the temperatures of deep oil–gas reservoirs vary from one type to another, such as the petroliferous basins in China, those in the east are mostly extensional basins that are typically hot basins with an average geothermal gradient of approximately 4 °C/100 m and oil reservoirs deeper than 4,500 m being hotter than 180 °C; the extrusion basins in the west and the craton basins in the center, to the contrary, have lower geothermal gradients and are typical cold basins (Liu et al. 2012) with an average geothermal gradient of approximately 2.5 °C/100 m. At the same depth of 4,500 m, the temperature in the center and west is less than 120 °C. The temperature difference is approximately 60 °C. Figure 33 compares the geothermal gradients of some of the representative basins in China as a function of time, from which we can observe that, at the same depth, the formation temperature tends to increase from west toward east, reflecting the eastward increase of the geothermal gradient or heat flux.

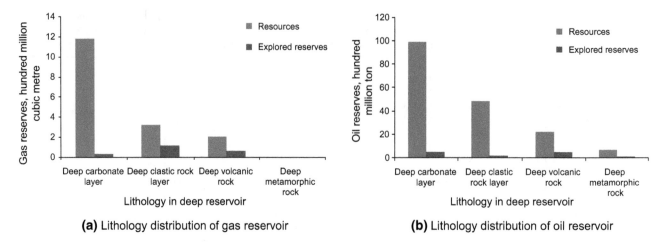

(a) Lithology distribution of gas reservoir   **(b)** Lithology distribution of oil reservoir

Fig. 25 Distribution of reservoir lithologies of deep oil–gas reservoirs in China

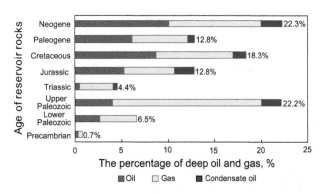

Fig. 26 The age distribution of deep hydrocarbon reservoirs in the world (Bai and Cao 2014)

4.9 The pressure of deep oil–gas reservoirs typically varies with the buried depth, genesis, and evolution period

Oil–gas reservoirs formed under high porosities and high permeabilities in middle–shallow petroliferous basins are typically buoyancy controlled, thereby generally displaying high pressures. Oil–gas reservoirs discovered in deep basins, however, have complex genesis, thereby displaying diverse pressures (Fig. 34). Statistics of the pressure records of 16,552 oil–gas reservoirs in the world revealed significant pressure differences among deep oil–gas reservoirs. The highest of the majority is 130 MPa, with a few outliers as high as 172 MPa, while the lowest is merely 8.4 MPa (Fig. 35).

Abnormal high pressures are generally contained in deep tight structural gas reservoirs and tight lithological gas reservoirs, like the Sichuan Xiaoquan gas reservoir which has abnormally high pressures with a pressure coefficient of 1.6–2.0 (Guan and Niu 1995), and the Fuyang tight oil reservoir in Songliao which also displays abnormally high pressures with a pressure coefficient larger

than 1.6. Deep oil–gas reservoirs with normal pressures also exist, like the 13 Lunnan buried hill oil reservoirs in the Tarim Basin that have the pressure coefficients between 1.03 and 1.14; and the Lunnan-17 well, Lunnan-30 well, Lunnan-44 well, and Jiefang-123 well oil reservoirs that have coefficients of 1.137, 1.130, 1.143, and 1.148, defining them as normal pressure oil reservoirs (Gu et al. 2001). Low-pressure oil–gas reservoirs are typically found in tight syncline sandstone gas reservoirs, like the tight sandstone hydrocarbon reservoirs discovered in Canada's Alberta Basin (Masters 1979), the tight sandstone hydrocarbon reservoirs discovered in the USA's Red Desert (Spencer 1989), and the tight sandstone hydrocarbon reservoirs in the USA's Green River Basin. The same occurs in the upper Paleozoic tight sandstone reservoirs in China's Erdos Basin and the tight sandstone gas reservoirs discovered in the Jurassic of the Tuha Basin (Fig. 36). In the main, deep oil–gas reservoirs have complex pressures; abnormally high, normal, and abnormally low-pressure hydrocarbon reservoirs can all be found in deep oil–gas reservoirs. So far, the complex hydrocarbon accumulation areas formed by coexistence of the three pressure categories are increasingly found in basins. Research (Pang et al. 2014a) shows that the pressure of deep oil–gas reservoirs is decided by their genesis mechanism and genesis process. Normal oil–gas reservoirs formed in free fluid dynamic fields in high-porosity, high-permeability media generally have high pressure, while unconventional oil–gas reservoirs formed in limited fluid dynamic fields in low-porosity, low-permeability media generally end up with negative pressure, though they appeared to show high pressure during hydrocarbon accumulation into reservoirs. Normal oil–gas reservoirs formed in the early years in petroliferous basins, as the depth increased, superimposed or compounded with the unconventional oil–gas reservoirs originated from the deep part before, eventually giving rise to

Table 5 Chronological distribution of deep hydrocarbon reservoir strata in China's petroliferous basins

Basin	Oil/gas field name	Target formation depth, m	Formation lithology	Proved oil reserve, 10^4 t	Proved natural gas reserve, 10^8 m^3	Target formation age
Tarim Basin	Central Tarim gas field	4,500–6,200	Carbonate	38,600	1,020	Ordovician
	Halahatang oilfield	5,900–7,100	Carbonate	20,812	–	Ordovician
	Donghetang oilfield	>6,000	Clastic rock	3,323	–	Carboniferous
	Kuche deep gas region	5,000–8,000	Clastic rock	–	6,448	Cretaceous
Junggar Basin	Xiazijie oilfield	4,800	Volcanic rock	1,548	34	Permian
Sichuan Basin	Longgang gas field	2,800–7,100	Carbonate	–	730	Permian
	Moxi Longwangmiao gas field	4,500–5,500	Dolomite	–	4,404	Sinian, Cambrian
Bohai Bay Basin	Qishan gas field	4,500–5,500	Clastic rock	–	95	Paleogene
	Xinglongtai gas field	4,500–5,500	Metamorphic rock	0.75	–	Archaeozoic
Songliao Basin	Changling gas field	4,500–5,000	Volcanic rock	–	640	Cretaceous

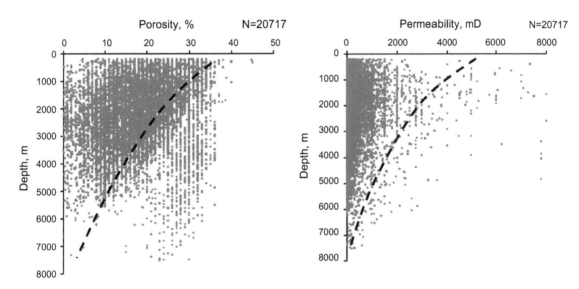

Fig. 27 Reservoir properties as a function of depth of petroliferous basins in the world

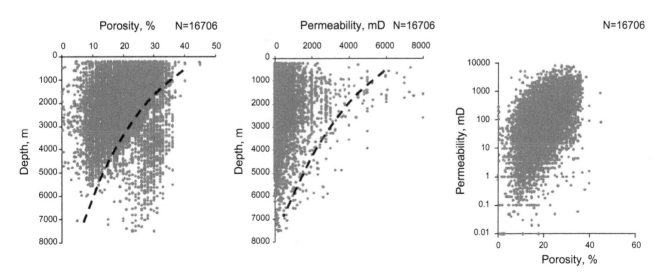

Fig. 28 Reservoir properties of clastic rocks as a function of buried depth of petroliferous basins in the world

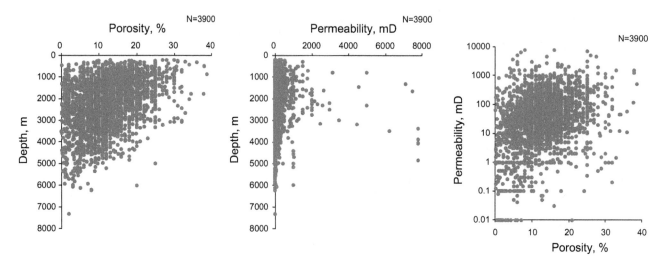

Fig. 29 Reservoir properties of carbonate as a function of buried depth of petroliferous basins in the world

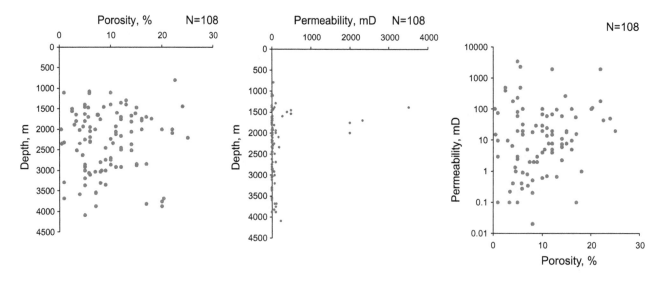

Fig. 30 Reservoir properties of volcanic and metamorphic rocks as a function of buried depth of petroliferous basins in the world

superimposing or compounded, continuous oil–gas reservoirs, among which there can be high-pressure petroliferous reservoirs coexisting with low-pressure petroliferous reservoirs.

4.10 Deep oil–gas reservoirs may exist
with or without a cap, and those without a cap are typically of unconventional genesis

Forming a hydrocarbon reservoir without relying on a cap is the unique geological feature of deep hydrocarbon entrapment. Conventional petroleum geological theory assumes that a cap is an indispensable geological element for the formation and preservation of any hydrocarbon reservoir; without a cap, it would be impossible for hydrocarbon to gather into a reservoir in any high-porosity, high-permeability media, since buoyancy would cause the

hydrocarbon to percolate upward to the basin surface. For deep hydrocarbon entrapment, however, as far as the reservoir media were commonly tight, a cap is not indispensable (Fig. 37). The realities underlying this phenomenon are the particular fluid dynamic fields and material equilibrium conditions of deep oil–gas reservoirs: (1) deep reservoirs, if commonly tight, had poor porosities and permeabilities and limited pore throat radii, so hydrocarbon had to overcome great capillary pressures and the hydrostatic pressures of the overlying water columns when it tried to charge into the reservoir; (2) for deep oil–gas reservoirs, buoyancy was no longer the primary drive for hydrocarbon migration; they expelled pore water and expanded their own area by relying on the hydrocarbon volume increase and finally became continuous oil–gas reservoirs; and (3) the sources of tight continuous oil–gas reservoirs formed from deep basins were close to the

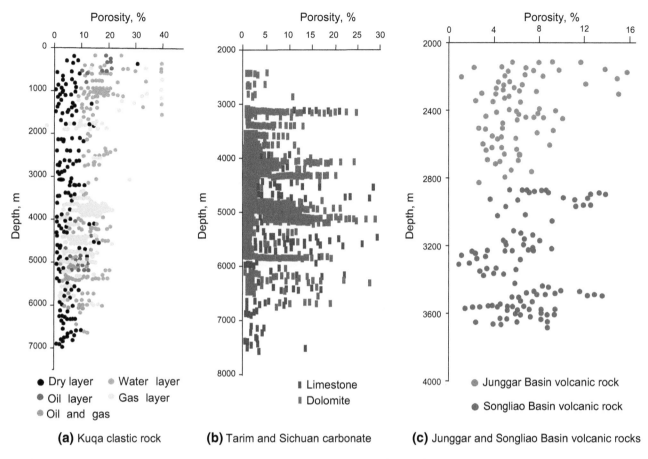

(a) Kuqa clastic rock **(b)** Tarim and Sichuan carbonate **(c)** Junggar and Songliao Basin volcanic rocks

Fig. 31 Porosities of target strata with different lithologies as a function of depth of petroliferous basins in China

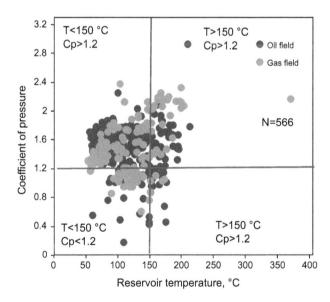

Fig. 32 Scatter diagram of temperatures and pressure coefficients of deep hydrocarbon fields of petroliferous basins in the world

reservoir, keeping the hydrocarbon under constant diffusion-accumulation equilibrium, thus enabling them to be preserved for a long time under structural stability.

Deep oil–gas reservoirs can be developed without a cap, but oil–gas reservoirs without a cap are generally found in areas with tight reservoirs and stable tectonism, like deep depressions and slope areas of the basin.

5 Major progress in deep hydrocarbon reservoir research

Deep hydrocarbon reservoirs are attracting the attention of more and more companies and scholars at home and abroad and a series of research achievements have been made in the past ten years. The major research achievements and progresses are described as follows:

5.1 Multiple deep hydrocarbon sources in petroliferous basins and the formation mechanisms

Traditionally, hydrocarbon researchers believed that hydrocarbon was formed from thermal degradation of organic matter under appropriate temperatures (60–135 °C) and pressure (burial depth 1,500–4,500 m) in sedimentary basins (Hunt 1979; Tissot and Welte 1978). But the formation mechanism of deep hydrocarbons in a high-

Fig. 33 Formation
temperatures of petroliferous
basins in different parts of
China as a function of buried
depth

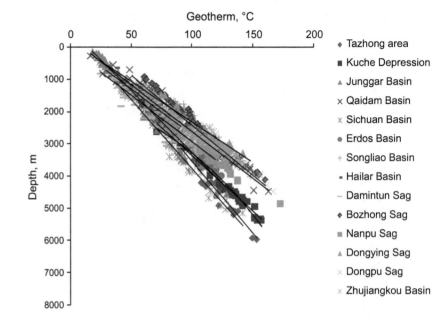

Fig. 34 Comparison of
pressures and genesis of
conventional gas reservoirs
versus deep basin oil–gas
reservoirs (Pang et al. 2006)

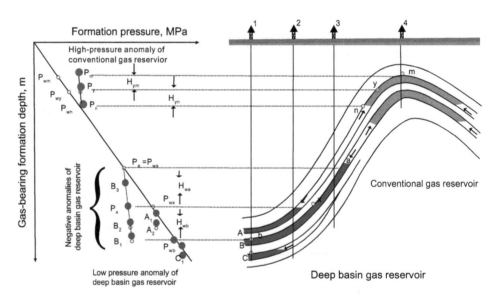

temperature and high-pressure environment is far more complicated than it was thought to be. So far, four hydrocarbon generation modes have been proposed by scholars.

5.1.1 Successive generation of gas from deep organic matter

The theory of successive generation of gas from organic matter solved the problem of generation and expulsion of natural gas from highly and over-mature source rocks deep down in hydrocarbon basins (Zhao et al. 2005). The "successive generation of gas" mechanism means the conversion of gas-generating matrix and succession of gas-generating time and contribution in the gas-generating process (Fig. 38), including two aspects: 1) the generation of gas from thermal degradation of kerogen and cracking of liquid hydrocarbon and soluble organic matter in coal, which occurs successively in terms of gas generation and contribution; and 2) part of the liquid hydrocarbon generated in the thermal degradation of kerogen is expelled from source rocks to form oil reservoirs, but most of the liquid hydrocarbon remains dispersed in the source rocks, where it is thermally cracked in the highly to over-mature areas so that the source rocks still have great gas-generating potential.

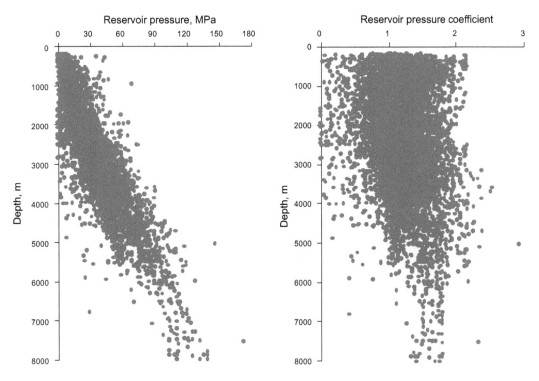

Fig. 35 Hydrocarbon reservoir pressures of petroliferous basins in the world as a function of buried depth

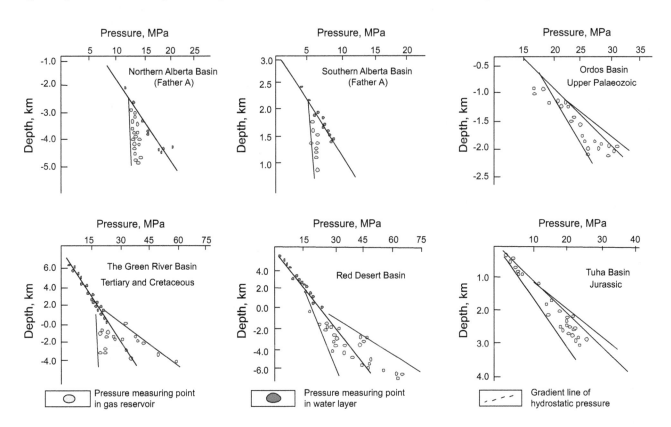

Fig. 36 Reservoir pressures of deep tight sandstone oil–gas reservoirs in representative petroliferous basins in China and elsewhere as a function of depth

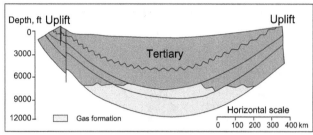

(a) American Red Desert Basin

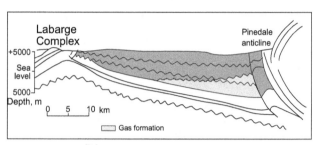

(b) American Green River Basin

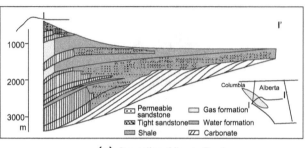

(c) Canadian Alberta Basin

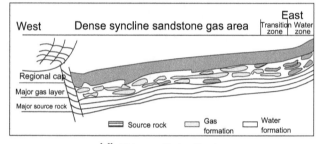

(d) Chinese Ordos Basin

Fig. 37 Reservoir pressures of a few known tight sandstone oil–gas reservoirs in China and worldwide

5.1.2 Hydrocarbon generation from hydrogenation of deep organic matter

The generation of hydrocarbons from thermal evolution of organic matter not only needs heat but also hydrogen. The injection of hydrogen-rich fluids into a sedimentary basin will definitely have a great impact on the generation of hydrocarbon. Jin et al. conducted a simulation test with olivine, zeolite, and source rocks, finding that the methane yield increased by 2–3 times after the source rocks interacted with the zeolite and olivine. According to analysis of hydrocarbon generated in low-maturity source rocks in the Dongying Sag and the Central Tarim region by Jin et al. (2002) by means of hydrogenation thermal simulation, the effect of hydrogenation on hydrocarbon generation for Type II$_2$ kerogen is distinct after the peak hydrocarbon-generating period, while the effect on humic-type kerogen is distinct in all periods. The effect of hydrogenation on source rocks with poor hydrogen kerogen is more distinct (Fig. 39).

5.1.3 Hydrocarbon generation and expulsion from deep asphalt cracking

Paleo-reservoirs, if damaged, may produce asphalt, and asphalt may be cracked under high temperature to generate lighter hydrocarbons.

Gong et al. (2004) collected asphaltic sand formed by bio-degradation in the Silurian from the Tarim Basin, and used a high-pressure reaction vessel to investigate the effects of heat on the compositions, isotopes, and physical properties of the sand (Fig. 40). The results indicated that heat had an effect on the compositions and structure of the Silurian asphaltic sand during later-stage burial in the Tarim Basin. The asphaltic sand produces gas at high temperature, and the cracked gas has lighter carbon isotopes: there are less compositions of C$_{6+}$ and above, and it is dominated by light oil; the gas yield is low at low temperature but increases substantially after 400 °C and reaches its peak at 550 °C.

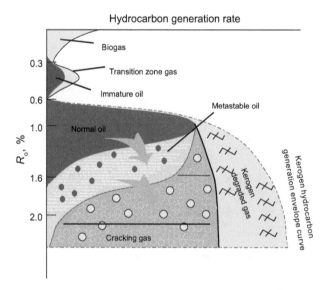

Fig. 38 Successive generation of gas from organic matter (Zhao et al. 2005)

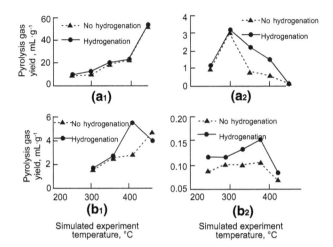

Fig. 39 Simulation test results comparison of different types of source rocks under hydrogenated and non-hydrogenated conditions (Jin et al. 2002). **a** Source rock from Sha-3 member at well Fan15 in the Dongying Sag (Type II_2 kerogen, water added) and **b** Carboniferous source rocks from Central Tarim well He6 (Type III kerogen, water not added)

Huang et al. (2012) obtained the yield curves of the asphalt degradation products (oil and gas) under different temperatures and further evolution process by using artificial and geological samples (Fig. 41). The hydrocarbon-generating process of asphaltic sandstone was divided into three stages: (1) evaporative fractionation stage ($R_o < 0.9$ %), during which the light components of crude oil escaped from damaged paleo-reservoirs to form heavy oil (or asphalt), which could not provide sufficient supply to new hydrocarbon reservoirs; (2) Cracking stage 1 ($R_o = 0.9$ %–1.8 %), during which oil precipitation increased abruptly and reached a peak, and the yield of hydrocarbon gas increased. It was an important hydrocarbon supply stage in the burial process of asphaltic sandstone (heavy oil); (3) Cracking stage 2 ($R_o > 1.8$ %), during which large amounts of large molecule hydrocarbons were decomposed and hydrocarbon gas was given off. With the rise of temperature, the amount of liquid hydrocarbon in the expelled hydrocarbon components decreased substantially and large amounts of hydrocarbon gas was generated, followed by a substantial decrease of the residual hydrocarbon, and the generation of large amounts of methane. It was an important gas supply stage.

5.1.4 Hydrocarbon generation and expulsion from deep source rocks with low TOC

Pang et al. (2014b) believed that deep-buried poor source rocks with low total organic carbon (TOC) concentration can be regarded as effective source rocks. They generate mass hydrocarbons in the long evolution of geological history, which makes a great contribution to hydrocarbon

accumulation. Pang et al. (2014b) obtained the evolutionary charts of TOC of carbonate source rocks based on the hydrocarbon expulsion threshold theory through simulation by using the material balance method (Fig. 42). As shown in the charts, a) the original TOC of source rocks decreases with an increase of R_o. The TOC has a minor increase first and then decreases substantially until a balance is reached. The sharp decrease occurs when $R_o = 0.5$ %–2.0 %, and the TOC of Type I, II, and III source rocks decreases by 62 %, 48 %, and 25 %, respectively. This is consistent with the period of generation and expulsion of large amounts of hydrocarbon during the thermal evolution of organic matter. b) The abrupt TOC decrease of different types of source rocks occurs at different times. For source rocks of Type I, II, and III organic matter, the abrupt TOC decrease occurs when $R_o = 0.5$ %, 0.7 %, and 0.9 %, respectively. This may be related to the increasing hydrocarbon expulsion threshold depth of the sources rocks. Many scholars studied the hydrocarbon yield of poor hydrocarbon carbonate source rocks through thermal experiments (Qin et al. 2005; Hao et al. 1993; Cheng et al. 1996; Fan et al. 1997; Xie et al. 2002; Hu et al. 2005; Liu et al. 2010b). According to the test results, the source rocks yielded large amounts of hydrocarbon. The maximum oil and gas yield was 40.4–482.6 kg/t TOC and 115–4,226 m^3/t TOC, respectively. This indicates the source rocks in the deep basins with low TOC concentration are the result of the mass generation and expulsion of hydrocarbons and can be taken as effective source rocks under certain conditions. Pang et al. (2014b) pointed out, for deep source rocks that have yielded large amounts of hydrocarbon in basins, if the regional hydrocarbon prospects are identified and evaluated by using residual organic abundance indices (TOC), it will lead to errors; for highly to over-matured source rocks buried deep in basins, more source rocks are poor in hydrocarbon and the errors will be more obvious. Therefore, residual hydrocarbon indices cannot be used directly to identify and evaluate deep-buried effective source rocks. If TOC is used as a basic index to identify and evaluate source rocks, comparisons should be made in the same geological conditions (Pang et al. 2014b). Figure 43 shows the charts of TOC recovery coefficient under different geological conditions in accordance with the material balance theory, we can conclude that the TOC recovery coefficient of argillaceous source rocks ($R_o > 2.0$ %) varies with different parent material types; the TOC recovery coefficient of parent material I, II, III type reaches 2.6, 1.75, 1.25 respectively; while for carbonate source rocks the TOC recovery coefficient can reach 2.65, 7.80, 1.30 respectively. The chart allows comparison of TOC of source rocks under different geological conditions on an equal footing. The criteria for evaluation of effective source rocks based on hydrocarbon expulsion thresholds of

Fig. 40 Curves of *n*-alkane carbon isotopes under different experimental temperatures (Gong et al. 2004)

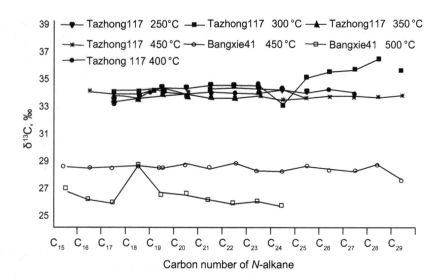

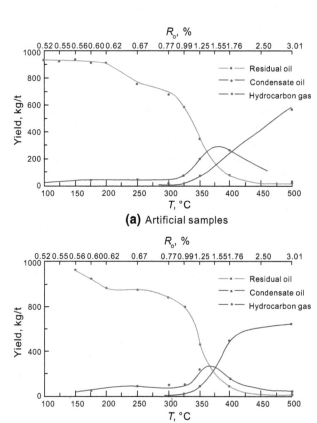

(a) Artificial samples

(b) Geological samples

Fig. 41 Hydrocarbon yield curves of asphaltic sand under different temperatures (Huang et al. 2012)

deep source rocks in basins are listed in Table 6. With the criteria, Pang et al. (2014b) identified and evaluated an effective Cambrian–Ordovician source rock in the Central Tarim region. According to the results, the thickness of the effective source rocks is 47 to 129 meters more than that evaluated using the TOC = 0.5 % criterion, and the scope

of distribution is increased by 7 km², total oil yield increased by 182,500 million tons, and total reserves increased by 9,300 million tons, accounting for 38.1 % of the total resource. This reflects the importance of research on effective source rocks with low TOC.

5.2 Deep-buried high-porosity and high-permeability reservoirs, the formation and distribution of which are related to structural changes and underground fluid activities

The porosity and permeability of deep hydrocarbon reservoirs in basins are usually low. The deeper the reservoirs are, the smaller the porosity and permeability will be. In actual geological conditions, the porosity and permeability of deep effective reservoirs vary significantly. Research results indicate that the formation of deep-buried high-porosity and high-permeability reservoirs is closely related to structural changes and evolution as well as fluid activities. In general, more faults and disconformities are developed in regions with stronger structural activities. Faults not only serve as pathways for hydrocarbon migration, but also improve the reservoir quality. Fault developed regions generally have favorable porosity and permeability. As shown in Fig. 44, most hydrocarbon reservoirs so far discovered in the Tarim region are distributed in fault regions. This is particularly true for carbonate reservoirs, which are developed near faults (Fig. 44a). Disconformities also serve as pathways for hydrocarbon migration, and improve the geological conditions of reservoirs by weathering and leaching. As shown in Fig. 45, reservoirs near disconformities have high porosity and permeability. High-quality carbonate reservoirs are usually distributed within 200–300 m from an unconformity. Sedimentary fluids have decisive effects on the granular

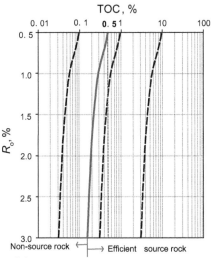

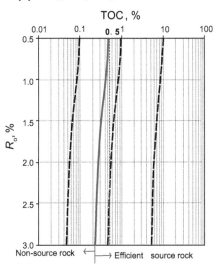

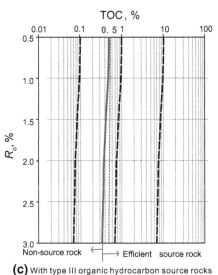

Fig. 42 Evolutionary charts of TOC of different carbonate source rocks of deep petroliferous basins (Pang et al. 2014b)

structure and composition of reservoirs, and influence the strain response and the formation of fractures and cavities in the rock-forming process. Surface fluids may change the physical properties of surface rocks, enabling the rocks to accept external fluids and providing conditions for solution-pore type reservoirs reconstructed by external fluids after being buried. Underground hot fluids are favorable for secondary pore formation in deep reservoirs and the accumulation of gas and oil may take place in the dissolution pores and cavities, fractures, and cracks.

The formation of deep effective reservoirs is usually affected by a range of factors. For example, organic matter generates organic acids in the process of thermal evolution. The organic acids dissolve some minerals in the surrounding rocks to form secondary pores and improve the porosity of deep reservoirs. In the fast subsidence–sedimentation of sedimentary basins, overpressure systems may develop in shallow and high-porosity strata to reduce the effective stress, and thus reducing the compaction and restraining the pressure dissolution. At the same time, the fluids in the overpressure systems have poor fluidity, which retards the formation and cementation of rocks. That is why deep-buried overpressure reservoirs usually have high porosities. In summary, the high porosities of deep-buried overpressure reservoirs in basins are the result of the following combined effects: reduced compaction under low effective stress, retarded cementation due to low-fluidity fluids, and dissolution of minerals by organic acids. Therefore, multi-factor integrated research is required for the evaluation of deep-buried effective reservoirs.

Deep-buried high-quality clastic reservoirs developed over different periods in basins in China are related to such factors as abnormally high pressure, early hydrocarbon charging, thermal convection, gypsum-salt effects, and sandstone and mudstone interbeds (Li and Li 1994; Gu et al. 1998; Zhong and Zhu 2003). Dissolution, early long-term shallow burial and late short-term deep burial, abnormally high pressure, and early hydrocarbon charging are key factors influencing the formation of deep-buried high-quality clastic reservoirs in China (Shi and Wang 1995; Yang et al. 1998; Li et al. 2001; Zhong et al. 2008). Many researchers realized that the formation and evolution of high-porosity and high-permeability carbonate reservoirs are related to the epidiagenesis, dissolution of organic acids, dolomitization, abnormally high pressure, modification of thermal fluids, and hydrocarbon charging (Surdam et al. 1989; Davies and Smith 2006; Fan 2005; Li et al. 2006; Zhu et al. 2006).

As research on deep-buried marine facies carbonate reservoirs goes on and exploration breakthroughs are continuously made in China, geologists have achieved a better understanding of the formation mechanisms of secondary pores in carbonate reservoirs. Previous researchers

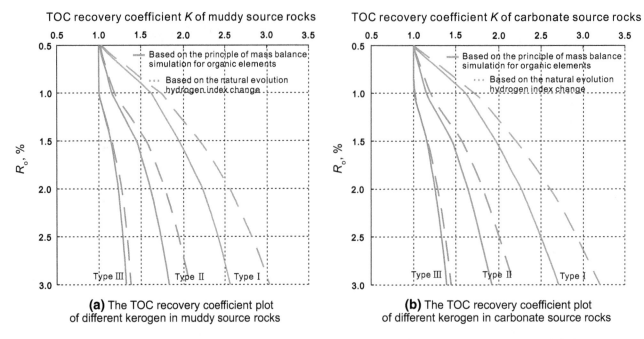

Fig. 43 Charts of restored TOC of source rocks under different geological conditions of petroliferous basins (Pang et al. 2014b)

Table 6 Revised TOC thresholds criterion for evaluation of effective source rocks

Evolutionary period	Muddy source rock			Carbonate source rock		
	I	II	III	I	II	III
Immature ($R_o < 0.5$ %)	0.50	0.50	0.50	0.50	0.50	0.50
Mature ($R_o = 0.5$ %–1.2 %)	0.35	0.45	0.47	0.30	0.40	0.45
Highly mature ($R_o = 1.2$ %–2.0 %)	0.25	0.35	0.45	0.20	0.30	0.40
Over-mature ($R_o > 2.0$ %)	0.20	0.30	0.40	0.15	0.25	0.35

insisted that secondary pores in deep-buried carbonate reservoirs were formed as a result of the dissolution of the carbonate rock exposed to the air, and many scholars focused on paleo-dissolution reservoirs, ignoring the role of karstification (dissolution) in improving the permeability and storage properties of deep-buried carbonate reservoirs (Zhu et al. 2006). In recent years, the fluid-rock interaction in deep-buried hydrocarbon-bearing carbonate reservoirs has attracted the attention of more and more scholars (Land and MacPherson 1992; Fisher and Boles 1990; Williams et al. 2001; Zhang et al. 2005b). As deep-buried karst is not controlled by the base level of erosion on the surface, many substances could induce water–rock reactions, such as acidic water and gases produced in the process of thermal evolution of organic matter; hot water produced by magmatic activities, compaction or diagenesis; acidic gases from deep strata in basins; and hydrogen sulfide gas produced in thermochemical or microbe reduction of sulfate-bearing carbonate rocks (Luo 2003). Pan et al. (2009) believed, based on samples from outcrops and drilling in the Central Tarim region, that hydrothermal karst

reservoirs may be an important type of reservoir, which has been ignored in the exploration for Lower Paleozoic carbonate hydrocarbon reservoirs in the Tarim Basin, and that there might be high-quality reservoirs along the faults, the main pathways for the migration of hydrothermal fluids, and fault-associated dissolution fractures and cavities. Li et al. (2010) studied the Mid-Lower Ordovician carbonate reservoirs in the Tahe region in the Tarim Basin. According to the research results, carbonate reservoirs might have experienced strong cementation in the normal process of deep burial diagenesis, but there were hardly any signs of dissolution. The modification of the reservoirs might be related to late epikarstification and structural-thermal fluid processes (Fig. 46). Generally, the epikarstification was followed by the structural-thermal fluid processes. The development and distribution of the latter might be related to the fault-fissure system and early epikarst system. According to the research results on the Ordovician carbonate reservoirs in the Tarim Basin by Lin et al. (2012), four dynamic mechanisms were involved in the structural modification of the reservoirs and the formation of fissures and cavities: (1) penecontemporaneous surface water karstification of carbonate rocks to form dissolution cavities, (2) raised surface fresh water karstification of the reservoir to form dissolution cavities, (3) deep hydrothermal fluid karstification of reservoir to form dissolution cavities, and (4) stress-induced faulting to form fissures (Fig. 47). Zhu et al. (2006) discovered in studying the deep-buried high-quality carbonate reservoirs in the Sichuan Basin that, besides the dolomitization and deep dissolution that controlled the formation of the porous oolitic dolomite in the Feixianguan Formation, strongly corrosive materials produced in the TSR (thermochemical sulfate reduction) boosted the dissolution of dolomite, leading to the formation of porous permeable spongy oolitic dolomite dissolution bodies, which played a constructive role in the formation of the reservoirs. The above research achievements indicated that the deep fluids in basins might be closely associated with the formation of high-quality carbonate reservoirs.

Fig. 45 Weathering crust and hydrocarbon distribution in the basin-platform region, Tarim Basin and the porosity–permeability features in reservoirs (Du et al. 2010)

5.3 Capillary pressure difference between inside and outside the target formation is the main force resulting in hydrocarbon accumulation in deep reservoirs

Forces involved in the formation of hydrocarbon reservoirs in hydrocarbon basins mainly include buoyancy, fluid pressure and dynamic force, expansive force of molecular volume, capillary force, molecule adsorption force, and intermolecular binding force (Pang et al. 2000). Under the action of buoyancy, natural gas migrates along migration pathways into traps in the upper part of a structure (White 1885). Under the action of the expansive force of gaseous molecules, natural gas migrates in a piston-type and accumulates in the lower part of a structure, with gas

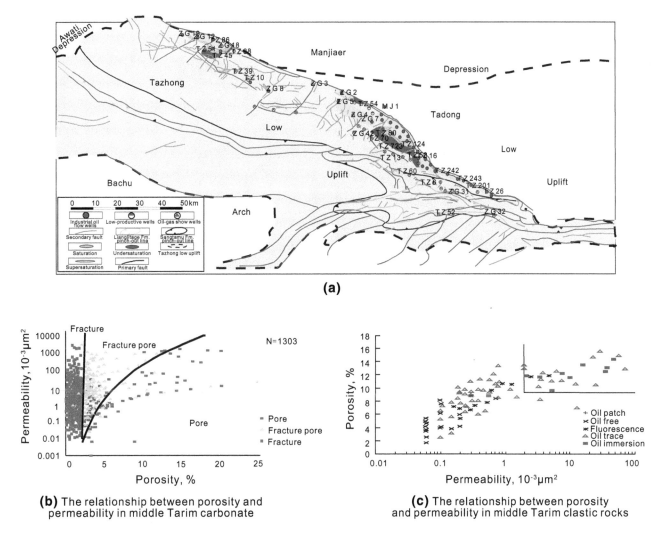

(a)

(b) The relationship between porosity and permeability in middle Tarim carbonate

(c) The relationship between porosity and permeability in middle Tarim clastic rocks

Fig. 44 Faults and hydrocarbon distribution in the Central Tarim Basin and the porosity–permeability features in reservoirs

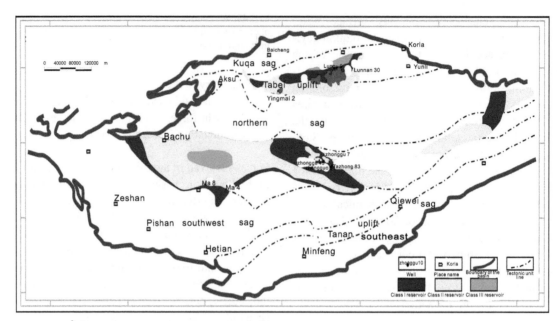

(a) Relationship between porosity distribution and distance to the weathering crust in the Yingshan Formation, Tazhong region

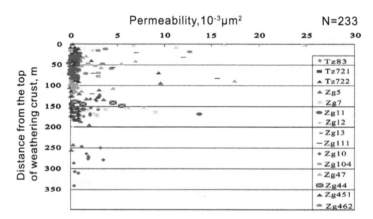

(b) The relationship between permeability of the Yingshan Formation and the distance from the top of weathering crust in Tazhong

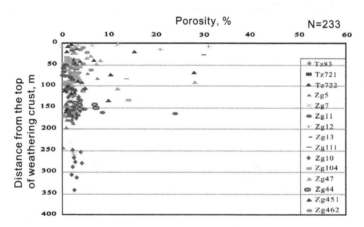

(c) Relationship between porosity distribution and the distance to the weathering crust in Yingshan Formation, Tazhong region

topped by water (Pang et al. 2003). Due to the capillary pressure difference between sandstone and mudstone, natural gas migrates into lens-type sandstones with larger pores, and accumulates at the top of the lens under the action of buoyancy to form a reservoir (Chen et al. 2004).

The temperature and pressure in the shallow strata of sedimentary basins are low and the rock porosity and permeability are relatively high, so the accumulation and migration of hydrocarbon are primarily controlled by buoyancy. But in the deep strata, the temperature and pressure are high and the rocks are relatively tight, so the accumulation and migration of hydrocarbon are controlled by a combination of forces. There are a variety of hydrocarbon reservoirs in the deep strata of sedimentary basins. In addition to conventional reservoirs in high-porosity and high-permeability strata, there are a large number of unconventional reservoirs in tight strata. Many hydrocarbon reservoirs have been discovered in the low-porosity carbonate and sandstone strata of the Junggar Basin and Tarim Basin (Fig. 48). With the ongoing research, geologists are gaining a better understanding of the formation mechanisms of deep hydrocarbon reservoirs. Capillary pressure that causes surface potentials has been considered to resist the flow of underground fluids. But since Magara (1978) puts forward the idea that capillary pressure is the primary power for hydrocarbon migration from source rocks to reservoirs, more and more scholars have realized that capillary pressure difference between mudstone and sandstone is an important factor in the formation of hydrocarbon reservoirs (Barker 1980; Magara 1978; Pang et al. 2000; Chen et al. 2004; Zhao et al. 2007; Li et al. 2007), and particularly that lithological traps such as sandstone lenses and pinch-outs are in a direct contact with deep mudstones in the part of basins. Therefore, capillary pressure difference might be one of the key forces for the formation of lithological hydrocarbon reservoirs included in sources rocks or in contact with source rocks. Surface potential is more significant to the formation of deep-buried lithological hydrocarbon reservoirs (Huo et al. 2014a). In the sand–mud contact zone, there is capillary pressure difference between sandstone and mudstone as

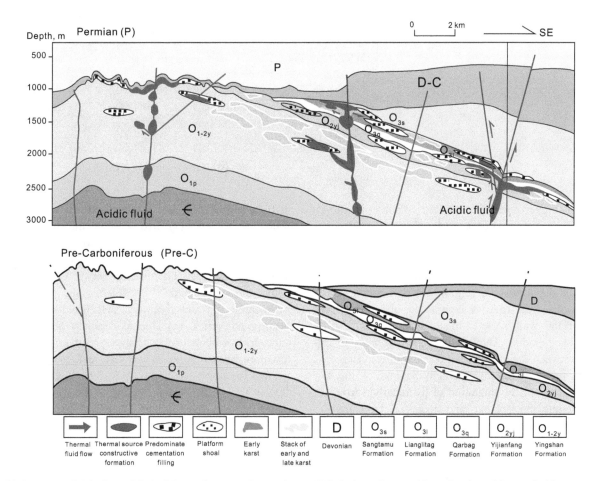

Fig. 46 Structures–fluid effect of Ordovician carbonate and controlling mode of reservoir distribution in the Tahe region, Tarim Basin (Li et al. 2010). Early karst: Mid-Late Ordovician–Silurian (Caledonian); Superposition of early and late-period karst: superposition of Mid-Late Ordovician–Silurian karst (Caledonian) + Late Devonian karst (Hercynian)

Fig. 47 Cavities and fissures formed as a result of structural changes of carbonate reservoirs in the Tarim Basin (Lin et al. 2012). **a** Dissolution cavities, **b** Fissures

(a)

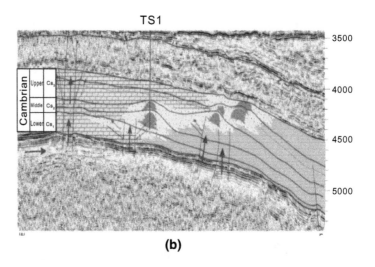

(b)

former has bigger pores than the latter. Pang et al. (2006) and Wang et al. (2013) through the physical experiments concluded that the greater the capillary pressure difference is between the inner sandstone and outer mudstone in a trap, the better hydrocarbon accumulation conditions will be in the sandstone (Fig. 49). Under strata conditions, as the wall rock is compacted, the pore throat radius of the wall rock is much smaller than that of isolated sand bodies. Due to the capillary pressure difference between them, hydrocarbon migrates inwards. So, capillary pressure difference might be one of the key factors that led to the migration of hydrocarbon from the wall rock into the isolated sand bodies. A large number of tight reservoirs were developed in the deep strata of basins. The research of Pang et al. (2006) showed that hydrocarbon was not affected or less affected by buoyancy in tight media. The surface potential or capillary pressure difference between reservoir and external strata is a key factor affecting the accumulation of hydrocarbon

in tight media (Fig. 50). The pressure difference has the following effects: (1) hydrocarbon migrates from the source rock into the reservoir due to capillary pressure difference. The greater the difference between the inner and outer capillary pressure or between the inner and outer surface potential, the higher the saturation of hydrocarbon in the reservoir, and the better the gas-bearing properties. (2) Due to capillary pressure difference, hydrocarbon in the tight reservoir migrates from a low-porosity and low-permeability rock to a high-porosity and high-permeability rock to form hydrocarbon-rich "sweet spots". (3) When faulting or fissuring occurs in a tight reservoir, the capillary force will decrease and the surface potential will drop, causing capillary pressure difference and the accumulation of hydrocarbon to fault zones and formation of "sweet spots". Exploration practice has proved that there are large numbers of hydrocarbon-rich "sweet spots" in the extensive, continuous tight sandstone reservoirs in the deep strata of basins, and most of them were formed due to faulting and fissuring in late periods.

5.4 Three dynamic boundaries of deep hydrocarbon reservoirs

The accumulation and migration of gas and oil are controlled by many geological conditions, including temperature, pressure, source of oil and gas, migration path, traps, etc. The formation and distribution of deep oil–gas reservoirs are mainly constrained by three force balance boundaries in terms of the dynamic mechanics. The first dynamic boundary is buoyancy-controlled threshold. It is the upper boundary of formation and distribution of unconventional tight reservoirs. The buoyancy-controlled threshold is the maximum depth of non-buoyancy-driven hydrocarbon migration in a hydrocarbon basin. The formation and distribution of hydrocarbon reservoirs in hydrocarbon basins are controlled by the dynamic boundary formed by different thresholds (Pang 2010; Pang et al. 2014a). The buoyancy-driven hydrocarbon accumulation threshold is a new geological concept in relation to buoyancy-driven hydrocarbon accumulation (White 1885). Pang et al. (2014a) maintained that the buoyancy-driven hydrocarbon accumulation threshold is the threshold beyond which buoyancy will have less effect on the accumulation of hydrocarbon in highly compacted strata as the burial depth increases. Generally, it is characterized by porosity,

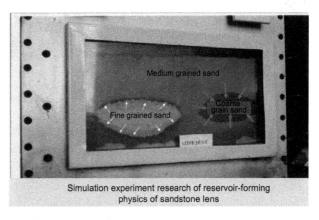

Simulation experiment research of reservoir-forming physics of sandstone lens

Fig. 49 Physical experiments simulating the formation of a sandstone lens reservoir (Pang et al. 2006; Wang et al. 2013)

pore throat radius, and permeability geological parameters of strata at certain burial depths (Fig. 51).

With respect to the causes of the buoyancy-controlled threshold, scholars have different views. Masters held that buoyancy-controlled threshold is caused by the relative variation of permeability of strata (Masters 1979); other scholars argued that geological conditions such as diagenetic difference (Cant 1986), fault seals (Cluff and Cluff 2004) and force equilibrium mechanisms (Berkenpas 1991) are the causes. These views may be used to explain the phenomena of non-buoyancy-controlled hydrocarbon accumulation in some basin areas, but cannot explain it in a

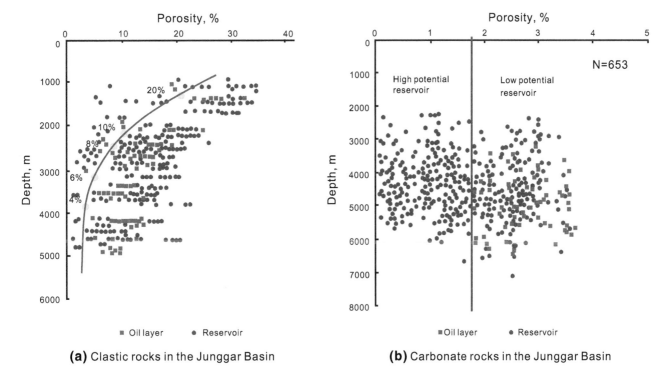

(a) Clastic rocks in the Junggar Basin

(b) Carbonate rocks in the Junggar Basin

Fig. 48 Diagram of effective porosity in different reservoirs changes as a function of burial depth

broader sense. Through physical simulation experiments (Fig. 52), Pang et al. (2013, 2014a) believed that there exists an equilibrium of forces affecting the migration of natural gas at the buoyancy-controlled threshold. Charge pressure is the dynamic force and capillary force and overlying water pressure are the resistances. The equilibrium between the driving force for upward migration of hydrocarbon (P_e) and capillary force in tight media (P_c) and overlying static water pressure (P_w) is the dynamic mechanism of buoyancy-driven hydrocarbon accumulation thresholds. Figure 52 shows the physical simulation experiment results of buoyancy-controlled threshold in a tapered glass tube; it can be quantified by the dynamic equilibrium equation $P_e = P_w + P_c$. Similar results have been obtained based on the physical simulation experiment using a thick tube filled with sandstone of different sizes (Pang et al. 2014a).

The fundamental reason why hydrocarbon in deep tight strata of hydrocarbon basins is not controlled by buoyancy is that the sum of rock capillary force and overlying static water pressure is greater than the pressure in the hydrocarbon reservoir. When the pressure in the hydrocarbon reservoir is greater than the sum of the two forces, buoyancy will cause hydrocarbons to migrate upwards to the surface or scatter. The equilibrium of forces is the dynamic mechanism of buoyancy-controlled threshold, which can be expressed in Eqs. (1)–(3):

$$P_e = P_w + P_c, \tag{1}$$

where P_e is oil reservoir pressure, which can be expressed as follows:

$$P_{eg} = \frac{z \times \rho_g}{M_g} \times R \times T \times 1.01 \times 10^2, \tag{2}$$

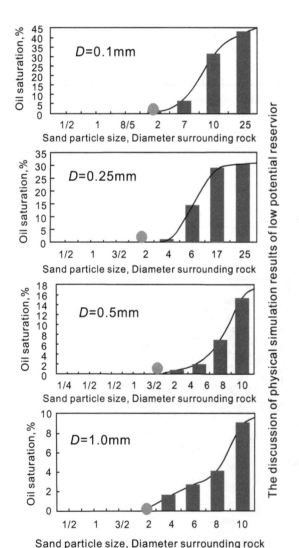

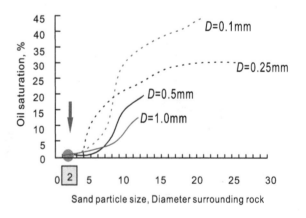

Petroliferous properties of sandstone under various conditions of wall rock ($D=0.5$)

No.	Sample No.	D/d	The sample size	Extraction of oil	Oil content	Oil saturation
1	30mesh-10mesh	1/4	125.1	5.3	0.0042	0.0261
2	30mesh-20mesh	1/2	113.8	4.7	0.0041	0.0255
3	30mesh-30mesh	1	82.4	4.9	0.0059	0.0367
4	30mesh-60mesh	2	167.6	21.2	0.0126	0.0780
5	30mesh-120mesh	4	140.3	180.4	0.1286	0.7929
6	30mesh-200mesh	6	152.4	452.0	0.2966	1.8290
7	30mesh-240mesh	8	142.2	1554.0	1.0905	6.7249
8	30mesh-300mesh	10	126.6	3143.8	2.4833	15.3134

Accumulation=Interfacial energy of wall rock≥2   ×Interfacial energy of reservoirs

Fig. 50 Physical experiment results of capillary pressure difference (surface potential) controlling hydrocarbon (Pang et al. 2006)

$$P_{eo} = \frac{RT}{V-b} - \frac{a}{V^2} = \frac{\rho_o RT}{M_o - \rho_o b} - \frac{\rho_o^2 \times a}{M_o^2}, \qquad (3)$$

where P_{eg} is the gas reservoir pressure, MPa; z is the gas deviation coefficient (compressibility factor), dimensionless; R is the universal gas constant, 0.008314 MPa m^3/ (kmol K); T is the absolute temperature of natural gas, K; M_g is natural gas molar mass, kg/kmol; ρ_g is the natural gas density under strata conditions, kg/m^3; P_{eo} is the oil reservoir pressure, MPa; ρ_o is the oil density under strata conditions, kg/m^3; M_o is the molar mass of oil, kg/kmol; and a, b are van der Waals constants.

The change of any geological parameter in the force equilibrium equations of the buoyancy-controlled threshold will lead to changes of the critical conditions of the buoyancy-controlled threshold. Factors include different driving forces, fluid physical and chemical properties, strata conditions, and the structural environment of basins. Driving forces include hydrocarbon reservoir pressure (P_e), overlying static water pressure (P_w), and reservoir media capillary forces (P_c). The change of any force will lead to change of the buoyancy-driven hydrocarbon accumulation threshold; fluid physical and chemical properties include hydrocarbon-water interface tension, contact angle, density, and temperature; strata media conditions include porosity, permeability, and pore throat radius; the structural environment of basins refers to structural changes of a basin that affect the distribution scope of the buoyancy-driven hydrocarbon accumulation threshold.

Under actual geological conditions, the buoyancy-controlled threshold is affected by a combination of the above factors. Research results indicate that the depth of the buoyancy-controlled threshold increases and the corresponding porosity, permeability, or pore throat radius decreases as the sand grain size increases under the

circumstances that all the conditions are favorable, and that the corresponding burial depth, porosity, permeability, or pore throat radius of the buoyancy-controlled threshold force equilibrium decreases as the sand grain sorting difficulty increases. According to statistics, generally, the buoyancy-controlled threshold of hydrocarbon basins is as follows: porosity <12 %, permeability <1 mD, and pore throat radius <2 μm. Although porosity is often considered a characteristic parameter of buoyancy-controlled threshold, in fact, whether hydrocarbon is controlled by buoyancy in a reservoir or not is mostly decided by the pore throat radius, because it has a direct effect on the capillary force affecting the migration of hydrocarbon. For carbonate reservoirs with smaller porosities, the migration of hydrocarbon is also affected by buoyancy in fissure developed areas. The general relations between porosity, permeability, and pore throat radius are shown in Fig. 53.

Buoyancy-controlled threshold distribution under actual geological conditions can be predicted based on the equilibrium equation (Eqs. (1)–(3)). Figure 54 shows the predicted buoyancy-controlled threshold of tight sandstone reservoirs of the Upper Paleozoic in the Ordos Basin in China based on the force equilibrium equations. The results indicated that they correspond to geological conditions of porosity <10 % and permeability <1 mD. Below the dynamic boundary, gas is filled in the tight reservoirs of adjacent source rocks; above the boundary, natural gas accumulates only in the upper part of the structure. Figure 55 shows the comparison of predicted buoyancy-driven oil accumulation thresholds with drilling results in the Putaohua reservoir in the Songliao Basin. As shown in Fig. 55, liquid oil also has a buoyancy-controlled threshold, and the force equilibrium boundary corresponds to the burial depth with a porosity between 10 % and 11 %. Above this boundary, liquid oil reservoirs have the features

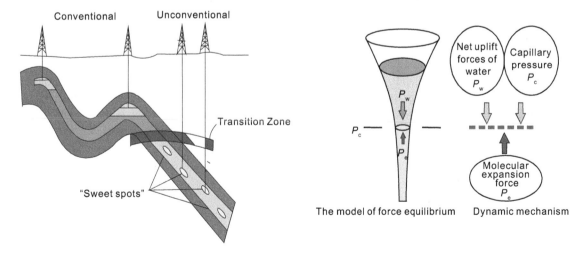

Fig. 51 Conceptual model of buoyancy-controlled thresholds in deep petroliferous basins and controlling the distribution of oil–gas

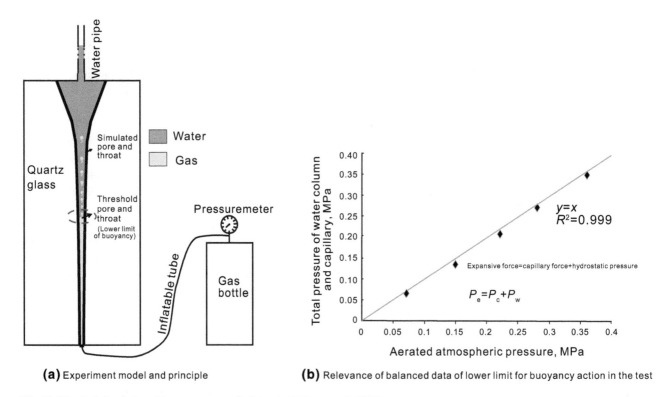

(a) Experiment model and principle   **(b)** Relevance of balanced data of lower limit for buoyancy action in the test

Fig. 52 Physical simulation of buoyancy-controlled threshold (Pang et al. 2013)

of high-point accumulation, high-porosity enrichment, high-point sealing, and high pressure; below the boundary, liquid oil reservoirs have the features of low-depression accumulation, low-stand inversion, low-porosity enrichment, and low-pressure stability. The consistency of the theoretic prediction results with the actual drilling results reflects the existence of a buoyancy-controlled threshold and the practicality and reliability of the prediction model for both natural gas and liquid oil. Their difference is small (Pang et al. 2013, 2014a).

The second dynamic boundary is the hydrocarbon accumulation limit. It is the maximum depth for hydrocarbon accumulation under geological conditions. The enrichment and accumulation of hydrocarbons require certain temperature and pressure conditions. Theoretically, each hydrocarbon basin may have a hydrocarbon accumulation threshold at a certain depth, below which hydrocarbon accumulations do not exist. The dynamic boundary corresponding to the lower limit of hydrocarbon accumulation is regarded as the hydrocarbon accumulation threshold. According to Pang et al. (2014a), the hydrocarbon accumulation limit is the maximum burial depth or corresponding critical geological condition for hydrocarbon accumulation in hydrocarbon basins, which can be characterized by the porosity, permeability, or pore throat radius of a reservoir (Fig. 56). Pang et al. (2014a) recognized through analysis that the corresponding critical

values of the hydrocarbon accumulation limit in clastic petroliferous basins are generally as follows: porosity is less than 2 %–2.4 %, permeability is less than 0.01 mD, pore throat radius is less than 0.01 μm, and burial depth is 5,000–8,000 m.

Many scholars discovered through research using different methods that there is a physical threshold for sedimentary basin oil–gas-bearing reservoirs, below which hydrocarbon does not accumulate or has no exploration significance (Wan et al. 1999; Guo 2004; Shao et al. 2008). Through analysis of actual mass exploration results, four different methods have been used to determine hydrocarbon accumulation limits in the deep part of hydrocarbon basins by Pang et al. (2014a): (1) As the burial depth increases, the bound water saturation in the reservoir will reach 100 %, resulting in the termination of hydrocarbon accumulation. When the bound water saturation reaches 100 % in the Fuyang sandstone reservoir in the Songliao Basin, the hydrocarbon accumulation limit porosity is 2.4 %–4 %; when the bound water saturation reaches 100 % in the hydrocarbon-bearing sandstone in the Kuqa Depression in the Tarim Basin, the hydrocarbon accumulation limit porosity is about 2.4 %. Figure 56 shows an example of this. (2) The hydrocarbon accumulation limit can be determined when the porosity and permeability as a function of buried depth do not allow the migration of oil and gas normally. As the burial depth increases, the

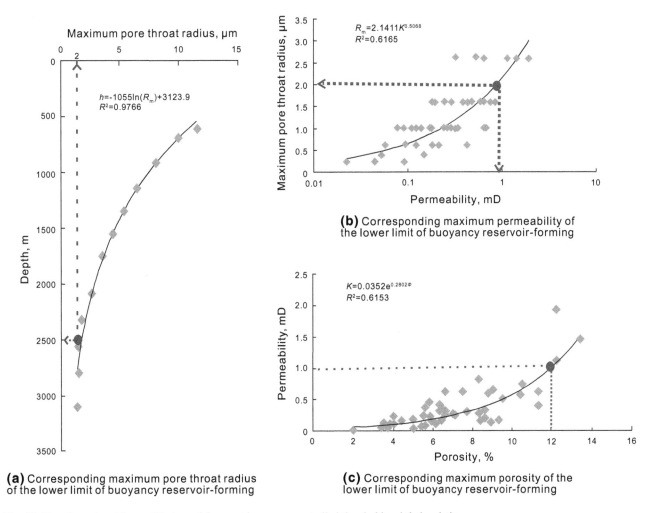

(a) Corresponding maximum pore throat radius of the lower limit of buoyancy reservoir-forming

(b) Corresponding maximum permeability of the lower limit of buoyancy reservoir-forming

(c) Corresponding maximum porosity of the lower limit of buoyancy reservoir-forming

Fig. 53 Key elements of the equilibrium of forces at buoyancy-controlled threshold and their relations

reservoir permeability under actual geological condition decreases as a result of compaction strengthening. Figure 57 shows an example. (3) The hydrocarbon accumulation limit can be determined when capillary pressure difference between inside and outside the reservoir or the potential difference tends to disappear as a function of buried depth. As the burial depth under actual geological conditions increases, the potential difference between inside and outside the reservoir will disappear, resulting in the termination of hydrocarbon accumulation. Figure 58 shows the hydrocarbon accumulation limits in terms of inner and outer potential difference of major target reservoirs in the Jiyang Depression in the Bohai Bay Basin in eastern China and the Kuqa Depression in the Tarim Basin in western China are at burial depths of 6,000 and 8,500 m, respectively (Fig. 58). (4) The hydrocarbon accumulation limit can be determined based on the exploration well data from a 100 % dry bed. When a 100 % dry bed was met in the exploration well during drilling for the purpose of

understanding the hydrocarbon and water distribution in the Central Tarim Basin region, the reservoir porosity was less than 2 %, so it is taken as the hydrocarbon accumulation limit (Fig. 59).

We highlight here that the hydrocarbon accumulation limit is not a threshold indicating the existence of hydrocarbon. When hydrocarbon exists in reservoirs below this limit, it was probably accumulated before entering the threshold, below which there may be exploration risks because the strata porosity is low and accumulation has terminated. There may be hydrocarbon (liquid oil) reservoirs of industrial value in strata with a burial depth of 6,000 m, and even with a burial depth exceeding 8,000 m in some basins in the world. The threshold of deep hydrocarbon accumulation in hydrocarbon basins is in a wide range and changes according to certain rules. As the sand grain size increases or the sand grain sorting difficulty decreases, the hydrocarbon accumulation limit becomes deeper.

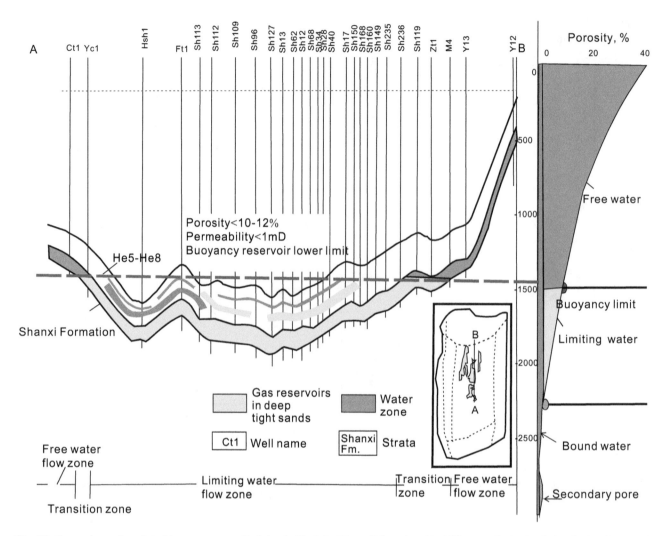

Fig. 54 Comparison of predicted buoyancy-controlled threshold of the Upper Paleozoic with drilling results in the Ordos Basin (Pang et al. 2014a)

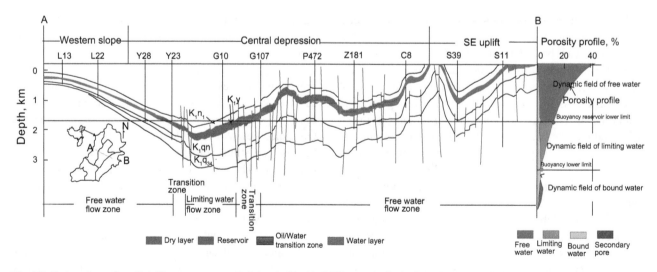

Fig. 55 Comparison of predicted buoyancy-controlled threshold with drilling results in the Putaohua reservoir in the Songliao Basin (Pang et al. 2014a)

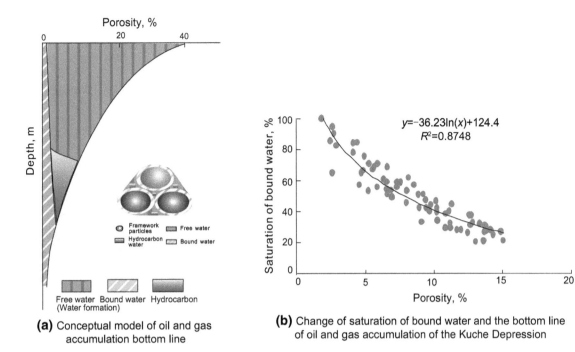

Fig. 56 Bound water saturation in an oil–gas-bearing target reservoir as a function of buried depth and hydrocarbon accumulation limits

The third dynamic boundary is the lower limit of hydrocarbon generation of source rocks. It usually refers to the critical geological condition that hydrocarbon output of organic-rich source rocks is less than 1 % of the total. Methods used to determine the lower limit of hydrocarbon generation mainly include the amount of residual hydrocarbon, organic element variation, hydrocarbon generation potential, and efficiency of hydrocarbon expulsion. The residual hydrocarbon capacity of source rocks is expressed by the amount of residual hydrocarbon in organic carbon (S_1/TOC) or chloroform asphalt ("A"/TOC). The residual hydrocarbon S_1/TOC or "A"/TOC of the source rocks increases at first followed by a gradual decrease as the depth or R_o increases, generating a "belly-shaped" curve. When S_1/TOC or "A"/TOC reaches a minimal value which is so small that change is hardly visible, it means that hydrocarbon is no longer generated and expelled from the organic matter, and the R_o value at such minimal value is the lower limit of hydrocarbon generation (Fig. 60a). The organic element variation method is described as follows: When H/C and O/C in the source bed reaches a minimal value which is so small that change is hardly visible, the source rock will no longer yield hydrogen-rich hydrocarbons. That is the lower limit of hydrocarbon generation. Theoretically, when hydrocarbon generation terminates, H/C will no longer change when reaching a minimal value. Huo et al. (2014b) determined the lower limit of hydrocarbon generation of carbonate rocks in the basin-platform region of the Tarim Basin, by supposing that the minimal value

H/C = 0.1 and the H/C value at R_o = 0.5 % as the maximum value (Fig. 60b). In the diagram of hydrocarbon generation potential changes of source rocks, $(S_1 + S_2)$/TOC is the current hydrocarbon generation potential index of the source rocks. Due to hydrocarbon generation and expulsion from the source rocks, $(S_1 + S_2)$/TOC increases first followed by a gradual decrease as the depth or R_o increases, generating a "belly-shaped" curve. When the hydrocarbon generation potential reaches a certain minimal value, the hydrocarbon generation potential stops changing. This means that the source rock stops generating hydrocarbon and the lower limit of hydrocarbon generation of the source rock is reached (Fig. 60c). The amount of hydrocarbon expelled from source rock increases at first followed by a gradual decrease as the depth or R_o increases. When the amount of expelled hydrocarbon drops to zero or a minimal value, it means that hydrocarbon expulsion stops, and hydrocarbon generation may also have stopped. The R_o value when the amount of expelled hydrocarbon is zero or reaches a minimal value is the lower limit of hydrocarbon generation. When the amount of expelled hydrocarbon is zero, the hydrocarbon expulsion rate is also zero, but the hydrocarbon expulsion efficiency increases to the maximum of nearly 100 %. The lower limit of hydrocarbon generation can be determined according to the hydrocarbon expulsion rate and efficiency instead of the amount of expelled hydrocarbon. According to the lower limit of hydrocarbon generation concept, the R_o value when F_e = 99 % is the lower limit of hydrocarbon generation (Fig. 60d).

Fig. 57 Porosity and
permeability changes and
hydrocarbon accumulation
limits in the burial process of a
hydrocarbon-bearing target
layer

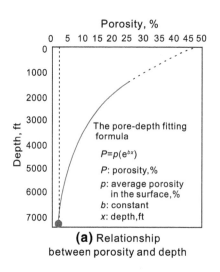

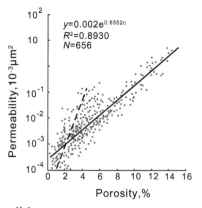

(a) Relationship
between porosity and depth

(b) Relationship between permeability
and effective porosity in Xu6,Chuanxi section

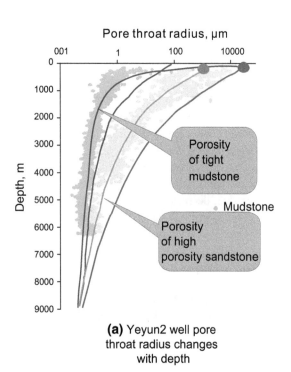

(a) Yeyun2 well pore
throat radius changes
with depth

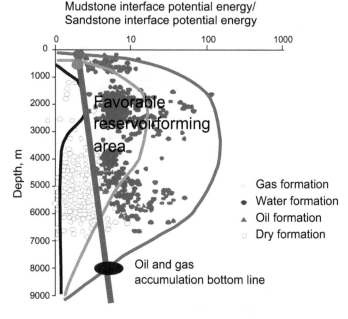

(b) Changes of potential difference of
inner-outer interface in Kuqa Depression,
Tarim as a function of depth

Fig. 58 Changes of inner–outer surface potential and capillary pressure difference and hydrocarbon accumulation limits

5.5 The formation and distribution of deep hydrocarbon reservoirs are controlled by three fluid dynamic fields

Research on fluid dynamic fields originated from fluid dynamics research in the domain of geodynamics. In 1953, Hubbert (1953) proposed the fluid potential concept in order to describe energy changes and migration rules of underground fluids. Ye et al. (1999) made a detailed description of the concept, pointing out that a fluid dynamic field is an integration of temperature, pressure, fluid potential, and structural stress fields in a sedimentary basin and their relations. Based on the above analysis that there is a buoyancy-controlled threshold and a hydrocarbon accumulation limit in the deep part of each hydrocarbon basin, Pang et al. (2014a) suggested that a hydrocarbon basin may be divided into three dynamic fields according to buoyancy-controlled threshold and hydrocarbon

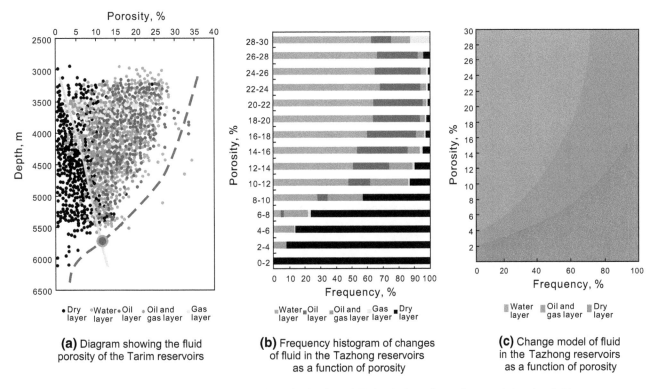

(a) Diagram showing the fluid porosity of the Tarim reservoirs

(b) Frequency histogram of changes of fluid in the Tazhong reservoirs as a function of porosity

(c) Change model of fluid in the Tazhong reservoirs as a function of porosity

Fig. 59 Distribution of oil–gas, water bearing, and dry layers in the Central Tarim Basin region and accumulation thresholds

accumulation limit. Strata above the buoyancy-controlled threshold are called the free fluid dynamic field; strata between the buoyancy-controlled threshold and hydrocarbon accumulation limit are in the limited fluid dynamic field; and strata below the hydrocarbon accumulation limit are in the bound fluid dynamic field (Fig. 61). The fluid dynamic field is referred to the strata field where the hydrocarbon has identical or similar media, migration–accumulation force and reservoir-forming rule.

The research conducted by Pang et al. (2014a) shows that deep hydrocarbon reservoirs are controlled by free and limited fluid dynamic fields. The free fluid dynamic field is where conventional hydrocarbon reservoirs develop, and where buoyancy has a dominant effect on the migration and accumulation of hydrocarbons, with the hydrocarbon distribution features of high-point sealing and accumulation, high-porosity enrichment, and high-pressure stability. The limited fluid dynamic field is where tight hydrocarbon reservoirs develop, and where buoyancy has less effect on the migration and accumulation of hydrocarbons. The sandstone porosity in this field is generally 2.4 %–12 %, permeability is 0.01–1 mD, and pore throat radius is 0.01–2 μm. In this formation area, three major tight hydrocarbon reservoirs are formed. (1) the conventional tight hydrocarbon reservoirs are developed from pre-existing conventional reservoirs under compaction, i.e., first forming reservoirs and then being compacted, such

reservoirs are characterized by the features of high-point accumulation, high-stand sealing, and high-pressure stability; (2) the tight deep basin gas reservoirs underwent the process of first compaction and then reservoir forming. They show the features of low-depression accumulation, low-stand inversion, low-porosity enrichment, and low-pressure stability. They are mainly formed by the expansion of the petroleum area caused by the volume expansion of hydrocarbon when migrating into the tight reservoirs; (3) the tight superimposed hydrocarbon reservoirs are formed by the combination of the two hydrocarbon reservoir types mentioned above, which are formed by compaction and molecule volume expansion. They underwent three stages, i.e., reservoir forming—compaction—reservoir forming. They generally show the features as follows: the coexistence of oil and gas of high and low points; the coexistence of oil and gas of high and low porosity; the coexistence of oil–gas-bearing zone of high and low pressure; the coexistence of oil–gas reservoirs of high and low production. A superimposed continuous hydrocarbon reservoir is formed by the superimposition of the tight hydrocarbon reservoirs mentioned above (Pang et al. 2014a). The bound fluid dynamic field is at the bottom of the basin, and the oil and gas in it were accumulated at an early stage and retained from that earlier time. The risk of exploration and development is huge in this field, as the target formation is characterized by deep-buried depth, low

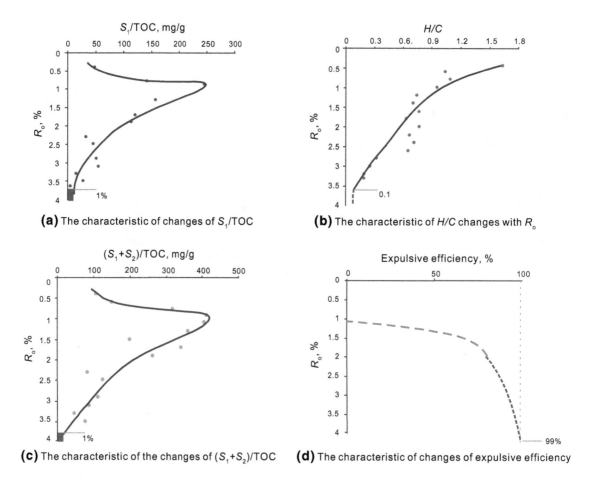

(a) The characteristic of changes of S_1/TOC

(b) The characteristic of H/C changes with R_o

(c) The characteristic of the changes of (S_1+S_2)/TOC

(d) The characteristic of changes of expulsive efficiency

Fig. 60 Changes of hydrocarbon generation and expulsion in the source rocks as a function of buried depth and the lower limit of hydrocarbon accumulation (*blue line*) (Huo et al. 2014b)

porosity and permeability, and lack of formation energy (Fig. 62). Later tectonism may damage and reconstruct the hydrocarbon reservoirs and develop fracture type, cave-type, or fracture-cave complex-type hydrocarbon reservoirs. They are tight reservoirs, but the porosity and permeability may be good in some parts of the area, which then show the geological features of conventional hydrocarbon reservoirs.

The distribution of fluid dynamic fields varies in different basins. It is affected by the following three factors in the deep strata of hydrocarbon basins: (1) The changing rate of porosity and permeability with the burial depth. The depth of a fluid dynamic field decreases as the changing rate increases, and vice versa; (2) The uplifting of deep strata as a result of erosion of overlying strata leads to the uplifting of tight strata as a whole and the limited and bound fluid dynamic fields in the basins uplift to a shallow formation or even to the surface; (3) Faults caused by structural changes may damage the fluid dynamic field boundary in hydrocarbon basins. Faults may change a limited fluid dynamic field into a free fluid dynamic field near the faults and damage some reservoirs. Overall

faulting may change the deep part into a free fluid dynamic field to form conventional hydrocarbon reservoirs under the effect of buoyancy.

5.6 Tight hydrocarbon resources in limited deep fluid dynamic fields and exploration prospects

Tight hydrocarbon reservoirs in limited deep fluid dynamic fields can be divided into three types by the development mode: conventional reservoirs, deep basin reservoirs, and composite reservoirs, and each has its own unique formation process. Conventional tight reservoirs were deep-buried conventional reservoirs formed under the action of buoyancy after long years of compaction and diagenesis, characterized by an "accumulation followed by tightening" process; deep basin tight reservoirs were formed by hydrocarbons expelled from source rocks and accumulated in adjacent tight rocks without being controlled by buoyancy forces. These reservoirs are characterized by a "tightening followed by accumulation" process; composite tight reservoirs are a combination of conventional tight reservoirs and deep basin tight reservoirs, characterized by

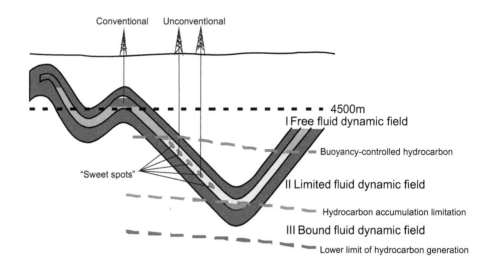

Fig. 61 Diagram showing the division of dynamic field boundary and fluid dynamic field of hydrocarbon migration and accumulation in deep petroliferous basins

an "accumulation-tightening-accumulation" process. Figure 63 shows the migration–accumulation force, source-reservoir matching, major controlling factors, and reservoir forming modes in the three tight hydrocarbon reservoirs.

Deep reservoirs in hydrocarbon basins have better forming and preserving conditions than shallow ones in the following aspects: (1) Reservoir tightness. Buoyancy has less effect on the migration and accumulation of hydrocarbons. This means that all hydrocarbons from source rocks during this period are not easily dispersed, and the preservation conditions are much better than those in the free fluid dynamic field. (2) Thermal maturity of source rocks. The geotemperature is high, between 100 and 200 °C, and R_o is 1.2 %–2.5 %. The hydrocarbon yield of per unit weight of parent matter is 1.0–2.2 t/tc, and hydrocarbon expulsion efficiency is 25 %–99 %, equivalent to 2 and 5 times those in the free fluid dynamic field, respectively. (3) Hydrocarbon migration and accumulation

efficiency. Hydrocarbons from source rocks accumulate in adjacent rocks to form reservoir resources. The migration and accumulation efficiency is 3–10 times that in the free fluid dynamic field. (4) The hydrocarbon targets in limited deep fluid dynamic fields have experienced the evolution period of free fluid dynamic fields, and the conventional reservoirs formed in early periods accumulated in limited fluid dynamic fields, where they were compacted and changed into conventional tight reservoirs as part of the hydrocarbon resources of limited fluid dynamic fields. The oil and gas resource evaluating results of limited fluid dynamic fields in several China major basins evaluated by Pang et al. (2014a) indicate the oil and gas resource accumulated in the field accounts for more than 84 % of the total basin resource.

The hydrocarbon resource potential is different in different types of tight reservoirs in deep petroliferous basins. Deep hydrocarbon reservoir distribution patterns are

Fig. 62 Distribution of deep fluid dynamic field-controlled hydrocarbons in hydrocarbon basins (Pang et al. 2014a)

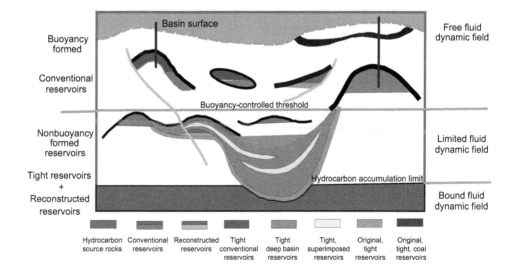

affected and controlled by single factor (Fig. 64). (1) under the control of terrestrial heat flow or geothermal gradient in the basin, "cold basins" characterized by low geothermal gradient and terrestrial heat flow values are favorable for deep oil and gas exploration. They can develop both free fluid dynamic fields and limited or bound fluid dynamic fields below a depth of 4,500 m with an extensive formation area for favorable exploration. Their resource potential is huge and a variety of hydrocarbon reservoir types are found in the basins with promising exploration prospects. "Hot basins" characterized by a high geothermal gradient and terrestrial heat flow values are unfavorable for deep oil and gas exploration, only developing a limited fluid dynamic field below the depth of 4,500 m with limited formation area favorable for exploration. Their resource potential and reservoir types are limited, with huge exploration risk. "Warm basins" whose geothermal gradient and terrestrial heat flow values are between "cold basins" and "hot basins" have an extensive formation area favorable for exploration and the resource potential and exploration prospects are between the two extremes. (2) When hydrocarbon source rocks are widely developed and hydrocarbon generation and expulsion amounts are large, the exploration potential is favorable. When a source-reservoir-cap rock combination is completed and the cap rock condition is good, the exploration potential is also favorable. (3) When intense tectonic movements occurred with fractures dominating, deep oil and gas reservoirs are susceptible to damage leading to low exploration potential. When a deep fluid dynamic field is uplifted to middle–shallow formation due to the denudation of overlying strata, the oil and gas exploration potential is weakened below a depth of 4,500 m. For example, the limited fluid dynamic field is currently above a depth of 4,500 m in the Chu-Saleisu Basin, Kazakhstan, resulting from the overall uplifting caused by the denudation of overlying strata. So the bound fluid dynamic field, which is unfavorable for oil and gas exploration, is currently developed below a depth of 4,500 m.

6 Geological research directions for hydrocarbons in deep petroliferous basins

6.1 Identification of deep hydrocarbon sources and relative contribution evaluation

The sources of deep hydrocarbons are complex. Some have single source and some have mixed sources. The hydrocarbon can originate from degradation of organic matter, from thermal cracking of asphalt sand or dispersed organic matter, or from the catalytic action of deep hydrothermal activity so it is very important to understand the origin of deep hydrocarbon and evaluate relative contributions. This will play a leading role in determining the hydrocarbon resource potential and favorable exploration directions. Scholars have done a lot of research into the identification of hydrocarbon sources and quantitative evaluation of hydrocarbon contributions using a variety of methods, laying a solid foundation for further work.

Methods for hydrocarbon source identification mainly include total hydrocarbon gas chromatographic

Classification		Mechanism of accumulation		Mode of accumulation			Example
Categories	Sub-categories	Power for migration	Matching of source rock and reservoir rock	Conceptual model	Factor combination	Regularities of distribution	At home and abroad
Tight oil and gas reservoirs in function of non-buoyancy	Conventional tight oil and gas oil reservoir	Static formation pressure	Accumulating first, and then becoming tight reservoir	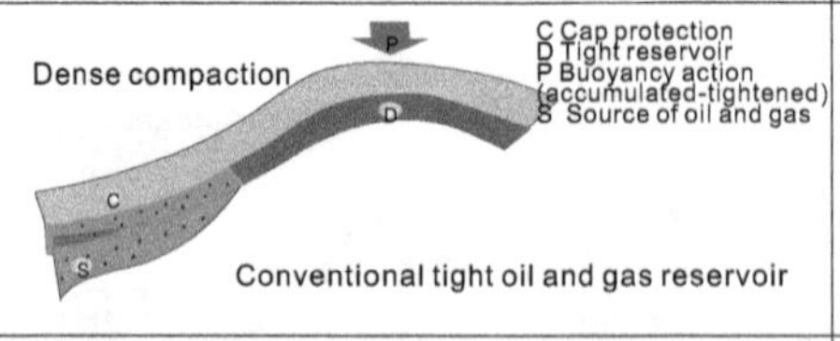 	T-CDPS	Stable tectonic zone in restricted dynamic field of liquid	Dina 2 and tight conventional oil and gas reservoirs currently discovered below 4500m
	Deep-basin tight oil and gas reservoir	Expansion stress of hydrocarbon; the differences of capillary force and concentration of hydrocarbon between the source rock and reservoir rock	Becoming tight reservoir first, and then accumulating	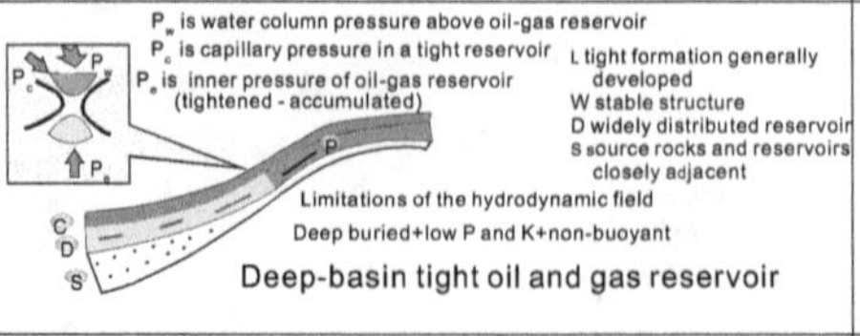 	T-LWDS		Yinan 2 and deep-basin reservoirs in the Alberta Basin
	Compound tight oil and gas reservoir	Static formation pressure; expansion stress of hydrocarbon; the difference between inner and outer	Accumulating first, becoming tight reservoir later, accumulating again	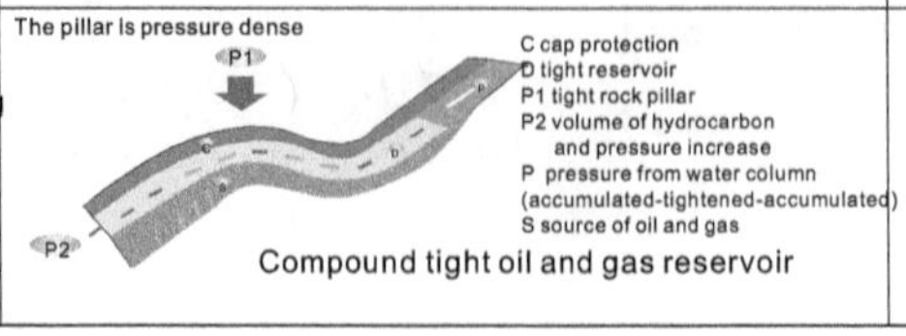 	T-CDPS + T-LWDS		Dibei Oil field structural high gas reservoir in Paleozoic tight gas reservoir of the Erdos Basin

Fig. 63 Three unconventional non-buoyancy tight hydrocarbon reservoirs in limited fluid dynamic field and basic modes

Fig. 64 Relations between the distribution of limited fluid dynamic fields and geothermal gradients in a deep petroliferous basin

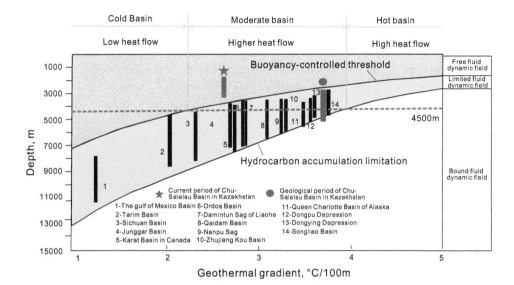

fingerprints, *n*- and iso-alkane hydrocarbon ratios, steroids/terpenoids, aromatics, non-aromatics, as well as family composition and carbon isotope studies (Gormly et al. 1994; Telnæs and Cooper 1991; Stahl 1978; Seifert 1978; Hirner et al. 1981); index comparison using aromatics and thiophene compounds (Michael et al. 1989; Mukhopadhyay et al. 1995; Jing 2005) and comprehensive crude oil light hydrocarbon analysis for multi-period and multi-source reservoirs (Chen et al. 2006; Philippi 1981; Odden et al. 1998; Chen et al. 2003).

Methods for quantitative evaluation of the relative contribution of mixed hydrocarbon sources mainly include special compound absolute concentration quantification (Zhang et al. 2005a), biomarker parameters (Li et al. 2002), carbon isotope ratios (Song et al. 2004), and chart matching (Wang et al. 1999, 2004).

In deep strata, there are multiple sets and multiple varieties of hydrocarbon sources (including shale, carbonate, paleo-reservoir asphalt, and inorganic hydrocarbons) and multiple hydrocarbon-generating points, which experienced a number of periods of hydrocarbon generation and expulsion and multiple hydrocarbon migration pathways (faults, unconformities, carrier systems, and combinations). For source rocks that experienced many periods of evolution, the results of comparison using a single biomarker will be seriously affected by the degree of maturity. For multiple sets of source rocks, the results of analysis using the carbon isotope method will be affected by the sedimentary environment and climate conditions. Therefore, integrated analysis is required for the determination of deep hydrocarbon sources, including geological (altitude, log response, color, and mud content) and geochemical (biomarker, carbon isotope) analysis. The determination of deep hydrocarbon sources and evaluation of relative

contributions are important to the evaluation of deep hydrocarbon resources and determination of hydrocarbon exploration fields.

6.2 Genetic mechanisms and preservation conditions of deep-buried high-quality reservoirs

Deep reservoirs include clastic, carbonate, volcanic, and metamorphic reservoirs. Research on the genesis and preservation of high-quality, high-porosity, and high-permeability reservoirs has drawn the attention of many geologists, but there is hardly any research done on the physical thresholds of deep-buried reservoirs. Although the limited fluid dynamic field mainly forms tight hydrocarbon and unconventional reservoirs, what concerns people is still "sweet point" formation in high-porosity and high-permeability reservoirs. The most important part of deep oil and gas exploration is to find these reservoirs under current conditions.

The genesis of deep-buried high-quality reservoirs mainly includes dissolution by organic acids (Surdam et al. 1984), effects of hydrocarbon charging and deep thermal fluids (Navon et al. 1988), clay mineral membranes (Ehrenberg 1993; Dolbier 2001), temperature and depth (Ezat 1997), faulting (Moretti et al. 2002), abnormal pressure (Wilkinson et al. 1997; Osborne and Swarbrick 1999), effects of fractures (Harris and Bustin 2002), sedimentary environment (Amthor and Okkerman 1998; Khidir and Catuneanu 2003; Pape et al. 2005; Rossi et al. 2001), and tectonics (Watkinson and Ward 2006). Preservation mechanisms mainly include early hydrocarbon charging (Gluyas et al. 1990; Robinson and Gluyas 1992; Rothwell et al. 1993), grain coating (Heald and Larese 1974; Ramm et al. 1997), and overpressure (Ramm et al. 1997; Osborne and Swarbrick 1999).

There is little research on physical thresholds of deep reservoirs. Deep clastic reservoirs are extensively distributed, ranging from 3,000 to 6,000 m, but most are distributed between 3,500 and 4,000 m (Wood and Hewett 1984; Surdam et al. 1984; Gaupp et al. 1993; Ehrenberg 1993; Gu 1996; Gu et al. 1998, 2001; Aase et al. 1996; Wilkinson et al. 1997). Deep carbonate reservoirs have developed secondary pores (such as leaching pores and dissolution pores), cracks, and fissures, which greatly improve the physical properties of the reservoirs even at considerable depths (Shi et al. 2005; Xie et al. 2009). Deep volcanic reservoirs are buried deep, with complicated geological conditions, which add to the particularity and uncertainty of the reservoirs (Zhang and Wu 1994; Liu et al. 2010a). The physical properties of deep metamorphic reservoirs are decided by the development of fractures and are affected by structural movements and faulting activities (Nelson 1985; Waples 1990; Walker and James 1992; Nelson 2000).

Due to the burial depth, long evolutional history, multiple formation mechanisms, and complicated distribution, geological and geophysical exploration for deep reservoirs is difficult, and limited by technical conditions. The accuracy of geophysical data is low, the number of deep wells and ultra-deep wells are limited, and the availability of original data cannot be guaranteed. Besides, there are intense diagenesis and structural volcanic activities and abnormal pressures. All of these add to the difficulty in understanding the genesis of deep-buried high-porosity and high-permeability reservoirs. Traditional petroleum and geological theories can neither explain why there are porosities of 20 % or more at such depths nor guide the exploration and development of deep reservoirs. Therefore, understanding the genesis and preservation mechanisms of deep-buried high-quality reservoirs and studying the physical thresholds of different reservoirs are of great significance to improving the success rate of deep effective reservoir exploration and reducing exploration risks.

6.3 Phase behavior and conversion mechanisms of deep hydrocarbon and its distribution prediction

Deep hydrocarbons are subject to complicated temperature and pressure conditions and are affected by multiple fluid compositions. Understanding the phase behavior of oil and gas accumulation in deep basins plays a leading role in revealing its genetic mechanism and distribution regularity.

The temperature and pressure environment of deep hydrocarbon reservoirs mainly include high temperature + high pressure, high temperature + low pressure, low temperature + high pressure, and low temperature + low pressure (Miao et al. 2000; Gu et al. 2001; Jiao et al. 2002; Ma et al. 2005; Meng et al. 2006; Zhang et al.

2008). Deep hydrocarbon reservoirs are primarily composed of gaseous hydrocarbons, oil gas mixtures, oil, water vapor, and water (Zhang 2006). Due to the unique temperature and pressure environment, deep hydrocarbons exist in three forms: free, dissolved, and adsorbed, and three phases: oil, gas, and mixed (Tuo 2002; Shi et al. 2005; Wu and Xian 2006; Huang et al. 2007).

The migration of deep hydrocarbons is affected by a variety of forces, such as buoyancy, (oil) gas molecular expansive force, capillary force, molecular adsorption, and binding forces, which jointly act on the accumulation and entrapment of hydrocarbons (Pang et al. 2007b). For deep conventional reservoirs, buoyancy is the main force on the migration of hydrocarbons; for deep unconventional reservoirs, gas molecular expansion is the main force (Gies 1984; Pang et al. 2003; Xiao et al. 2008; Ma et al. 2009; Jiang et al. 2010; Zhu et al. 2010).

Affected by the widely different temperature and pressure conditions and complicated fluid compositions, deep hydrocarbon reservoirs occur in different forms, which are difficult to predict. Besides, due to the unique porosity and permeability conditions, the fluid moving forces are so complicated that it is impossible to characterize the dynamic mechanism that controls the hydrocarbons. In addition, the hydrocarbon driving forces are related to the conditions in a complicated manner. Different hydrocarbon phases have different forces, and the forces are affected by the hydrocarbon phases. Therefore, the formation of deep hydrocarbon reservoirs is not only related to fluid composition and occurrence conditions, but also is jointly controlled by a range of forces including buoyancy. Prediction of the occurrence conditions of deep fluids and characterization of their dynamic mechanism are important to the scientific prediction of deep hydrocarbon resources and the determination of favorable exploration directions.

6.4 Economic feasibility evaluation for deep-buried tight hydrocarbon exploration and development

The exploration and development of deep hydrocarbons is difficult, and the investment return is low. It not only relies on the further understanding of related theories, but also is affected by technical (well drilling and completion techniques) and economic feasibilities.

Economic feasibility is a key to the development of deep hydrocarbon exploration and can decide the exploration direction. It is related to development costs and profits. Profits are affected by the international oil price. Oil exploration is a system, in which drilling costs account for about 50 %–70 % of the total production costs and drilling costs increase considerably as the drilling depth increases (Du and Yao 2001). According to research by some scholars, drilling costs increase exponentially, instead of linearly, as the depth increases (Guan et al. 2012). Profits are mainly affected by oil price and output. The higher the oil price is, the more output there will be, and the more profit it will bring. Output is controlled by the market as well as geological and engineering factors. The higher the international oil price is, the more output there will be. Rapid increase of output will in turn restrain the price rise.

According to the 2014 International Energy Agency, the United States will close 2 % of its shale wells if the international oil price drops below $80/bbl; it will close 18 % of its shale wells if the international oil price drops below $60/bbl. This suggests that production profits are the motive force of hydrocarbon exploration. Cost and break-even analysis and economic feasibility research are important factors for the prediction of deep hydrocarbon exploration prospects.

Acknowledgments The paper is completed based on the National Basic Research Program of China (973 Program, 2011CB201100) "Complex hydrocarbon accumulation mechanism and enrichment regularities of deep superimposed basins in Western China" and National Natural Science Foundation of China (U1262205) under the guidance of related department heads and experts. Here, the authors express gratitude and appreciation for their contributions. In addition, we thank the postgraduates including Li-Ming Zhou, Hua Bai, Jing Bai, Lu-Ya Wu, Qian-Wen Li, and Rui Yu for their help in data gathering and sorting.

References

Aase NE, Bjørkum PA, Nadeau PH. The effect of grain coating micro quartz on preservation of reservoir porosity. AAPG Bull. 1996;80(10):1654–73.

Amthor JE, Okkerman J. Influence of early diagenesis on reservoir quality of Rotliegende sandstones, northern Netherlands. AAPG Bull. 1998;82(12):2246–65.

Bai GP, Cao BF. Characteristic and distribution patterns of deep petroleum accumulations in the world. Oil Gas Geol. 2014;01:7–19 (in Chinese).

Barker C, Takach NE. Prediction of natural gas composition in ultradeep sandstone reservoirs. AAPG Bull. 1992;76(12):1859–73.

Barker C. Calculated volume and pressure changes during the thermal cracking of oil to gas in reservoirs. AAPG Bull. 1990;74:1254–61.

Barker C. Primary migration: The importance of water organic mineral matter interaction in the source rock. Tulsa: AAPG Studies in Geology; 1980.

Baskin DK. Atomic ratio H/C of kerogen as an estimate of thermal maturity and organic matter conversion. AAPG Bull. 1997;81(9):1437–50.

Berkenpas PG. The Milk River shallow gas pool: role of the up dip water trap and connate water in gas production from the pool. SPE. 1991;229(22):371–80.

Bluokeny TB. Translated by Shi D. Development properties under deep abnormally high pressure. Nat Gas Geosci. 2001;12(4–5):61–4 (in Chinese).

Burruss RC. Stability and flux of methane in the deep crust a review. In: The future of energy gases. US Geological Survey Professional Paper; 1993. p. 21–29.

Cant DJ. Diagenetic traps in sandstones. AAPG Bull. 1986;70(2):155–60.

Chen DX, Pang XQ, Qiu NS, et al. Accumulation and filling mechanism of lenticular sand body reservoirs. Earth Sci. 2004;29(4):483–8 (in Chinese).

Chen JP, Liang DG, Wang XL, et al. Mixed oils derived from multiple source rocks in the Cainan oilfield, Junggar Basin, Northwest China. Part II: artificial mixing experiments on typical crude oils and quantitative oil-source correlation. Org Geochem. 2003;34(7):911–30.

Chen SJ, Zhang Y, Lu JG, et al. Limitation of biomarkers in the oil-source correlation. J Southwest Petrol Inst. 2006;28(5):11–4 (in Chinese).

Cheng KM, Wang ZY, Zhong NN, et al. Carbonate hydrocarbon generation theory and practice. Beijing: Petroleum Industry Press; 1996. p. 290–303 (in Chinese).

Cluff RM, Cluff SG. The origin of Jonah field, Northern Green River Basin, Wyoming. In: Robinson JW and Shanley KW (eds.) Jonah field: case study of a tight-gas fluvial reservoir. AAPG Stud Geol. 2004;52:215–41.

Dai JX, Qin SF, Tao SZ, et al. Developing trends of natural gas industry and the significant progress on natural gas geological theories in China. Nat Gas Geosci. 2005;16(2):127–42 (in Chinese).

Dai JX. Characteristic of abiogenic gas resource and resource perspective. Nat Gas Geosci. 2006;17(1):1–6 (in Chinese).

Dai JX. Enhance the studies on natural gas geology and find more large gas field in China. Nat Gas Geosci. 2003;14(1):3–14 (in Chinese).

Darouich T, Behar F, Largeau C. Thermal cracking of the light aromatic fraction of Safaniya crude oil—experimental study and compositional modelling of molecular classes. Org Geochem. 2006;37:1130–54.

Davies GR, Smith LB. Structurally controlled hydrothermal dolomite reservoir facies: an overview. AAPG Bull. 2006;90(11):1641–90.

Dickey PA. Possible primary migration of oil from source rock in oil phase. AAPG Bull. 1975;59(2):337–45.

Dolbier RA. Origin, distribution, and diagenesis of clay minerals in the Albian Pinda Formation, offshore Cabinda, Angola. Reno: University of Nevada; 2001.

Dominé F, Dessort D, Brevart O. Towards a new method of geochemical kinetic modelling: implications for the stability of crude oils. Org Geochem. 1998;28:576–612.

Du CG, Hao F, Zou HY, et al. Process and mechanism for oil and gas accumulation, adjustment and reconstruction in Puguang gas field, Northeast Sichuan Basin, China. Sci China. 2009;52(12):1721–31 (in Chinese).

Du JH, Wang ZM, Li QM, et al. Oil and gas exploration of cambrian-ordovician carbonate in Tarim Basin. Beijing: Petroleum Industry Press; 2010. p. 1–4 (in Chinese).

Du XD, Yao C. The necessary need of deep oil and gas exploration. Mar Oil and Gas Geol. 2001;6(1):1–5 (in Chinese).

Durand B. Understanding of HC migration in sedimentary basins (present state of knowledge). Org Geochem. 1988;13(1–3):445–59.

Dyman TS, Cook TA. Summary of deep oil and gas wells in the United States through 1998, Chapter B. In: Dyman TS, Kuuskraa BN, editors. Geologic Studies of Deep Natural Gas Resources. U.S. Geological Survey Digital Data Series DDS-67, Version 1.00; 2001. p. B1–B9.

Dyman TS, Crovelli RA, Bartberger CE, et al. Worldwide estimates of deep natural gas resources based on the US Geological Survey

World Petroleum Assessment 2000. Nat Resour Res. 2002;11(6):207–18.

Dyman TS, Litinsky VA, Ulmishek GF. Geology and natural gas potential of deep sedimentary basins in the Former Soviet Union, Chapter C. In: Dyman TS, Kuuskraa VA, editors. U. S. Geological Survey studies of deep natural gas resources. U. S. Geological Survey Digital Data Series DDS-67, Version 1.00; 2001. p. C1–C29.

Dyman TS, Spencer CW, Baird JK, et al. Geologic and production characteristics of deep natural gas resources based on data from significant fields and reservoirs, Chapter C. In: Dyman TS, Rice DD, Westcott WA, editors. Geologic controls of deep natural gas resources in the United States. U.S. Geological Survey Bulletin 2146-C; 1997. p. 19–38.

Ehrenberg SN. Preservation of anomalously high porosity in deeply buried sandstones by grain-coating chlorite: examples from the Norwegian Continental Shelf. AAPG Bull. 1993;77(7):1260–86.

Ezat H. The role of burial diagenesis in hydrocarbon destruction and H_2S accumulation, Upper Jurassic Smackover Formation, Black Creek Field, Mississippi. AAPG Bull. 1997;81(1):26–45.

Fan JS. Characteristics of carbonate reservoirs for oil and gas field in the world and essential controlling factors for their formation. Earth Sci Front. 2005;13(3):23–30 (in Chinese).

Fan SF, Zhou ZY, Jie QD. Characteristics and simulation experimental study of generation and conservation of deep carbonate oils and gases. Acta Sedimentol Sin. 1997;2:114–7 (in Chinese).

Feng ZQ. Exploration of large Qingshen gas field in the Songliao Basin. Nat Gas Ind. 2006;26(6):1–5 (in Chinese).

Fisher JB, Boles JR. Water rock interaction in tertiary sandstones, San Joaquin Basin, California, USA: diagenetic controls on water composition. Chem Geol. 1990;82:83–101.

Gaupp R, Matter A, Platt J, et al. Diagenesis and fluid evolution of deeply buried Permian (Rotliegende) gas reservoir, Northwest Germany. AAPG Bull. 1993;77(7):1111–28.

Gies RM. Case history for a major Alberta deep basin gas trap: the Cadomin Formation. AAPG Mem. 1984;38:115–40.

Gluyas JG, Leonard AJ, Oxtoby NH. Diagenesis and oil emplacement: the race for space Ultra Trend, North Sea. In: 13th International sedimentological congress. International Association of Sedimentologists; 1990. p. 193.

Gong S, Peng PA, Lu YH, et al. The model experiments for secondary pyrolysis of biodegraded bituminous sandstone. Chin Sci Bull. 2004;49:39–47 (in Chinese).

Gormly JR, Buck SP, Chung HM. Oil-source rock correlation in the North Viking Graben. Org Geochem. 1994;22(3–5):403–13.

Gu JY. Sedimentary environment and reservoir characters of the Carboniferous Donghe sandstone in the Tarim Basin. Acta Geol Sinica. 1996;70(2):153–61 (in Chinese).

Gu JY, Ning CQ, Jia JH. High-quality reservoir features and generate analysis of clastic rocks in the Tarim Basin. Geol Rev. 1998;44(1):83–9 (in Chinese).

Gu JY, Zhou XX, Liu WL. Buried-hill Karst and the distribution of hydrocarbon in Lunnan Area, Tarim Basin. Beijing: Petroleum Industry Press; 2001 (in Chinese).

Guan DS, Niu JY. Unconventional oil and gas geology in China. Beijing: Petroleum Industry Press; 1995. p. 60–5 (in Chinese).

Guan D, Luo Y, Zhang XD, et al. Research on prediction methods and analysis of cost influential factors for offshore drilling. Drill Prod Technol. 2012;35(4):41–9 (in Chinese).

Guo R. Supplement to determining method of cut-off value of net play. Petrol Explor Dev. 2004;31(5):140–4 (in Chinese).

Hao F, Zou HY, Huang BJ. Gas generation model and geo-fluid response, Yinggehai Basin. Science China Ser D. 2002;32(11):889–96 (in Chinese).

Hao SS, Gao G, Wang FY. Highly-over mature marine source rocks. Beijing: Petroleum Industry Press; 1996. p. 98–109 (in Chinese).

Hao SS, Liu GD, Huang ZL, et al. Dynamic equilibrium model of migration and accumulation for natural gas resource evaluation. Petrol Explor Dev. 1993;20(3):16–21 (in Chinese).

Harris RG, Bustin RM. Diagenesis, reservoir quality, and production trends of Doig Formation sand bodies in the Peace River area of Western Canada. Bull Can Pet Geol. 2002;48(4):339–59.

He DF, Jia CZ, Zhou XY, et al. Control principles of structures and tectonics over hydrocarbon accumulation in multi-stage super imposed basins. Acta Petrol Sin. 2005;26(3):1–9 (in Chinese).

He ZL, Wei XC, Qian YX, et al. Forming mechanism and distribution prediction of quality marine carbonate reservoirs. Oil Gas Geol. 2011;32(4):489–98 (in Chinese).

Heald MT, Larese RE. Influence of coating on quartz cementation. J Sediment Petrol. 1974;44(4):1269–74.

Hirner A, Graf W, Hahn-Weinheimer P. A contribution to geochemical correlation between crude oils and potential source rocks in the eastern Molasse Basin (Southern Germany). J Geochem Explor. 1981;15(1–3):663–70.

Hu GY, Xiao ZY, Luo X, et al. Light hydrocarbon composition difference between two kinds of cracked gases and its application. Nat Gas Ind. 2005;25(9):23–5 (in Chinese).

Huang JW, Gu Y, Chen QL, et al. Thermal simulation of Silurian bituminous sandstone of hydrocarbon supply in northern Tarim Basin. Petrol Geol Exp. 2012;34(4):445–50 (in Chinese).

Huang J, Zhu RK, Hou DJ, et al. The new advances of secondary porosity genesis mechanism in deep clastic reservoir. Geol Sci Technol Inf. 2007;26(6):76–82 (in Chinese).

Hubbert MK. Entrapment of petroleum under hydrodynamic conditions. AAPG Bull. 1953;37(8):1954–2026.

Hunt JM. Petroleum geochemistry and geology. 1st ed. New York: Freeman; 1979. p. 261–73.

Huo ZP, Pang XQ, Fan K, et al. Application of facies-potential coupling reservoir-controlling effect in typical lithologic reservoirs in Jiyang depression. Petrol Geol Exp. 2014a;36(5):574–83 (in Chinese).

Huo ZP, Pang XQ, Ouyang XC, et al. Upper limit of maturity for hydrocarbon generation in carbonate source rocks in the Tarim Basin, China. Arab J Geosci. 2014b;. doi:10.1007/s12517-014-1408-9.

Ingersoll RV. Tectonics of sedimentary basins with revise nomenclature. In: Busby CJ, Perez AA, editors. Tectonics of sedimentary basins: recent advances. Cambridge, MA: Blackwell Science; 1995. p. 1–151.

Jemison RM. Geology and development of Mills Ranch complex—world's deepest field. AAPG Bulletin. 1979;63(5):804–9.

Jia CZ, Zhou XY, Wang ZM, et al. Petroleum geological features of Kela-2 gas field. Chin Sci Bull. 2002;47(S1):90–6 (in Chinese).

Jiang FJ, Pang XQ, Wu L. Geologic thresholds and its gas controlling function during forming process of tight sandstone gas reservoir. Acta Petrol Sin. 2010;31(1):49–54 (in Chinese).

Jiao HS, Fang CL, Niu JY, et al. Deep petroleum geology of Eastern China. Beijing: Petroleum Industry Press; 2002. p. 24–162 (in Chinese).

Jin ZJ, Zhang LP, Yang L, et al. Primary study of geochemical features of deep fluids and their effectiveness on oil-gas reservoir formation in sedimentary basins. Earth Sci. 2002;27(6):659–65 (in Chinese).

Jing GL. Aromatic compounds in crude oils and source rocks and their application to oil-source rock correlations in the Tarim Basin, NW China. J Asian Earth Sci. 2005;25(2):251–68.

Jones RW, Edison TA. Microscopic observations of kerogen related to geochemical parameters with emphasis on thermal maturation. In: Oltz DF, editor. Low temperature metamorphism of kerogen and clay minerals. Los Angeles: Pacific Section, SEPM; 1978. p. 1–12.

Kang YZ. Geological conditions of the formation of the large Tahe oilfield in the Tarim Basin and its prospects. Geol China. 2003;30(3):315–9 (in Chinese).

Khidir A, Catuneanu O. Sedimentology and diagenesis of the Scollard sandstones in the Red Deer Valley area, central Alberta. Bull Can Pet Geol. 2003;51(1):45–69.

Kutcherov V, Lundin A, Ross RG, et al. Theory of abyssal abiotic petroleum origin: challenge for petroleum industry. AAPG Eur Reg Newslett. 2008;3:2–4.

Land LS, MacPherson GL. Origin of saline formation waters, Cenozoic section, Gulf of Mexico sedimentary basin. AAPG Bull. 1992;76(9):1344–62.

Li GY, Lü MG. Atlas of petroliferous basins in China. Beijing: Petroleum Industry Press; 2002 (in Chinese).

Li GY. Atlas of world petroliferous basins. Beijing: Petroleum Industry Press; 2009 (in Chinese).

Li HX, Ren JH, Ma JY, et al. Abnormal high pore fluid pressure and clastic rock deep reservoir. Petrol Explor Dev. 2001;28(6):5–8 (in Chinese).

Li L, Ren ZW, Sun HB. An integrated evaluation on petroleum geology of the deep reservoirs in the west depression, Liaohe Basin, China. Acta Petrol Sin. 1999;20(6):9–15 (in Chinese).

Li MC, Shan XQ, Ma CH, et al. Dynamics of sand lens reservoir. Oil Gas Geol. 2007;28(2):209–15 (in Chinese).

Li SM, Pang XQ, Jin ZJ, et al. Geochemical characteristics of the mixed oil in Jinhu Sag of Subei Basin. J China Univ Petrol. 2002;26(1):11–5 (in Chinese).

Li Z, Li HS. An approach to genesis and evolution of secondary porosity in deeply buried sandstone reservoirs, Dongpu depression. Sci Geol Sin. 1994;29(3):267–75 (in Chinese).

Li Z, Chen JS, Guan P. Scientific problems and frontiers of sedimentary diagenesis research in oil-gas-bearing basins. Acta Petrol Sin. 2006;22(8):2113–22 (in Chinese).

Li Z, Huang SJ, Liu JQ, et al. Buried diagenesis, structurally controlled thermal-fluid process and their effect on Ordovician carbonate reservoirs in Tahe, Tarim Basin. Acta Sedimentol Sin. 2010;28(5):969–74 (in Chinese).

Lin CS, Li H, Liu JY. Major unconformities, tectonostratigraphic framework, and evolution of the superimposed Tarim Basin, northwest China. J Earth Sci. 2012;23(4):395–407.

Liu JQ, Meng FC, Cui Y, et al. Discussion on the formation mechanism of volcanic oil and gas reservoirs. Acta Petrol Sin. 2010a;26(1):1–13 (in Chinese).

Liu QY, Jin ZJ, Gao B, et al. Characterization of gas pyrolysates from different types of Permian source rocks in Sichuan Basin. Nat Gas Geosci. 2010b;21(5):700–4 (in Chinese).

Liu WH, Chen MJ, Guan P, et al. Ternary geochemical-tracing system in natural gas accumulation. Sci China Ser D. 2007a;37(7):908–15 (in Chinese).

Liu WH, Zhang JY, Fan M, et al. Gas generation character of dissipated soluble organic matter. Petrol Geol Exp. 2007b;29(1):1–6 (in Chinese).

Liu Z, Zhu WQ, Sun Q, et al. Characteristics of geotemperature-geopressure systems in petroliferous basins of China. Acta Petrol Sin. 2012;33(1):1–17 (in Chinese).

Luo P, Qiu YN, Jia AL, et al. The present challenges of Chinese petroleum reservoir geology and research direction. Acta Sedimentol Sin. 2003;21(1):142–7 (in Chinese).

Luo XR, Yu J, Zhang FQ, et al. Numerical migration model designing and its application in migration studies in the 8th segment in the Longdong area of Ordos Basin, China. Sci China Ser D. 2007;37(S1):73–82 (in Chinese).

Luo XR. Review of hydrocarbon migration and accumulation dynamics. Nat Gas Geosci. 2003;14(5):337–46 (in Chinese).

Ma WM, Wang XL, Ren LY, et al. Over-pressures and secondary pores in Dongpu Depression. J Northwest Univ. 2005;35(3):325–30 (in Chinese).

Ma XZ, Pang XQ, Meng QY, et al. Hydrocarbon expulsion characteristics and resource potential of deep source rocks in the Liaodong Bay. Oil Gas Geol. 2011;32(2):251–9 (in Chinese).

Ma YS, Chu ZH. Building-up process of carbonate platform and high-resolution sequence stratigraphy of reservoirs of reef and oolitic shoal facies in Puguang gas field. Oil Gas Geol. 2008;29(5):548–56 (in Chinese).

Ma YS, Guo TL, Zhao XF, et al. Formation mechanism of high porosity and permeability dolomites deeply buried in Puguang gas field. Sci China Ser D. 2007;37:43–52 (in Chinese).

Ma YS, Zhang SC, Guo TL, et al. Petroleum geology of the Puguang sour gas field in the Sichuan Basin, SW China. Mar Petrol Geol. 2008;25(4–5):357–70.

Ma ZZ, Wu HY, Dai GW. Progress of study on reservoir forming mechanism of deep basin gas. Petrol Geol Oilfield Dev Daqing. 2009;28(6):57–61 (in Chinese).

Magara K. Compaction and fluid migration: practical petroleum geology. London: Elsevier; 1978. p. 319.

Mango FD. The stability of hydrocarbon under the time/temperature conditions of petroleum genesis. Nature. 1991;352:146–8.

Manhadieph NR. Translated by Shi D. Geothermal conditions of deep hydrocarbon-bearing layers. Nat Gas Geosci. 2001;12(2):56–60 (in Chinese).

Masters JA. Deep basin gas trap, Western Canada. AAPG Bull. 1979;63(z2):152–81.

Melienvski BN. Translated by Shi D. Discussion on the deep zonation of the oil-gas formation. Nat Gas Geosci. 2001;12(4–5):52–5 (in Chinese).

Meng YL, Gao JJ, Liu DL, et al. Diagenetic facies analysis and anomalously high porosity zone prediction of the Yuanyanggou area in the Liaohe depression. J Jilin Univ. 2006;36(2):227–33 (in Chinese).

Miao JY, Zhu ZQ, Liu WR, et al. Relationship between temperature, pressure and secondary pores of deep reservoirs in eogene in the Jiyang depression. Acta Petrolei Sinica. 2000;21(3):36–40 (in Chinese).

Michael GE, Lin LH, Philp RP, et al. Biodegradation of tar-sand bitumens from the Ardmore/Anadarko Basins, Oklahoma—II correlation of oils, tar sands and source rocks. Org Geochem. 1989;14(6):619–33.

Ministry of Land and Resources. Resource evaluation of the new round national oil and gas resources. Beijing: China Land Press; 2005 (in Chinese).

Moretti I, Labaume P, Sheppard SMF, et al. Compartmentalization of fluid migration pathways in the sub-Andean zone, Bolivia. Tectonophysics. 2002;348(1–3):5–24.

Mukhopadhyay PK, Wade JA, Kruge MA. Organic facies and maturation of Jurassic/Cretaceous rocks, and possible oil-source rock correlation based on pyrolysis of asphaltenes, Scotian Basin, Canada. Org Geochem. 1995;22(1):85–104.

Navon O, Hutcheon ID, Rossman GR, et al. Mantle-derived fluids in diamond micro-inclusions. Nature. 1988;335:784–9.

Nelson RA. Production characteristics of the fractured reservoirs of the La Paz field, Maracaibo Basin, Venezuela. AAPG Bull. 2000;84(11):1791–809.

Nelson RA. Geologic analysis of naturally fractured reservoirs. Houston: Gulf Publishing Company; 1985. p. 304–51.

Odden W, Patience RL, Van Graas GW. Application of light hydrocarbons (C_4-C_{13}) to oil/source rock correlations: a study of the light hydrocarbon compositions of source rocks and test fluids from offshore Mid-Norway. Org Geochem. 1998;28(12):823–47.

Osborne MJ, Swarbrick RE. Diagenesis in North Sea HPHT clastic reservoirs— consequences for porosity and overpressure prediction. Mar Pet Geol. 1999;16:337–53.

Pan WQ, Liu YF, Dickson J, et al. The geological model of hydrothermal activity in outcrop and the characteristics of carbonate hydrothermal karst of Lower Paleozoic in Tarim Basin. Acta Sedimentol Sin. 2009;27(5):983–94 (in Chinese).

Pang XQ. Key challenges and research methods of petroleum exploration in the deep of superimposed basins in western China. Oil Gas Geol. 2010;31(5):517–41 (in Chinese).

Pang XQ, Chen DX, Zhang J, et al. Concept and categorize of subtle reservoir and problems in its application. Lithol Reserv. 2007a;19(1):1–8 (in Chinese).

Pang XQ, Chen DX, Zhang J, et al. Physical simulation experimental study on mechanism for hydrocarbon accumulation controlled by facies—potential—source coupling. Journal of Palaeogeography. 2013;5:575–92 (in Chinese).

Pang XQ, Gao JB, Meng QY. A discussion on the relationship between tectonization and hydrocarbon accumulation and dissipation in the platform-basin transitional area of Tarim Basin. Oil Gas Geol. 2006;27(5):594–603 (in Chinese).

Pang XQ, Gao JB, Lü XX, et al. Reservoir accumulation pattern of multi-factor recombination and procession superimposition and its application in Tarim Basin. Acta Petrol Sin. 2008;29(2):159–66 (in Chinese).

Pang XQ, Jiang ZX, Huang HD, et al. Genetic mechanism, development mode and distribution forecast of overlapping continuous oil and gas reservoir. Acta Petrol Sin. 2014a;35(5):1–34 (in Chinese).

Pang XQ, Jin ZJ, Zuo SJ. Dynamics models and classification of hydrocarbon accumulations. Earth Sci Front. 2000;7(4):507–14 (in Chinese).

Pang XQ, Jin ZJ, Jiang ZX, et al. Critical condition for gas accumulation in the deep basin trap and physical modeling. Nat Gas Geosci. 2003;14(2):207–14 (in Chinese).

Pang XQ, Li QW, Chen JF, et al. Recovery method of original TOC and its application in source rocks at high mature—over mature stage in deep petroliferous basin. J Palaeogeogr. 2014b;16(6):769–88 (in Chinese).

Pang XQ, Li SM, Jin ZJ, et al. Geochemical evidence of hydrocarbon expulsion threshold and its application. Earth Sci. 2004;29(4):384–9 (in Chinese).

Pang XQ, Luo XR, Jiang ZX, et al. Advancements and problems on hydrocarbon accumulation research of complicated superimposed basins in Western China. Adv Earth Sci. 2007b;22(9):879–87 (in Chinese).

Pape H, Clauser C, Iffland J, et al. Anhydrite cementation and compaction in geothermal reservoirs: interaction of pore-space structure with flow, transport, P-T conditions, and chemical reactions. Int J Rock Mech Min Sci. 2005;42(7–8):1056–69.

Perry W Jr. Structural settings of deep natural gas accumulations in the conterminous United States. In: Dyman TS, Rice DD and Westcott PA (eds.), geologic controls of deep natural gas resources in the United States. USGS Bull. 1981;2146-D:41–6.

Philippi GT. Correlation of crude oils with their oil source formation, using high resolution GLC C_6-C_7 component analyses. Geochim Cosmochim Acta. 1981;45(9):1495–513.

Qiao HS, Zhao CL, Tian KQ, et al. Petroleum geology of deep reservoirs in the eastern China. Beijing: Petroleum Industry Press; 2002. p. 254–8 (in Chinese).

Qin JZ, Jin JC, Liu BQ. Thermal evolution pattern of organic matter abundance in various marine source rocks. Oil Gas Geol. 2005;26(2):177–84 (in Chinese).

Ramm M, Forsberg AW, Jahren J S. Porosity-depth trends in deeply buried Upper Jurassic reservoirs in the Norwegian Central Graben: an example of porosity preservation beneath the normal economic basement by grain-coating microquartz. In: AAPG Memoir 69: reservoir quality prediction in sandstones and carbonates; 1997. p. 177–199.

Robinson A, Gluyas J. Duration of quartz cementation in sandstones, North Sea and Haltenbanken Basins. Mar Pet Geol. 1992;9:324–7.

Rodrenvskaya МИ. Translated by Shi D. Hydrocarbon potential in deep layers. Nat Gas Geosci. 2001;12(4–5):49–51 (in Chinese).

Rossi C, Marfil R, Ramseyer KR, et al. Facies related diagenesis and multiphase siderite cementation and dissolution in the reservoir sandstones of the Khatatba Formation, Egypt's Western Desert. J Sediment Res. 2001;71(3):459–72.

Rothwell NR, Sorensen A, Peak JL, et al. Gyda: recovery of difficult reserves by flexible development and conventional reservoir management. In: SPE26778, Proceedings of the offshore Europe conference; 1993. p. 271–280.

Samvelov PГ. Translated by Guan F X Deep oil-gas reservoir forming characteristics and distribution. Northwest Oil Gas Explor. 1997;9(1):52–7 (in Chinese).

Seifert WK. Steranes and terpanes in kerogen pyrolysis for correlation of oils and source rocks. Geochim Cosmochim Acta. 1978;42(5):473–84.

Shao CX, Wang YZ, Cao YZ. Determine the effective reservoir lower limit—two new methods and application to Paleogene deep clastic reservoirs in Dongying Sag as an example. J Oil Gas Technol. 2008;30(2):414–6 (in Chinese).

Shi JA, Wang Q. A discussion on main controlling factors on the properties of clastic gas reservoirs. Acta Sedimentol Sin. 1995;13(2):128–39 (in Chinese).

Shi X, Dai JX, Zhao WZ. Analysis of deep oil and gas reservoir exploration prospect. China Petrol Explor. 2005;10(1):1–10 (in Chinese).

Simmons MR. The world's giant oilfields. Hubbert Center Newslett. 2002;1:1–62.

Song CC, Peng YM, Qiao YL, et al. Direction for future oil & gas exploration in SINOPEC's exploration area in the Junggar Basin. Oil Gas Geol. 2008;29(4):453–9 (in Chinese).

Song FQ, Zhang DJ, Wang PR, et al. A contribution evaluation method for biodegraded mixed oils. Petrol Explor Dev. 2004;31(2):67–70 (in Chinese).

Spencer CW. Review of characteristics of low-permeability gas reservoirs in western United States. AAPG Bull. 1989;73(5):613–29.

Stahl WJ. Source rock-crude oil correlation by isotopic type-curves. Geochim Cosmochim Acta. 1978;42(10):1573–7.

Sugisaki R. Deep-seated gas emission induced by the earth tide: a basic observation for geochemical earthquake prediction. Science. 1981;212(4500):1264–6.

Sun LD, Fang CL, Li F, et al. Petroleum exploration and development practices of sedimentary basins in China and research progress of sedimentology. Petrol Explor Dev. 2010;37(4):385–96 (in Chinese).

Sun LD, Zou CN, Zhu RK, et al. Formation, distribution and potential of deep hydrocarbon resources in China. Petrol Explor Dev. 2013;40(6):641–9 (in Chinese).

Surdam RC, Boese SW, Crossey LJ. The chemistry of secondary porosity. AAPG Mem. 1984;37:127–49.

Surdam RC, Crossey LJ, Hagen ES, et al. Organic-inorganic and sandstone diagenesis. AAPG Bull. 1989;73(1):1–23.

Svetlakova EA. Model of formation regularities of distribution of hydrocarbon pools in the North Caspian Basin. In: Krylov NA, Nekhrikova NA, editors. Petroleum potential of the North Caspian Basin and Adjacent Areas (Neftegazonosnost Prikaspiyskoy vpadinyi sopredelnykh rayonov). Moscow: Nauka; 1987. p. 151–4.

Telnæs N, Cooper BS. Oil-source rock correlation using biological markers, Norwegian continental shelf. Mar Pet Geol. 1991;8(3):302–10.

Tissot BP, Welte DH. Petroleum formation and occurrence: a new approach to oil and gas exploration. Berlin: Springer; 1978.

Tissot B, Durand B, Espitalé J, et al. Influence of the nature and diagenesis of organic matter in the formation of petroleum. AAPG Bull. 1974;58:499–506.

Tuo JC. Research status and advances in deep oil and gas exploration. Adv Earth Sci. 2002;17(4):565–71 (in Chinese).

Tuo JC. Secondary hydrocarbon generation in carbonate rocks. Nat Gas Geosci. 1994;3:9–13 (in Chinese).

Tuo JC, Huang XZ, Ma WY. The lagging phenomenon of the petroleum generation in carbonate rocks. Petrol Explor Dev. 1994;21(6):1–5 (in Chinese).

Tuo JC, Wang XB, Zhou SX. Prospect of oil-gas resources in the Bohai Bay Basin. Nat Gas Geosci. 1999a;06:27–31 (in Chinese).

Tuo JC, Wang XB, Zhou SX, et al. Exploration situation and research development of deep hydrocarbon. Nat Gas Geosci. 1999b;06:1–8 (in Chinese).

Walker RG, James NP. Facies models: response to sea level change. Geol Assoc Can. 1992;4(5):153–70.

Wan L, Sun Y, Wei GQ. A new method used to determine the lower limit of the petrophysical parameters for reservoir and its application: a case study on Zhongbu Gas Field in Ordos Basin. Acta Sedimentol Sin. 1999;17(3):454–7 (in Chinese).

Wang GQ, Chen YD, Zhou YH. Difficulty analysis and solution discussion for deep and ultra-deep exploration well. Oil Drill Prod Technol. 1998;20(1):1–17 (in Chinese).

Wang Q, Song YJ, Zhang SM. The study on the methods used to delineate the deep tight gas bearing sands in Daqing oil field. J Daqing Petrol Inst. 1994;18(2):22–6 (in Chinese).

Wang TG, Wang CJ, He FQ, et al. Determination of double filling ratio of mixed crude oils in the Ordovician oil reservoir, Tahe Oilfield. Petrol Geol Exp. 2004;26(1):74–9 (in Chinese).

Wang T. Deep Basin Gas in China. Beijing: Petroleum Industry Press; 2002 (in Chinese).

Wang WJ, Song N, Jiang NH, et al. The mixture laboratory experiment of immature oils and mature oils, making and using the mixture source theory plat and its application. Petrol Explor Devel. 1999;26(4):34–7 (in Chinese).

Wang XB, Tuo JC, Yan H, et al. The formation mechanism of deep oil and gas. The meeting abstract on Chinese sedimentology in 2001. Wuhan: China Univ Geosci; 2001. p. 163–4 (in Chinese).

Wang YM, Niu JY, Qiao HS, et al. An approach to the hydrocarbon resource potential of Bohai Bay Basin. Petrol Explor Dev. 2002;29(2):21–5 (in Chinese).

Wang YS, Pang XQ, Liu HM, et al. Low-potential reservoir-controlling characteristics and dynamics mechanism and its role in oil and gas exploration. J Earth Sci. 2013;38(1):165–72 (in Chinese).

Wang Y, Su J, Wang K, et al. Distribution and accumulation of global deep oil and gas. Nat Gas Geosci. 2012;23(3):526–34 (in Chinese).

Wang ZM, Wang QH, Wang Y. Poll formation condition and controlling factor for Hetian gas field in the Tarim Basin. Mar Oil Gas Geol. 2000;5(1):124–32 (in Chinese).

Waples DW. Time and temperature in petroleum formation: application and detachment free deformation. J Struct Geol. 1990;12(3):355–81.

Watkinson AJ, Ward EMG. Reactivation of pressure solution seams by a strike slip fault sequential, dilational jog formation and fluid flow. AAPG Bull. 2006;90(8):1187–200.

White IC. The geology of natural gas. Science. 1885;6(128):42–4.

Wilkinson M, Darby D, Haszeldine RS, et al. Secondary porosity generation during deep burial associated with overpressure leak-off: Fulmar Formation, United Kingdom Central Graben. AAPG Bull. 1997;81(5):803–13.

Williams LB, Hervig RL, Wieser ME, et al. The influence of organic matter on the boron isotope geochemistry of the gulf coast sedimentary basin, USA. Chem Geol. 2001;174:445–61.

Wood JR, Hewett TA. Reservoir diagenesis and convective fluid flow. AAPG Mem. 1984;37:99–110.

Wu FQ, Xian XF. Current state and countermeasure of deep reservoirs exploration. Sediment Geol Tethyan Geol. 2006;26(2):68–71 (in Chinese).

Wu YX, Wang PJ, Bian WH, et al. Reservoir properties of deep volcanic reservoirs in the Songliao Basin. Oil Gas Geol. 2012;33(4):236–47 (in Chinese).

Xiao ZH, Zhong NN, Huang ZL, et al. A study on hydrocarbon pooling conditions in tight sandstones through simulated experiments. Oil Gas Geol. 2008;29(6):721–5 (in Chinese).

Xie JL, Huang C, Wang XX. Distribution of proved reserves of carbonate oil and gas pools in China. Mar Orig Petrol Geol. 2009;14(2):24–30 (in Chinese).

Xie ZY, Jiang ZS, Zhang Y, et al. Novel method of whole rock thermal simulation and application to the evaluation of gas source rock. Acta Sedimentol Sin. 2002;20(3):510–4 (in Chinese).

Yang XH, Ye JR, Sun YC, et al. Diagenesis evolution and preservation of porosity in the member2, Shahejie Formation, Qinan fault step zone. China Offshore Oil Gas (Geol). 1998;12(4):242–8 (in Chinese).

Ye JR, Wang LJ, Shao R. Fluid dynamic fields in hydrocarbon accumulation dynamics. Oil Gas Geol. 1999;20(2):86–9 (in Chinese).

Zappaterra E. Source-rock distribution model of the Periadriatic region. AAPG Bull. 1994;78(3):333–54.

Zhang DY. Fluid characteristics and reservoir forming condition in deep oil and gas reservoirs. Petrol Geol Recovery Effic. 2006;13(5):45–9 (in Chinese).

Zhang HF, Fang CL, Gao XZ, et al. Petroleum geology. Beijing: Petroleum Industry Press; 1999 (in Chinese).

Zhang WC, Li H, Li HJ, et al. Genesis and distribution of secondary porosity in the deep horizon of Gaoliu area, Nanpu Sag. Petrol Explor Dev. 2008;35(3):308–12 (in Chinese).

Zhang XD, Tan XC, Chen JS. Reservoir and controlling factors of the secondary member of the Jialing River Formation in the hit and the south transition strip of Sichuan. Nat Gas Geosci. 2005a;16(3):338–42 (in Chinese).

Zhang YW, Jin ZJ, Liu GC, et al. Study on the formation of unconformities and the amount of eroded sedimentation in Tarim Basin. Earth Sci Front. 2000;7(4):449–57 (in Chinese).

Zhang ZS, Wu BH. Research on volcanic reservoir and exploration technology home and abroad. Nat Gas Explor Dev. 1994;1:1–26 (in Chinese).

Zhang ZY, Shao LY, Zhang SH, et al. Distribution of mixed crude in western arc structural belt of Taibei Depression in Turpan-Hami Basin. Acta Petrol Sin. 2005b;26(2):15–20 (in Chinese).

Zhao MJ, Zeng Q, Zhang BM. The petroleum geological condition and prospecting orientation in Tarim Basin. Xinjiang Petrol Geol. 2001;22(2):93–6 (in Chinese).

Zhao WZ, Wang ZY, Zhang SC, et al. Successive generation of natural gas from organic materials and its significance in future exploration. Petrol Explor Devel. 2005;32(2):1–7 (in Chinese).

Zhao WZ, Zou CN, Gu ZD, et al. Preliminary discussion on accumulation mechanism of sand lens reservoirs. Petrol Explor Dev. 2007;34(3):273–83 (in Chinese).

Zhong DK, Zhu XM. Pore evolution and genesis of secondary pores in Paleogene clastic reservoirs in Dongying Sag. Oil Gas Geol. 2003;24(3):281–5 (in Chinese).

Zhong DK, Zhu XM, Wang HJ. Characteristics and formation mechanism analysis of deep buried sandstone reservoirs in China. Sci China. 2008;38(S1):11–8 (in Chinese).

Zhou J, Pang XQ. A method for calculation the quantity of hydrocarbon generation and expulsion. Petrol Explor Dev. 2002;29(1):24–7 (in Chinese).

Zhou SX, Wang XB, Tuo JC, et al. Advances in geochemistry of deep hydrocarbon. Nat Gas Geosc. 1999;10(6):9–15 (in Chinese).

Zhu GY, Zhang SC. Hydrocarbon accumulation conditions and exploration potential of deep reservoirs in China. Acta Petrol Sin. 2009;30(6):793–802 (in Chinese).

Zhu GY, Zhang SC, Liang YB, et al. Dissolution and alteration of the deep carbonate reservoirs by TSR: an important type of deep-buried high-quality carbonate reservoirs in Sichuan Basin. Acta Petrol Sin. 2006;22(8):2182–94 (in Chinese).

Zhu ZQ, Zeng JH, Wang JJ. Microscopic experiment on oil flowing through porous media in low permeability sandstone. J Southwest Petrol Univ Sci Technol. 2010;32(1):16–20 (in Chinese).

Zou CN. Unconventional oil and gas geology. Beijing: Geological Publishing House; 2011. p. 50–2 (in Chinese).

4

Nonlinear dynamic analysis and fatigue damage assessment for a deepwater test string subjected to random loads

Kang Liu[1] · Guo-Ming Chen[1] · Yuan-Jiang Chang[1] · Ben-Rui Zhu[1] ·
Xiu-Quan Liu[1] · Bin-Bin Han[1]

Abstract The deepwater test string is an important but vulnerable component in offshore petroleum exploration, and its durability significantly affects the success of deepwater test operations. Considering the influence of random waves and the interaction between the test string and the riser, a time-domain nonlinear dynamic model of a deepwater test string is developed. The stress-time history of the test string is obtained to study vibration mechanisms and fatigue development in the test string. Several recommendations for reducing damage are proposed. The results indicate that the amplitude of dynamic response when the string is subjected to random loads gradually decreases along the test string, and that the von Mises stress is higher in the string sections near the top of the test string and the flex joints. In addition, the fatigue damage fluctuates with the water depth, and the maximum damage occurs in string sections adjacent to the lower flex joint and in the splash zone. Several measures are proposed to improve the operational safety of deepwater test strings: applying greater top tension, operating in a favorable marine environment, managing the order of the test string joints, and performing nondestructive testing of components at vulnerable positions.

Keywords Deepwater test string · Pipe-in-pipe model · Random wave · Nonlinear vibration · Fatigue damage

1 Introduction

In recent years, the deepwater drilling unit HYSY981, which is operated by the China National Offshore Oil Corporation (CNOOC), has drilled several exploration wells in the South China Sea (Xu et al. 2013). Petroleum exploration in China has gradually progressed from near-shore to deepwater locations (Chen et al. 2013; Hu et al. 2013). The deepwater test string is a crucial component used for assessing the formation fluid characteristics and potential production of wells during the early stage of offshore petroleum exploration. However, for the complexity of testing process, marine environment and structure of the string, vibration and fatigue that occur in a deepwater test string are not fully understood. Therefore, the study of nonlinear vibration dynamics and fatigue damage in deepwater test strings is of great scientific and engineering significance.

A deepwater test string is inevitably subjected to rapid currents and waves resulting in wear and fatigue due to friction and collision with the riser. However, previous studies have primarily focused on wave-induced fatigue (Nazir et al. 2008; Khan and Ahmad 2010; Li and Low 2012) and vortex-induced fatigue (Yang et al. 2007; Tognarelli et al. 2010; Song et al. 2011) for a single-layer deepwater string, such as a drilling riser and production riser. There has been little research on the coupling dynamics of deepwater test strings and risers at present. Only a few studies have focused on extrusion and contact analysis of offshore double-layer strings in construction and installation (Wang et al. 2009; Chen and Chia 2010).

✉ Guo-Ming Chen
offshore@126.com

Kang Liu
lkzsww@163.com

[1] Centre for Offshore Engineering and Safety Technology,
China University of Petroleum, Qingdao 266580, Shandong,
China

Edited by Yan-Hua Sun

Currently, studies of nonlinear contact analysis of down-hole oil/drill pipe (Pang et al. 2009; Dong et al. 2012) and multistring analysis (Liu et al. 2014b) are available for reference. In the study of test strings, some researchers have focused on deepwater job safety analysis and risk control techniques (Stomp et al. 2005; Chen et al. 2008; Mogbo 2010; Wendler and Scott 2012), while others have focused on mechanical analysis, design optimization, axial deformation, and other aspects of the downhole string (Zeng et al. 2010; Li 2012; Cheng et al. 2014). Liu et al. (2014a) studied the limits of platform offset for a deep-water test string, but its static and dynamic mechanical behavior have not been evaluated. Xie et al. (2011) studied the dynamics of a deepwater test string based on the top boundary simulation, but the interaction between the test strings and the riser and fatigue damage were not investigated.

An equivalent composite model is often used to analyze deepwater double-layer hydrocarbon strings. In this model, it is assumed that the pipes move together uniformly under external and internal loading, and the equivalent pipe bending and tension are shared equally based on the stiff-ness and the bearing area of the pipes. This model can be used to analyze motion, but is likely to be inaccurate in fatigue damage evaluation (Harrison and Helle 2007). To study the interaction between the test string and the riser, and the structural response of a deepwater test string in various operating modes, a model of the nonlinear dynamics for a deepwater test string is developed. Vibra-tion and damage in deepwater test strings are studied to provide guidelines for structural design and operation management.

2 Nonlinear dynamic model

Deepwater well testing is generally conducted on a floating platform, as illustrated in Fig. 1. The top tension of the riser and the test string are provided by a tensioner and a hook, respectively. The riser and the conductor comprise the outer string system with connection of the wellhead, forming a circulation channel for testing fluid. The inner string system is composed of oil tube/drill pipe, centralizer, subsea test tree, and fluted hanger (Bavidge 2013), forming a passage for hydrocarbons from the sea floor to the platform, which can be used to measure and control the test parameters. The entire deepwater test string system is not only subjected to various external loads (e.g., waves, currents, and platform movement) but also random contact and collision between the test string and the riser.

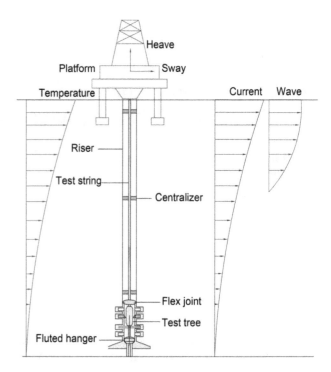

Fig. 1 Schematic diagram for the deepwater test string system

The dynamics of the deepwater test string and the riser are expressed using the partial differential equation of a beam in a vertical plane (Park and Jung 2002)

$$\frac{\partial^2}{\partial z^2}\left(EI\frac{\partial^2 y}{\partial z^2}\right) - \frac{\partial}{\partial z}\left(T\frac{\partial y}{\partial z}\right) + c\frac{\partial y}{\partial t} + m\frac{\partial^2 y}{\partial t^2} = F(z,\,t), \quad (1)$$

where z is the vertical height, m; y is the horizontal dis-placement, m; E is the elastic modulus, Pa; I is the moment of inertia, m^4; T is the effective tension, N; c is the damping factor of the structure, N s/m; m is the mass of the string per unit length, kg; t is time, s; and F is the trans-verse load per unit length, including environmental loads and contact forces between the test string and the riser, N.

The test string makes contact with riser due to the action of waves and currents in deepwater operations. Thus, a pipe-in-pipe model is used to simulate the interaction between the test string and the riser. In Fig. 2, the blue area represents the riser, the red area represents the test string, and the green area represents the gap element that con-strains the riser and the test string.

The displacement vector at any point on the test string is denoted by

$$U_i = [w_i, v_i, \theta_i], \quad (2)$$

where w_i and v_i represent linear displacements in the local coordinate system, m; and θ_i represents the angular dis-placement, rad.

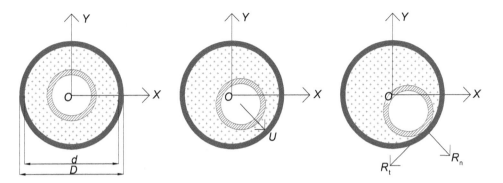

Fig. 2 The contact model for test string and riser

Applying the law of conservation of momentum and the theory of tube-string mechanics, the contact–impact model for the test string and the riser is given by

$$R_{ni} = \begin{cases} m(1-e)\sqrt{\dot{w}_i^2 + \dot{v}_i^2}/\Delta t & \left(\sqrt{\dot{w}_i^2 + \dot{v}_i^2} > \varepsilon, \text{ collision}\right) \\ -T_{ni} & \left(\sqrt{\dot{w}_i^2 + \dot{v}_i^2} < \varepsilon, \text{ contact}\right) \end{cases},$$
$$R_{ti} = \mu R_{ni}$$
$$M_i = d_i R_{ti}/2$$
$$(3)$$

where R_{ni} and R_{ti} represent the normal force and the friction force between the test string and the riser at the ith gap element, N; M_i is the moment of the friction, N m; m is the mass of the string per unit length, kg; e is the restitution coefficient of collision, which is obtained from experiment, and we set $e = 0.58$ in this paper; Δt is the duration of the collision, s; T_{ni} is the reaction force, N; ε is the speed threshold value of the collision and contact, m/s; μ is the friction coefficient, and we set $\mu = 0.25$ in this paper; and d_i is the inner diameter of the riser, m.

Based on these equations, a finite element model of the nonlinear dynamics for the deepwater test string system is obtained

$$\begin{cases} \mathbf{M}_t \ddot{\mathbf{U}}_t + \mathbf{C}_t \dot{\mathbf{U}}_t + \mathbf{K}_t \mathbf{U}_t = \mathbf{F}_t + \mathbf{f}(\mathbf{U}_t, \mathbf{U}_r) \\ \mathbf{M}_r \ddot{\mathbf{U}}_r + \mathbf{C}_r \dot{\mathbf{U}}_r + \mathbf{K}_r \mathbf{U}_r = \mathbf{F}_r - \mathbf{f}(\mathbf{U}_t, \mathbf{U}_r) \end{cases}, \quad (4)$$

where $\mathbf{M}_t$ and $\mathbf{M}_r$ represent the mass matrixes of the test string and the riser, respectively, kg; $\mathbf{C}_t$ and $\mathbf{C}_r$ represent the damping matrixes of the test string and the riser, respectively, N s/m; $\mathbf{K}_t$ and $\mathbf{K}_r$ represent the stiffness matrixes of the test string and the riser, respectively, N/m; $\mathbf{U}_t$ and $\mathbf{U}_r$ represent the global displacement vectors of the test string and the riser, respectively, m; $\mathbf{F}_t$ is the force vector of the test string including the effective weight of the test string and the internal fluid, N; $\mathbf{F}_r$ is the force vector of the riser including both the effective weight of riser, annular fluid, and auxiliary lines and the transverse

loads of random waves and currents, N; $\mathbf{f}(\mathbf{U}_t, \mathbf{U}_r)$ represents the contact or impact load of the test string and the riser, N.

3 Nonlinear vibration induced by waves

3.1 Wave-induced vibration mechanism

Waves and the motion of floating platform are the main loads that drive the dynamic response of a deepwater test string. The boundary condition is given by the wave frequency and the low-frequency motion of the platform due to waves. Currents act on the riser directly, and loads are transmitted to the test string via the gap element, which primarily affects the time-invariant portion in the dynamic response of the test string. Waves affect the dynamic response of a test string in two ways: (1) they produce hydrodynamic loads on the riser, which are transmitted to the test string via the gap element; (2) they cause the floating platform to move, which is modeled through the response amplitude operator (RAO), and form the moving boundary condition at the top of the test string. Figure 3 shows the effects of waves on the deepwater test string system.

The wave spectrum is typically used to simulate ocean waves. In this paper, the P–M spectrum is used. The loads on the deepwater test string system caused by waves are calculated using the Morison equation (Morooka et al. 2005). Simulation of the motion of the floating platform is an essential part of the vibration analysis of the deepwater test string system. Including the effect of long-term drift, the motion of the floating platform is represented as follows (Chang et al. 2008):

$$S(t) = S_0 + S_1 \sin\left(\frac{2\pi t}{T_1} - \alpha_1\right)$$
$$+ \sum_{n=1}^{N} S_n \cos(k_n x - \omega_n t + \varphi_n + \alpha_n), \quad (5)$$

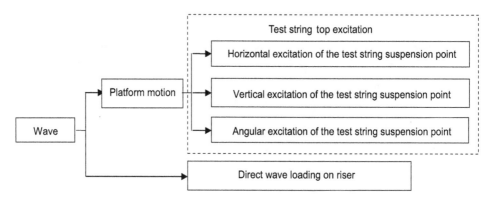

Fig. 3 Effects of waves on the deepwater test string system

where $S(t)$ is the time-dependent platform motion response, m; S_0 is the mean platform offset, m; S_1 is the single amplitude of the platform drift, m; T_1 is the period of platform drift motion, s; α_1 is a phase angle difference between the drift motion and the wave, rad; S_n is the amplitude of the nth wave component, m; k_n, ω_n, and φ_n are the wave number, frequency, and phase angle of the nth wave component, respectively, m^{-1}, Hz, rad; and α_n is the phase angle of the RAO, rad.

Assuming a platform drift amplitude of 10 m, a long-term platform drift period of 250 s, and a significant wave height of 4 m, the time history of floating platform motion is obtained using the method described previously; the results are shown in Fig. 4.

3.2 Dynamic response analysis

A deepwater well at a depth of 1500 m in the South China Sea was studied as an example. The configurations of the outer and inner string systems are presented in Tables 1 and 2, respectively. The inner string system includes top,

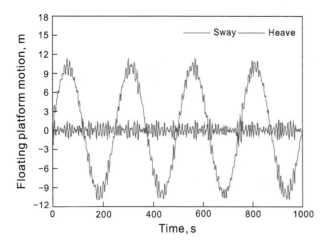

Fig. 4 Time history of floating platform motion

middle, and bottom centralizers used to protect the test sting, a lubricator valve, a subsea test tree, and other critical equipment. The rotational stiffnesses of the upper and lower flex joints in the outer string system are 8.8 and 127.4 kN m/(°), respectively.

Using the motion of the floating platform as a boundary condition, the analysis software ABAQUS Standard and Python were used to simulate the dynamic response of the test string system in the deepwater. The time histories of the flex joint rotation angle, and the von Mises stress envelope of the test string and the riser are shown in Figs. 5 and 6, respectively. Figure 5 shows that the magnitude of the rotation angle for the lower flex joint is much smaller than that for the upper flex joint, which is expected because the rotational stiffness of the former is larger. Moreover, both flex joints rotate in a random pattern, whereas the frequency of the joint rotation remains consistent with the long-term drift motion of the platform. It can be concluded that the platform offset is the dominant factor affecting the flex joint rotation angle.

In Fig. 6, a relative position of 1.0 represents the top of the test string and riser (near the surface of the water), and a relative position of 0 represents the level of the mud line.

Figure 6 shows that (1) the mean value and the amplitude of the von Mises stress tend to decrease gradually along the test string and riser from top to bottom, indicating that the transmission of the vibration at the top of the test string and riser decrease with depth; (2) the von Mises stress of the test string exhibits a sudden decline at points where the cross-sectional areas change, such as at the centralizers, and the stress levels in the corresponding positions of the flex joints increase sharply due to the change of bending moment; (3) the rate of decrease of the von Mises stress in the riser at depths of 120–1216 m generally remains constant with the buoyancy joints; the von Mises stress increases suddenly at 520 m due to the change in wall thickness; the von Mises stress in the riser at 1216–1400 m decreases rapidly because eight slick joints are used; (4) the string sections near the top of the test

Table 1 Configuration of the outer string system

Components	Number of joints	Outside diameter, m	Internal diameter, m	Length of joints, m
Outer barrel of telescopic joint	1	0.6604	0.6096	–
Slick joint 1	2	0.5334	0.4826	22.86
Buoyancy joint 1	20	0.5334	0.489	22.86
Buoyancy joint 2	32	0.5334	0.4826	22.86
Slick joint 2	8	0.5334	0.489	22.86
Buoyancy joint 3	2	0.5334	0.489	22.86
LMRP/BOP	1	–	–	16.288
Wellhead	1	–	–	2
Conductor	1	0.9144	0.87	–

Table 2 Configuration of the inner string system

Components	Outside diameter, m	Internal diameter, m	Length, m
Oil tube	0.1143	0.0857	21.798
Centralizer	0.4064	0.0762	1.29
Lubricator valve	0.3175	0.0762	3.42
Oil tube	0.1143	0.0857	732.48
Centralizer	0.4064	0.0762	1.29
Oil tube	0.1143	0.0857	732.48
Centralizer	0.4064	0.0762	0.97
Retainer valve	0.3175	0.0762	3.73
Tubing	0.1143	0.0857	2.72
Subsea test tree	0.3493	0.0762	1.29
Tubing	0.1143	0.0857	1.55
Fluted hanger	0.3302	0.0762	1.77

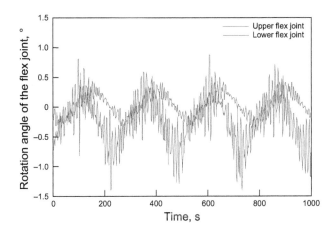

Fig. 5 Time history of the upper and lower flex joints

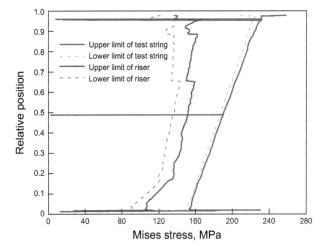

Fig. 6 The von Mises stress envelope of the deepwater test string and riser

string and flex joints are most prone to damage, so these areas should be given special attention during inspection and maintenance.

The mean contact force between the test string and the riser in a service cycle is displayed in Fig. 7. The following observations regarding the contact force were made: (1) the

deepwater test string and riser make contact in a random pattern, and the contact forces are relatively larger in the locations corresponding to the flex joints and the

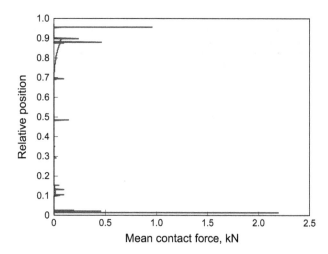

Fig. 7 The mean contact force between the deepwater test string and the riser

centralizers, among which the maximum contact force occurs at the position of the lower flex joint; (2) the mean contact forces are nearly zero in the figure because the gap elements generally do not have a continuous effect on the test string; (3) the contact forces are negligible at the lubricator valve, the retainer valve, and the subsea test tree, because of the effective protection of the centralizers; (4) the part of the test string located in the splash zone bears higher wave loads, whereas the lower part bears greater bending loads, resulting in contact primarily occurring in two areas (depths of 150–450 and 1200–1400 m) apart from the flex joints and the centralizers. Hence, centralizers should be added in those areas to reduce contact. These results have been successfully applied in a deepwater testing operation and have guided preliminary hazard analyses and maintenance for the deepwater test string in the South China Sea.

4 Fatigue analysis induced by waves

4.1 Fatigue damage assessment

The deepwater testing operation runs periodically and typically consists of four stages: deployment/retrieval mode, pressure perforation mode, flowing test mode, and shut-in well mode. These four stages are treated as a service cycle in the fatigue analysis of the deepwater test string system.

Based on the vibration response of the test string under combined loads, a rain-flow counting method (Khosrovaneh and Dowling 1990) was implemented in the MATLAB programming environment to analyze the fatigue damage of the test string in various stages. According

to the linear accumulated damage criterion, the fatigue damage of the test string in a service cycle is calculated as follows:

$$D = \sum_{i,j}^{i=4} \frac{n_{ij}}{N_{ij}} = \sum_{i,j}^{i=4} \frac{n_{ij} S_{ij}^m}{K}, \qquad (6)$$

where D is the fatigue damage, a^{-1}; $i = 1$ represents the deployment/retrieval mode, $i = 2$ represents the pressure perforation mode, $i = 3$ represents the flowing test mode, and $i = 4$ represents the shut-in well mode; S_{ij} and n_{ij} represent the fatigue stress amplitude and the cycle index for the corresponding operation modes, respectively; K, m are the S–N curve coefficients, in this paper, we set $m = 3$, and $K = 4.16 \times 10^{11}$.

In deepwater testing, it is common to use high rates of output flowback to prevent the formation of hydrate before downhole sampling and finding production by changing the flow (Triolo et al. 2013). To evaluate wave-induced fatigue damage in the deepwater test string, it was assumed that a service cycle contains: deployment/retrieval for 12 h, perforating for 1 h, flowback testing at 2.92×10^4 m^3/h for 8 h, flowing sampling at 1.25×10^4 m^3/h for 10 h, flowing test at 5.00×10^4 m^3/h for 10 h and at 6.25×10^4 m^3/h for 8 h, and shut-in well for 30 h. The change in flow in flowing testing mode was realized by adjusting the choke, and the shut-in time in transition was negligible.

The fatigue damage to the deepwater test string in the various stages of operation was calculated, as shown in Figs. 8 and 9. The results show that under the influence of waves and floating platform motion, the fatigue damage fluctuates with the water depth. The damage varies considerably with the position, and the maximum damage occurs adjacent to the lower flex joint. For this example,

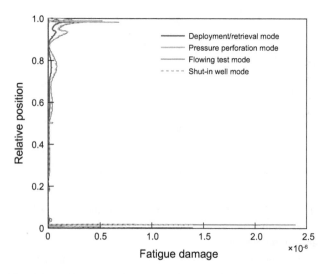

Fig. 8 Fatigue damage in the deepwater test string in different modes

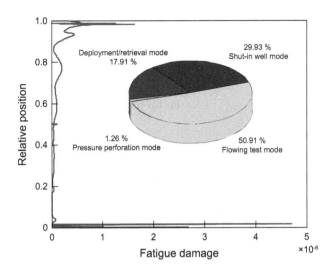

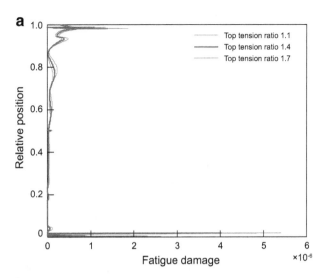

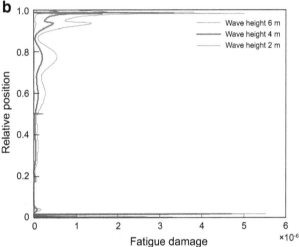

Fig. 9 Fatigue damage in the deep water test string in a typical service cycle

the fatigue damage in the deepwater test string in a service cycle is 4.706e−6, including 8.427e−7 (17.91 %) in the deployment/retrieval mode, 5.910e−8 (1.26 %) in the pressure perforation mode, 2.396e−6 (50.91 %) in the flowing test mode, and 1.409e−6 (29.93 %) in the shut-in well mode.

There are two locations at which the deepwater test system is most susceptible to fatigue: in the vicinity of the splash zone and the lower flex joint. These locations should be given special attention in managing the fatigue life of the deepwater test string. The fatigue damage in the sections located at the top of the test string in the splash zone primarily results from wave loads and the platform heaving motion, whereas the sections adjacent to the lower flex joint are mainly influenced by the offset motion of the floating platform. When the rotational stiffness of the flex joints is lower than that of the riser joints, the bending deformation and the vibration amplitude of the riser and the test string at this position are more significant.

The number of joints in a test string in each service cycle depends on the water depth; thus, the number of joints and their positions will vary. Therefore, the arrangement of the test string should be managed to avoid having the same joints located adjacent to the flex joints in more than one cycle. Moreover, nondestructive testing of the weakness positions should be conducted after each operation, and the damaged joints should be serviced or replaced to ensure that subsequent test operations can be conducted without failure.

4.2 Analysis of contributing factors

Fatigue damage in the deepwater test string is affected most by the top tension and dynamic wave loads. The fatigue resulting from these two factors is shown in Fig. 10.

Fig. 10 Fatigue damage in the deepwater test string for two main contributing factor. **a** Fatigue damage for various ratios of top tension. **b** Fatigue damage for various wave heights

As Fig. 10a shows, the fatigue damage tends to decrease with increasing top tension in the riser and the test string, and the fatigue damage in sections adjacent to the flex joints is significant. Higher levels of top tension could decrease the lateral displacement of the test string and the rotation angle of the flex joints, which would reduce the vibration amplitude. Thus, greater top tension could reduce fatigue damage in the test string due to wave action. As Fig. 10b shows, the fatigue damage tends to increase with wave height. The tendency in the upper sections of the test string is quite notable, a consequence of the flow velocity being an exponential function of the water depth. Furthermore, the platform motion caused by waves has a significant influence on the fatigue in the upper sections of the test string. Based on the results of this study, it can be concluded that applying greater top tension and operating in a favorable marine environment would markedly reduce fatigue damage in the test string..

5 Conclusions

(1) A nonlinear dynamic model of pipe-in-pipe configuration for a deepwater test string is established to study the vibration mechanism of a deepwater test string in waves. The results show that the mean value and the amplitude of von Mises stress gradually decrease along the test string. The von Mises stress is lower in locations with larger cross-sectional areas such as the centralizers, whereas the stress is higher in the sections near the top of the test string and the flex joints. The deepwater test string and the riser contact in a random pattern, and the contact loads are relatively larger in the points corresponding to the flex joints and the centralizers. The maximum contact load occurs in the sections adjacent to the lower flex joint.

(2) A fatigue analysis method and procedure are established for analyzing fatigue damage in deepwater test strings subjected to wave action. The results show that the fatigue damage in the deepwater test string fluctuates with the water depth, and damage varies considerably with position. The damage in the sections adjacent to the splash zone and the lower flex joint are much greater. The fatigue damage in the deepwater test string tends to increase as the top tension decreases and with wave height. Moreover, the damage in the flex joint areas is sensitive to the top tension, whereas the damage in the upper sections of the test string is sensitive to the wave height.

(3) Several measures are proposed to improve the operation safety of the deepwater test string. These measures include applying greater top tension and operating in a favorable marine environment to reduce fatigue damage, managing the arrangement of the test string to prevent the test string joints from being placed adjacent to the flex joints in more than one cycle, and performing nondestructive testing of components at vulnerable positions after test operations.

Acknowledgments This work is supported by the National Key Basic Research Program of China (973 Program, Grant No. 2015CB251203) and the Fundamental Research Funds for the Central Universities (14CX06119A). The authors are also grateful to the editors and the reviewers for their valuable comments.

References

Bavidge M. Husky Liwan deepwater subsea control system. In: Offshore technology conference, 6–9 May, Houston; 2013. doi:10.4043/23960-MS.

Chang YJ, Chen GM, Sun YY, et al. Nonlinear dynamic analysis of deepwater drilling risers subjected to random loads. China Ocean Eng. 2008;22(4):683–91.

Chen GM, Liu XQ, Chang YJ, et al. Advances in technology of deepwater drilling riser and wellhead. J China Univ Pet (Ed Nat Sci). 2013;37(5):129–39 (in Chinese).

Chen Q, Chia HK. Pipe-in-pipe walking: understanding the mechanism, evaluating and mitigating the phenomenon. In: ASME 2010 29th international conference on ocean, offshore and arctic engineering, 6–10 June, Shanghai; 2010. doi:10.1115/OMAE2010-20058.

Chen SM, Gong WX, Antle G. DST design for deepwater wells with potential gas hydrate problems. In: Offshore technology conference, 5–8 May, Houston; 2008. doi:10.4043/19162-MS.

Cheng WH, Wang LJ, Li GH. An optimal design of a DST string for high temperature, high pressure and sour gas fields on the Right Bank of the Amu Darya River, Turkmenistan. Nat Gas Ind. 2014;34(4):76–82 (in Chinese).

Dong SM, Zhang WS, Wang Q, et al. Mechanism of eccentric wear between rod string and tubing string of a surface driving screw pump lifting system in vertical wells. Acta Pet Sin. 2012;32(3):304–9 (in Chinese).

Harrison RI, Helle Y. Understanding the response of pipe-in-pipe deepwater riser systems. In: The seventeenth international offshore and polar engineering conference, 1–6 July, Lisbon; 2007.

Hu WR, Bao JW, Hu B. Trend and progress in global oil and gas exploration. Pet Explor Dev. 2013;40(4):439–43. doi:10.1016/S1876-3804(13)60055-5.

Khan RA, Ahmad S. Probabilistic fatigue safety analysis of oil and gas risers under random loads. In: ASME 2010 29th international conference on ocean, offshore and arctic engineering, 6–11, June, Shanghai; 2010. doi:10.1115/OMAE2010-20464.

Khosrovaneh AK, Dowling NE. Fatigue loading history reconstruction based on the rainflow technique. Int J Fatigue. 1990;12(2):99–106. doi:10.1016/0142-1123(90)90679-9.

Li FZ, Low YM. Fatigue reliability analysis of a steel catenary riser at the touchdown point incorporating soil model uncertainties. Appl Ocean Res. 2012;38(1):100–10. doi:10.1016/j.apor.2012.07.005.

Li ZF. Mechanical analysis of tubing string in well testing operation. J Pet Sci Eng. 2012;90(91):61–9. doi:10.1016/j.petrol.2012.04.019.

Liu K, Chen GM, Chang YJ, et al. Warning control boundary of platform offset in deepwater test string. Acta Pet Sin. 2014a;35(6):1204–10 (in Chinese).

Liu XQ, Chen GM, Chang YJ, et al. Multistring analysis of wellhead movement and uncemented casing strength in offshore oil and gas wells. Pet Sci. 2014b;11(1):131–8. doi:10.1007/s12182-014-0324-7.

Mogbo O. Deepwater DST design, planning and operations-Offshore Niger Delta experience. In: Production and operations conference and exhibition, 8–10 June, Tunis; 2010. doi:10.2118/133772-MS.

Morooka CK, Coelho FM, Ribeiro EJB, et al. Dynamic behavior of a vertical riser and service life reduction. In: ASME 2005 24th international conference on offshore mechanics and arctic engineering, 12–17, June, Halkidiki; 2005. doi:10.1115/OMAE2005-67294.

Nazir M, Khan F, Amyotte P. Fatigue reliability analysis of deep water rigid marine risers associated with Morison-type wave loading. Stoch Environ Res Risk Assess. 2008;22(3):379–90. doi:10.1007/s00477-007-0125-2.

Pang DX, Liu QY, Meng QH, et al. Solving method for nonlinear contact problem of drill strings in 3D curved borehole. Acta Pet Sin. 2009;30(1):121–4 (in Chinese).

Park HI, Jung DH. A finite element method for dynamic analysis of long slender marine structures under combined parametric and forcing excitations. Ocean Eng. 2002;29(11):1313–25. doi:10.1016/S0029-8018(01)00084-1.

Song J, Lu L, Teng B, et al. Laboratory tests of vortex-induced vibrations of a long flexible riser pipe subjected to uniform flow. Ocean Eng. 2011;38(11):1308–22. doi:10.1016/j.oceaneng.2011.05.020.

Stomp RJ, Fraser GJ, Actis SC, et al. Deepwater DST planning and operations from a DP vessel. J Pet Technol. 2005;57(6):55–7. doi:10.2118/90557-MS.

Tognarelli M, Fontaine E, Beynet P, et al. Reliability-based factors of safety for VIV fatigue using field measurements. In: ASME 2010 29th international conference on ocean, offshore and arctic engineering, 6–11 June, Shanghai; 2010. doi:10.1115/OMAE20 10-21001.

Triolo D, Mosness T, Habib R K. The Liwan gas project: a case study of South China Sea deepwater drilling campaign. In: The international petroleum technology conference, 26–28 March, Beijing; 2013. doi:10.2523/16722-MS.

Wang H, Sun J, Jukes P. FEA of a laminate internal buckle arrestor for deep water pipe-pipe flowlines. In: ASME 2009 28th international conference on ocean, offshore and arctic engineering, 31 May–5 June, Honolulu; 2009. doi:10.1115/OMAE2009-79520.

Wendler C, Scott M. Testing and perforating in the HPHT deep and ultra-deep water environment. In: The SPE Asia Pacific oil & gas conference and exhibition, 22–24 Oct, Perth; 2012. doi:10.2118/158851-MS.

Xie X, Fu JH, Zhang Z, et al. Mechanical analysis of deep water well testing strings. Nat Gas Ind. 2011;31(1):77–9 (in Chinese).

Xu LB, Jiang SQ, Zhou JL. Challenges and solutions for deep water drilling in the South China Sea. In: Offshore technology conference, 6–9 May, Houston; 2013. doi:10.4043/23964-MS.

Yang J, Liu CH, Liu HB, et al. Strength and stability analysis of deep sea drilling risers. Pet Sci. 2007;4(2):60–5. doi:10.1007/BF03187443.

Zeng ZJ, Hu WD, Liu JC, et al. A mechanical analysis of the gas testing string used in HTHP deep wells. Nat Gas Ind. 2010;30(2):85–7 (in Chinese).

5

Optimization of dispersed carbon nanoparticles synthesis for rapid desulfurization of liquid fuel

Effat Kianpour[1] · Saeid Azizian[1]

Abstract Stringent regulations and environmental concerns make the production of clean fuels with low sulfur content compulsory for the petroleum refining industry. Because of ease of operation without high energy consumption, the adsorption of sulfur compounds seems the most promising process. Central composite design was used to optimize parameters influencing the synthesis of dispersed carbon nanoparticles (CNPs), a new class of sorbents, in order to obtain an excellent adsorbent for desulfurization of liquid fuel. The optimized dispersed CNPs, which are immiscible in liquid fuel, can effectively adsorb different benzothiophenic compounds. Equilibrium adsorption was achieved within 2 min for benzothiophene, dibenzothiophene, and 4,6-dimethyldibenzothiophene with removal efficiency values of 75 %, 83 %, and 52 %, respectively. The rate of desulfurization by the prepared CNPs in the present work is seven times higher than the previously reported CNPs. Optimized CNPs were characterized by different techniques. Finally, the effect of the mass of CNPs on the removal efficiency was studied as well.

Keywords Carbon nanoparticles · Liquid fuel · Adsorption · Central composite design · Desulfurization

✉ Saeid Azizian
sazizian@basu.ac.ir; sdazizian@yahoo.com

[1] Department of Physical Chemistry, Faculty of Chemistry, Bu-Ali Sina University, Hamedan 65167, Iran

Edited by Xiu-Qin Zhu

1 Introduction

Sulfur compounds which are naturally present in unrefined oils and transportation fuels severely deactivate most catalysts, such as those used in automotive emissions control, petrochemicals production, and fuel cells (Shi et al. 2011). Furthermore, sulfur-containing fuels produce SO_X during combustion, and this seriously affects the environment. Therefore, stringent regulations are imposed on the presence of sulfur in fuel and have caused a growing demand for ultra-low sulfur fuels (<10 ppmw) (Liu et al. 2012). To produce low sulfur-containing fuels, various processes including hydrodesulfurization (HDS) (Song and Ma 2003; Farag et al. 1998; Lara et al. 2005), oxidation (Jiang et al. 2003; Yang et al. 2007), extraction (Bösmann et al. 2001; Zhang et al. 2004), and adsorption (Blanco-Brieva et al. 2011; Hernandez et al. 2010; Fallah and Azizian 2012a, b; Fallah et al. 2012a, b, 2014; Song et al. 2013) have been proposed. HDS, the major technology in current industries not only operates in difficult and costly conditions but also is inefficient in removing the more refractory sulfur compounds such as benzothiophene (BT), dibenzothiophene (DBT), and 4,6-dimethyldibenzothiophene (DMDBT) (Liu et al. 2012). Because of straightforward operation without high energy consumption, the adsorption of organosulfur compounds is an efficient alternative to the conventional HDS process. Recently, dispersed carbon nanoparticles (CNPs) in an aqueous phase was reported (Fallah and Azizian 2012a; Fallah et al. 2012) as a new class of sorbents in removing of BT, DBT, and DMDBT from liquid fuel, and it was shown that the adsorption of different organosulfur compounds on this adsorbent was rapid and had a good selectivity. The synthesis of CNPs is affected by different parameters (volume of water (W), time of irradiation (*t*), polyethylene glycol (PEG) volume, and weight of glucose (G)) which may have an effect on adsorption and removal efficiency. The main objective of the present work is to optimize these factors to enhance the desulfurization efficiency of CNPs.

Traditionally, a one-variable-at-a-time methodology, which involves changing one variable while keeping the other factors constant, has been used in assessing the effect of factors on the experimental response (Bezerra et al. 2008). This is time consuming and ignores the interactions among various parameters, so it is not an effective optimization technique (Puri et al. 2002). In order to overcome this problem, multivariable optimization methods should be carried out. Response Surface Methodology (RSM), which includes factorial design and regression analysis, is among the most relevant techniques (Bezerra et al. 2008; Puri et al. 2002). This approach has been recently applied for optimization of the production of microporous-activated carbon from various resources (Asenjo et al. 2011). Before using the RSM methodology, we needed to select an experimental design and to fit an adequate mathematical function for estimating the optimum points. Central composite design (CCD) is a symmetrical second-order experimental design most utilized for the improvement of procedures (Bezerra et al. 2008). In this work, central composite design was selected to study simultaneously the effects of four synthesis variables on the desulfurization efficiency as an experimental response and to get the optimum conditions. Based on this method, we arrived at the conclusion and conditions where desulfurization of liquid fuel can be done easily at the highest rate (within less than 2 min).

2 Experimental

2.1 Materials

+D-Glucose, PEG-200, and heptane (purity >99%) were provided by Merck Co. (USA). Extra-pure reagents of BT, DBT, and DMDBT (mass fraction purity of >98 %) were purchased from Sigma-Aldrich Co. (USA). CNPs were prepared following literature procedure (Zhu et al. 2009) but with changes in the amounts of reactants and period of

irradiation time. After dissolving +D-glucose and PEG-200 in distilled water, the clear solution was heated in a microwave oven for different times at 400 W and 2450 Hz. At the end, the produced dark suspension of CNPs was used for desulfurization experiments without any further processing. In order to take transmission electron microscope (TEM) images of the solid phase of adsorbent, suspended CNPs were separated from solution as follows: a saturated solution of NaCl was added, and after agglomeration of the CNPs, the solid phase was separated from the liquid phase by centrifugation. After drying the sample at 110 °C overnight, it was washed three times with distilled water in order to remove the NaCl and was dried again at the same condition.

2.2 Design of experiment

Variables optimization for the synthesis of CNPs was performed using a CCD, a standard RSM design. This approach limits the number of actual experiments performed while considering possible interaction between factors as well as individual variables and was suitable for fitting a quadratic surface (Asenjo et al. 2011; Kalavathy et al. 2009; Tan et al. 2008a). The selected variables in the present study are irradiation time (X_1), water volume (X_2), glucose weight (X_3), and volume of polyethylene glycol (X_4). The range and coded levels employed for the four factors in three levels, according to CCD design, are categorized in Table 1. Twenty-eight experimental runs, consisting of 18 cube points (for a full factorial), 8 star or axial points, and 4 center points, were generated by the principle of CCD (listed in Table 2) (Kalavathy et al. 2009). The order of these experiments was randomized in order to minimize the effect of uncertain variability in the monitored responses due to irrelevant factors (Asenjo et al. 2011). The quadratic polynomial model (Eq. 1) was chosen for fitting the experimental data:

$$Y = b_0 + \sum_{i=1}^{n} b_i X_i + \left(\sum_{i=1}^{n} b_{ii} X_i \right)^2 + \sum_{i=1}^{n-1} \sum_{j=1}^{n} b_{ij} X_i X_j, \quad (1)$$

where Y is the response variable (i.e., removal efficiency (%) of DBT), b_0, b_i, b_{ii}, and b_{ij} represent the intercept, the linear, quadratic, and interaction coefficients, respectively.

Table 1 Experimental range and levels of the independent variables used for CCD

Variable name	Range and coded factor levels				
	−2 (low)	−1	0	+1	+2 (high)
Irradiation time (t), X_1, min	7	9	11	13	15
Water volume (W), X_2, mL	7	13	19	25	31
Glucose weight (G), X_3, g	0.2	1.4	2.6	3.8	5
Polyethylene glycol volume (PEG), X_4, mL	8	13	18	23	28

Table 2 List of experiments according to the CCD

Experiment number	Factors levels				Response: removal, %
	t, min	W, mL	G, g	PEG, mL	
1	9	13	1.4	13	77.9
2	13	13	1.4	13	78.3
3	9	25	1.4	13	76.1
4	13	25	1.4	13	77.8
5	9	13	3.8	13	71.1
6	13	13	3.8	13	75.3
7	9	25	3.8	13	69.2
8	13	25	3.8	13	73.9
9	9	13	1.4	23	80.1
10	13	13	1.4	23	81.2
11	9	25	1.4	23	79.5
12	13	25	1.4	23	80.6
13	9	13	3.8	23	76.4
14	13	13	3.8	23	77.9
15	9	25	3.8	23	76.4
16	13	25	3.8	23	78.7
17	7	19	2.6	18	71.2
18	15	19	2.6	18	78.9
19	11	7	2.6	18	79.5
20	11	31	2.6	18	76.9
21	11	19	0.2	18	80.1
22	11	19	5	18	75.4
23	11	19	2.6	8	71.7
24	11	19	2.6	28	78.7
25[a]	11	19	2.6	18	77.8
26[a]	11	19	2.6	18	77.3
27[a]	11	19	2.6	18	78.0
28[a]	11	19	2.6	18	77.6

[a] Center points

A statistical package, 'Design Expert' software (Version 7.1.3, Stat-Ease Inc., Minneapolis, USA), was used to analyze and calculate the polynomial coefficients of the quadratic equation and to evaluate the statistical significance of the equation.

2.3 Batch adsorption measurements

In all experiments, a glass bottle containing CNPs and model fuel, prepared by dissolving sulfur-containing compounds in heptane with an initial concentration of 500 ppmw, was placed in a stirred thermostated batch system at 25.0 ± 0.1 °C. At first, desulfurization performance of sorbents (synthesized under different conditions) was evaluated in order to determine the optimum synthesis situation of CNPs using RSM. For this purpose, 2.5 g of

model fuel containing DBT 500 ppmw, as a representative sulfur compound because of having a typical structure, was added to 5 g of CNPs solution. After 5 min of high-speed magnetic stirring, a sample was taken from the organic phase to evaluate the residual DBT content and estimate the removal efficiency. Afterward, the best adsorbent synthesized under optimum conditions was chosen for further investigations. The procedure used for kinetic tests was as follows: 10 g of CNPs suspension and 5 g of model fuel were mixed well under high-speed magnetic stirring, and then the sampling was done at desired time intervals. To study the effect of the mass of CNPs on adsorbent performance, 2.5 g of the DBT containing solutions, with an initial concentration of 1000 ppmw, was added to bottles containing different amounts of CNPs, from 0.1 to 8 g, and they were mixed well under stirring and left to equilibrate for 20 min. Afterward, the removal efficiency of adsorbent was estimated by $\%R = (C_i - C_f) \times 100/C_i$, where C_i and C_f are the initial and final concentrations of sulfur compound (ppmw), respectively, and $\%R$ is the removal efficiency. These experiments were repeated for other sulfur compounds (i.e., BT and DMDBT). In all the tests, the sulfur compound was dissolved in heptane, and then the initial and residual concentration of it was determined using UV–Visible spectrophotometry at the related λ_{max} (Fallah and Azizian 2012b).

2.4 Characterization methods

Carbon nanoparticles synthesized under the optimum condition were characterized by using a transmission electron microscope (Philips EM208), photoluminescence (PL) (LS50 spectrofluorimeter, Perkin Elmer), Fourier transform infrared spectroscope (FTIR) (Perkin Elmer), and UV–Visible spectrophotometer (Spekol 2000 spectrophotometer, Analytik Jena).

3 Results and discussion

3.1 Development of regression model equation

The experimental conditions, including value of each parameter, are presented in Table 2 along with the observed response. Analysis of variance (ANOVA) for the response surface quadratic model is summarized in Table 3. The suitability of the models was justified through ANOVA. From the ANOVA, the model F value of 18.54 implied that the model was statistically significant. Values of 'Prob > F' less than 0.0500 indicated that the model terms were significant and those greater than 0.1000 were not. In this case, $X_1, X_2, X_3, X_4, X_1X_3, X_3X_4, X_1^2$ and X_4^2 were significant. Therefore, the final regression equation for removal efficiency in terms of actual factors after ignoring the insignificant terms is shown in Eq. (2):

$$R = 51.72 + 3.70X_1 - 0.08X_2 - 5.45X_3 + 1.03X_4 \\ + 0.22X_1X_3 + 0.09X_3X_4 - 0.16X_1^2 - 0.02X_4^2 \quad (2)$$

A positive sign in front of the terms indicated a synergistic effect, whereas a negative sign showed antagonistic results. The quality of the model equation was further estimated using the correlation coefficients (r^2) (Tan et al. 2008b). r^2 of this model with the value of 0.934 demonstrated a good agreement between experimental data and the model prediction. Figure 1 shows the predicted versus actual (experimental) removal efficiency. As mentioned above and represented in Fig. 1, this model was adequate to predict response.

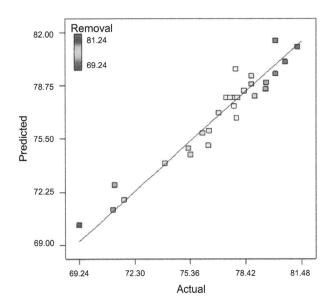

Fig. 1 Comparison plot between the experiments and the model predicted for DBT removal efficiency

3.2 Optimization of CNPs synthesis for the removal of DBT

According to Table 3, the volume of polyethylene glycol (PEG) (X_4), weight of glucose (G) (X_3), and irradiation time (t) (X_1) showed the large F value of 92.40, 79.11, and 47.79, respectively, indicating that these factors had more significant effects on adsorbent performance in removing of DBT. From Eq. (2), t and PEG were found to have synergistic effects on the removal efficiency of DBT in contrast with the effect of G and water volume (W). Besides, the quadratic effect of irradiation time (X_1^2) and PEG volume (X_4^2) on the DBT uptake was relatively significant with antagonist effects. So, it was expected that an increase in irradiation time could increase efficiency and then decrease it after reaching a maximum value. This is very clear from the one factor plot of varying responses with changing irradiation time (the other variables were constant at central levels) (Fig. 2). Similar behavior was observed for PEG volume. Equation (2) suggests that the weight of glucose variable interacted with the time of irradiation and volume of PEG. But these interactions (($X_1 X_3$) and ($X_3 X_4$)) and also the volume of water (X_2) have less significant effects as shown by the low F value. Isoresponse counter plots (three dimensional response surface contour plots), displayed in Fig. 3a–c, are the graphical representations of the regression equation. The main object of the response surface was finding the optimum values of the factors, so the response variable was maximized. Each contour curve represents a combination of irradiation time (t) and polyethylene glycol (PEG) with the volume of water (W) and weight of glucose

Table 3 Analysis of variance (ANOVA) for response surface quadratic model

Source	Sum of square	Degree of freedom	Mean square	F value	probe > F
Model	239.36	14	17.10	18.54	<0.0001
X_1	44.06	1	44.06	47.79	<0.0001
X_2	5.15	1	5.15	5.59	0.0343
X_3	72.94	1	72.94	79.11	<0.0001
X_4	85.20	1	85.20	92.40	<0.0001
$X_1 X_2$	0.42	1	0.42	0.46	0.5103
$X_1 X_3$	4.35	1	4.35	4.71	0.0490
$X_1 X_4$	1.55	1	1.55	1.68	0.2173
$X_2 X_3$	0.10	1	0.10	0.11	0.7443
$X_2 X_4$	1.66	1	1.66	1.80	0.2021
$X_3 X_4$	4.60	1	4.60	4.99	0.0437
X_1^2	8.56	1	8.56	9.28	0.0094
X_2^2	0.93	1	0.93	1.06	0.3223
X_3^2	0.13	1	0.13	0.14	0.7159
X_4^2	7.48	1	7.48	8.11	0.0137

Significant at "probe>F" less than 0.05

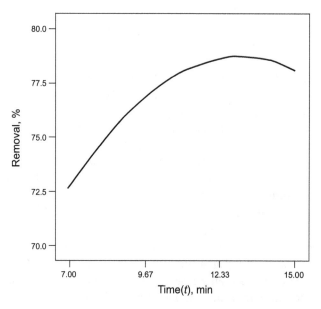

Fig. 2 One factor plot showing the effect of irradiation time on removal efficiency of DBT (the other factors were maintained at their central points)

Therefore, the optimal experimental conditions for obtaining optimized dispersed CNPs were found to be $t = 11$ min, PEG = 20 mL, $G = 0.2$ g and $W = 7$ mL. The predicted maximum removal efficiency of DBT by CNPs was 83 %.

3.3 Characterization of the prepared CNPs at optimum conditions

The optimum conditions for preparation of CNPs with maximum removal efficiency of DBT from fuel were found in the above section. The prepared CNPs were characterized by different methods, which will be explained below. Although the photoluminescence (PL) spectra for CNPs in Fig. 4 are similar to those of the reported data (Fallah and Azizian 2012a), but the UV–Visible spectrum (Fig. 5) of CNPs synthesized in optimum conditions is different from the reports in literature (Fallah and Azizian 2012a; Fallah et al. 2012). In this work, maximum absorption was observed at a wavelength of 220 nm (Fig. 5), while in the previous reports, $\lambda_{\max}$ was at 280 nm. According to the report by Tian et al. (2010), the UV–Visible spectra of carbon nanoparticles contain a major absorption peak at around 200 nm (peak A) and another broad peak at approximately 300 nm (peak B). The first absorption peak A is attributed to the $\pi–\pi^*$ electronic transitions of internal (sp^2) graphitic carbons, whereas the second absorption peak B is most likely related to $n–\pi^*$ electron transfer of carbonyl (C=O) groups (e.g., aldehydes and ketones) on the CNPs surface. Upon heating, the peak A remained, whereas the peak B diminished (Tian et al. 2010). The temperatures of CNPs solutions (synthesized in the present (CNPs II) and previous (CNPs I) (Fallah et al. 2012) were

(G) maintained at different levels. The predicted maximum response was indicated by the surface confined in the smallest ellipse in the contour diagram. Elliptical contours are obtained when there is a perfect interaction between the independent variables (Tanyildizi et al. 2005). As can be seen from Fig. 3a–c, the removal percentage of DBT increased with the increase in t and PEG but decreased with the increase of W and G. The highest removal of DBT was obtained when both latter variables (W and G) were at the minimum point within the range studied and the response surface turned into an ellipse in the last figure (Fig. 3c).

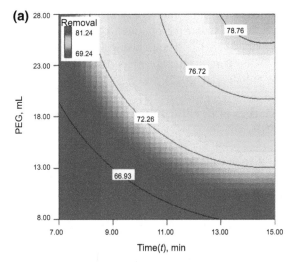

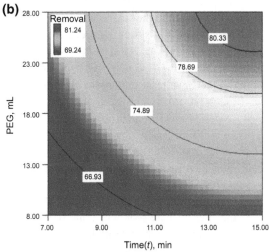

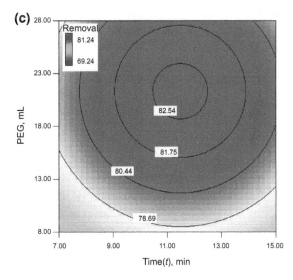

Fig. 3 Isoresponse counter plot showing the effect of PEG volume, irradiation time, and their mutual effect on removal efficiency of DBT by CNPs at different levels of water volume (W) and glucose weight (G) (**a** W and G at high levels, **b** W at low level and G at high level, **c** W and G at low levels)

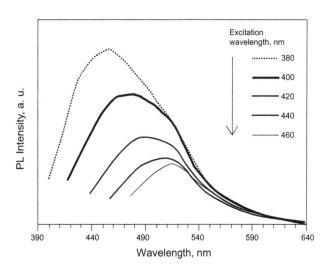

Fig. 4 Photoluminescence (PL) spectra of the prepared CNPs with different excitation wavelengths

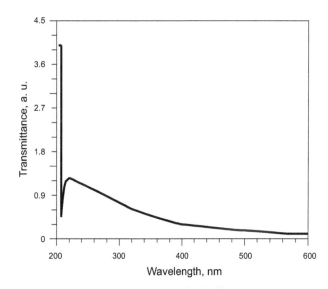

Fig. 5 UV/Visible spectra of the prepared CNPs in suspension

measured to be 265 °C and 240 °C, respectively, using a thermocouple immediately after bringing the samples out of the microwave. The increase in temperature may cause the surface functional groups containing oxygen to be decomposed. In fact, the shift in the UV–Visible spectrum of CNPs from 280 nm (Fallah et al. 2012) to 220 nm in the present work suggests the removal of carbonyl groups from the nanoparticles surface. The above interpretation was further justified by the result from FTIR measurements (Fig. 6). It can be seen that the overall FTIR spectrum of solid CNPs (Fig. 6a) prepared in this work was similar to that of the carbon nanoparticles prepared in the previous study (Fig. 6b) (Fallah et al. 2012), and only some differences in the intensity ratio of a number of peaks were

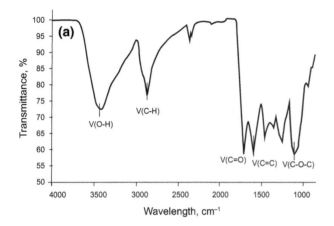

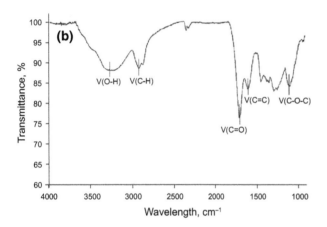

Fig. 6 FTIR spectra of the solid CNPs prepared in **a** the present study and **b** previous study (Fallah et al. 2012)

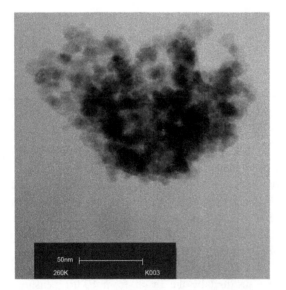

Fig. 7 TEM image of the solid CNPs synthesized under optimal conditions in this work

TEM images were obtained for further characterization of CNPs prepared under the optimum conditions. As can be seen from Fig. 7, CNPs are spherical less than 10 nm in size, smaller than the previous reports (Fallah and Azizian 2012a; Fallah et al. 2012).

3.4 Kinetics of desulfurization by the prepared CNPs

Figure 8 shows the effect of agitation time on removal of BT, DBT, and DMDBT by the synthesized CNPs. It can be seen that the concentrations of BT, DBT, and DMDBT decreased to 126, 86, and 241 ppmw and equilibrium was achieved within 90 s for BT and DBT and 120 s for DMDBT with removal efficiency values of 75 %, 83 % and 52 %, respectively. The rate of desulfurization by the prepared CNPs is much higher than the previously reported data for CNPs where the system approached equilibrium after 15 min (Fallah and Azizian 2012a). This observation was probably due to the size of CNPs synthesized in different conditions (synthesis conditions of this work: $t = 11$ min, PEG = 20 mL, $G = 0.2$ g and $W = 7$ mL and previously reported work (Fallah and Azizian 2012a): $t = 9$ min, PEG = 10 mL, $G = 2$ g and $W = 10$ mL). The smaller particles lead to an increase in the surface area of adsorbent, and therefore, the uptake rate of adsorbate on the sorbent has been increased as a consequence of increasing the surface area of adsorbent. As noted in Sect. 3.3, the present CNPs of less than 10 nm are finer than previously reported CNPs (Fallah and Azizian 2012a; Fallah et al. 2012). The experimental kinetic data of sulfur compounds' uptake on CNPs were best fitted using a first-order kinetic model, $\ln(1 - F) = -kt$, where

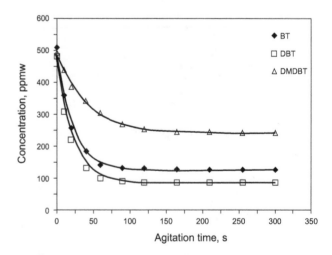

Fig. 8 Concentration variation of different sulfur compounds with time using CNPs. *Symbols* are the experimental values and lines are the predicted values from the first-order kinetic model. Experimental conditions: mass ratio of CNP/model oil = 1:2, sulfur compound concentration 500 ppmw, $T = 25$ °C

observed. The ratio of the C=O to C=C absorbance peak decreased for CNPs II in comparison to CNPs I. This observation is in agreement with the above discussions.

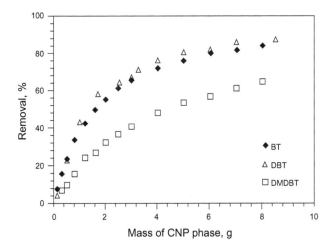

Fig. 9 Effect of CNP mass on BT, DBT, and DMDBT removal. Sulfur compound concentration 1000 ppmw, $T = 25\,°C$

$F = (C_0 - C_t)/(C_0 - C_e)$; C_0, C_e, and C_t are concentrations at initial, equilibrium, and at any time, respectively. Solid lines in Fig. 8 represent the predicted values by this model. The obtained rate constants (k) for BT, DBT, and DMDBT at 25 °C are 0.046, 0.045, and 0.025 s^{-1}, respectively, which confirm the higher rate of BT and DBT removal.

3.5 Effect of the mass of CNPs

The effect of the mass of CNPs on the removal performance was investigated, and the results are presented in Fig. 9. It is clear that removal efficiency values of BT, DBT, and DMDBT were increased rapidly with increasing the mass of CNPs up to 5 g and then increased very slowly.

4 Conclusion

A central composite design (CCD) was conducted to optimize the process conditions for the synthesis of CNPs in order to obtain an adsorbent with higher removal efficiency for sulfur compounds. The optimum conditions for preparation of CNPs with the highest rate of desulfurization was $t = 11$ min, PEG = 20 mL, $G = 0.2$ g, and $W = 7$ mL. Kinetic studies revealed that the rate of the sulfur removal process was very rapid and equilibrium was achieved within about 2 min. The experimental kinetic data were best fitted with a first-order kinetic model. The rate of desulfurization of liquid fuel by the prepared CNPs in the present work was seven times higher than that of the previously reported CNPs. Therefore, the prepared CNPs, which was synthesized by a simple and environmentally friendly method and which can be removed simply from the operation medium because of immiscibility with organic phase (fuel), is one of the best sorbents for the desulfurization of fuel.

Acknowledgments The financial support from Bu-Ali Sina University was gratefully acknowledged.

References

Asenjo NG, Botas C, Blanco C, et al. Synthesis of activated carbons by chemical activation of new anthracene oil-based pitches and their optimization by response surface methodology. Fuel Process Technol. 2011;92(10):1987–92.

Bezerra MA, Santelli RE, Oliveira EP, et al. Response surface methodology (RSM) as a tool for optimization in analytical chemistry. Talanta. 2008;76(5):965–77.

Blanco-Brieva G, Campos-Martin JM, Al-Zahrani SM, et al. Effectiveness of metal-organic frameworks for removal of refractory organo-sulfur compound present in liquid fuels. Fuel. 2011;90(1):190–7.

Bösmann A, Datsevich L, Jess A, et al. Deep desulfurization of diesel fuel by extraction with ionic liquids. Chem Commun. 2001;23:2494–5.

Fallah RN, Azizian S. Rapid and facile desulfurization of liquid fuel by carbon nanoparticles dispersed in aqueous phase. Fuel. 2012a;95:93–6.

Fallah RN, Azizian S. Removal of thiophenic compounds from liquid fuel by different modified activated carbon cloths. Fuel Process Technol. 2012b;93(1):45–52.

Fallah RN, Azizian S, Reggers G, et al. Effect of aromatics on the adsorption of thiophenic sulfur compounds from model diesel fuel by activated carbon cloth. Fuel Process Technol. 2014;119:278–85.

Fallah RN, Azizian S, Reggers G, et al. Selective desulfurization of model diesel fuel by carbon nanoparticles as adsorbent. Ind Eng Chem Res. 2012;51(44):14419–27.

Farag H, Whitehurst DD, Mochida I. Synthesis of active hydrodesulfurization carbon-supported Co–Mo catalysts. Relationships between preparation methods and activity/selectivity. Ind Eng Chem Res. 1998;37(9):3533–9.

Hernandez SP, Fino D, Russo N. High performance sorbents for diesel oil desulfurization. Chem Eng Sci. 2010;65(1):603–9.

Jiang Z, Liu Y, Sun X, et al. Activated carbons chemically modified by concentrated H_2SO_4 for the adsorption of the pollutants from wastewater and the dibenzothiophene from fuel oils. Langmuir. 2003;19(3):731–6.

Kalavathy MH, Regupathi I, Pillai MG, et al. Modelling, Analysis and optimization of adsorption parameters for H_3PO_4 activated rubber wood sawdust using response surface methodology (RSM). Colloids Surf, B. 2009;70(1):35–45.

Lara G, Escobar J, De Los Reyes JA, et al. Dibenzothiophene HDS over sulphided CoMo on high-silica USY zeolites. Can J Chem Eng. 2005;83(4):685–94.

Liu B, Zhu Y, Liu S, Mao J. Adsorption equilibrium of thiophenic sulfur compounds on the Cu- BTC metal-organic framework. J Chem Eng Data. 2012;57(4):1326–30.

Puri S, Beg QK, Gupta R. Optimization of alkaline protease production from bacillus sp. by response surface methodology. Curr Microbiol. 2002;44(4):286–90.

Shi F, Hammoud M, Thompson LT. Selective adsorption of dibenzothiophene by functionalized metal organic framework sorbents. Appl Catal B. 2011;103(3–4):261–5.

Song CS, Ma XL. New design approaches to ultra-clean diesel fuels by deep desulfurization and deep dearomatization. Appl Catal B. 2003;41(1–2):207–38.

Song H, Wan X, Sun X. Preparation of Agy zeolites using microwave irradiation and study on their adsorptive desulphurisation performance. Can J Chem Eng. 2013;91(5):915–23.

Tan IAW, Ahmad AL, Hameed BH. Optimization of preparation conditions for activated carbons from coconut husk using response surface methodology. Chem Eng J. 2008a;137(3):462–70.

Tan IAW, Ahmad AL, Hameed BH. Preparation of activated carbon from coconut husk: optimization study on removal of 2,4,6-trichlorophenol using response surface methodology. J Hazard Mater. 2008b;153(1–2):709–17.

Tanyildizi MS, Ozer D, Elibol M. Optimization of α-amylase production by Bacillus sp. using response surface methodology. Process Biochem. 2005;40(7):2291–6.

Tian L, Song Y, Changa X, et al. Hydrothermally enhanced photoluminescence of carbon nanoparticles. ScriptaMaterialia. 2010;62(11):883–6.

Yang Y, Lu H, Ying P, et al. Selective dibenzothiophene adsorption on modified activated carbons. Carbon. 2007;45(15):3042–4.

Zhang S, Zhang Q, Zhang ZC. Extractive desulfurization and denitrogenation of fuels using ionic liquids. Ind Eng Chem Res. 2004;43(2):614–22.

Zhu H, Wang X, Li Y, et al. Microwave synthesis of fluorescent carbon nanoparticles with electrochemiluminescence properties. Chem Commun. 2009;34:5118–20.

6

Improving slurryability, rheology, and stability of slurry fuel from blending petroleum coke with lignite

Jun-Hong Wu · Jian-Zhong Liu · Yu-Jie Yu ·
Rui-Kun Wang · Jun-Hu Zhou · Ke-Fa Cen

Abstract Petroleum coke and lignite are two important fossil fuels that have not been widely used in China. Petroleum coke–lignite slurry (PCLS), a mixture of petroleum coke, lignite, water, and additives, efficiently utilizes the two materials. In this study, we investigate the effects of the proportion (α) of petroleum coke on slurryability, rheological behavior, stability, and increasing temperature characteristics of PCLSs. The results show that the fixed-viscosity solid concentration (ω_0) increases with increasing α. The ω_0 of lignite-water slurry (LWS, $\alpha = 0$) is 46.7 %, compared to 71.3 % for the petroleum coke–water slurry (PCWS, $\alpha = 100$ %), while that of PCLS is in between the two values. The rheological behavior of PCLS perfectly fits the power-law model. The PCWS acts as a dilatant fluid. As α decreases, the slurry behaves first as an approximate Newtonian fluid, and then turns into a pseudo-plastic fluid that exhibits shear-thinning behavior. With increasing α, the rigid sedimentation and water separation ratio (WSR) increase, indicating a decrease in the stability of PCLS. When α is 60–70 %, the result is a high-quality slurry fuel for industrial applications, which has high slurryability ($\omega_0 = 57$–60 %), good stability (WSR < 2 %), and superior pseudo-plastic behavior ($n \approx 0.9$).

Keywords Petroleum coke · Lignite · Coal–water slurry · Rheological characteristics · Slurry stability

J.-H. Wu · J.-Z. Liu (✉) · Y.-J. Yu · R.-K. Wang ·
J.-H. Zhou · K.-F. Cen
State Key Laboratory of Clean Energy Utilization, Zhejiang
University, Hangzhou 310027, Zhejiang, China
e-mail: jzliu@zju.edu.cn

Edited by Yan-Hua Sun

1 Introduction

Petroleum coke is the main byproduct of the petroleum refining process, and it can be used for electrodes or fuel. Owing to a continuous demand for crude oil, the production of petroleum coke in China is increasing steadily (Zhou et al. 2012). Petroleum coke is a cheap fuel traditionally used as an important feedstock for circulating fluidized bed combustors because of its high calorific value, low volatile content, and high sulfur content (He et al. 2011). Therefore, achieving efficient and clean utilization of petroleum coke has become an urgent issue. To this end, some utilization technologies are worthy of being explored in depth (Wu et al. 2009). One of the most efficient ways of using petroleum coke is by mixing it with water to make a slurry fuel, which is then pumped into boilers and gasifiers as feedstock for combustion and gasification (Milenkova et al. 2003; Wang et al. 2006).

One of the major problems with slurry fuels is in simultaneously maintaining the highest possible solid concentration stably and at a viscosity that is optimal for industrial application. Petroleum coke–water slurry (PCWS) has good slurryability but inferior stability because of high hydrophobic properties of petroleum coke, which can be overcome by co-slurrying of petroleum coke with lignite. Rich reserves of lignite in China—which account for up to 17 % of all Chinese coal resources—have attracted increasing attention as raw material for gasification (Zhan et al. 2011). The presence of significant amounts of inherent moisture and high oxygen content in lignite renders the preparation of lignite–water slurry (LWS) difficult. In contrast, such high hydrophilicity due to the extensive numbers of oxygen functional groups is beneficial for maintaining a good and stable slurry (Yu et al. 2012). Therefore, the blending of petroleum coke with lignite to prepare petroleum coke–lignite slurry (PCLS) can produce synergistic effects. PCLS with high-solid concentration, high calorific value, good fluidity, and excellent stability is desirable for gasification and combustion.

Abundant reports are available on the influence of blending methods on the co-combustion and co-gasification characteristics of petroleum coke and lignite. As lignite has a low calorific value and petroleum coke has low reactivity due to its low number of pores and alkali species, blending the two may potentially overcome their individual drawbacks by combining their advantages (Wang et al. 2004; Wu et al. 2009; Zhan et al. 2011). Extensive studies have been carried out to determine the factors that affect their rheological properties, such as additive types, temperature, and particle-size distribution (PSD), in order to obtain slurries with high-solid concentration, good fluidity, and sufficient stability against sedimentation of the solid particles (Goudoulas et al. 2010; Zhan et al. 2010; Gao et al. 2010, 2012). Some significant studies involving co-slurrying of low-rank coal and petroleum coke have been carried out. Vitolo et al. (1996) reported that the addition of fine pulverized petroleum coke to coal water slurry improved its rheological properties at room temperature. Recently, Yang et al. (2008) and Xu et al. (2008) have qualitatively evaluated the co-slurrying ability of petroleum coke and low-rank coal. Most efforts in the past were directed toward enhancing the slurryability of coal by mixing it with petroleum coke. However, on further examination, it seems likely that the effect of mass ratio of petroleum coke to lignite on slurrying properties has not been adequately investigated. For the sake of convenience, the mass ratio of petroleum coke to the total blend of petroleum coke and lignite samples has been defined as the petroleum coke-mixing proportion (α). In the present study, the effect of variable values of α on co-slurrying of petroleum coke and lignite was studied with respect to slurrying ability, rheological behavior, slurry stability, and the effects of increased temperature. This study aims to obtain an optimal range of α that is suitable for industrial application, that is, to simultaneously attain the highest solid concentration and stability at a given viscosity. The findings of this study are of great importance for the promotion of industrial applications of PCLS.

2 Experimental

2.1 Materials

XiMeng (XM), a type of Chinese lignite found in Xilingol, Inner Mongolia, China, was used for all the experiments performed in the present study. Petroleum coke samples procured from Jinshan Petrochemical Company (Shanghai,

China) were chosen for mixing with the XM lignite. The results from proximate and ultimate analysis of test samples are shown in Table 1.

Thus, the proximate and ultimate analysis results (Table 1) show huge differences between XM lignite and petroleum coke. Petroleum coke has lower oxygen content, indicating a paucity of oxygen functional groups such as hydroxyl and carboxyl, as opposed to XM lignite. Consequently, petroleum coke is hydrophobic in contrast with lignite which is hydrophilic.

Chemical additives are an important component of slurry fuels. Prior studies have demonstrated that the copolymer of methylene naphthalene–styrene sulfonate–maleate (NDF) had the best dispersing effect among the most common additives. The chemical structure of NDF additives is shown in Fig. 1. Thus, NDF was added to the PCLS preparation in this study. The additive dosage was fixed at 0.8 wt% based on dry coal–coke samples.

2.2 Methods

2.2.1 Fourier transform infrared spectroscopy (FTIR) analysis

FTIR analysis of the petroleum coke and lignite were carried out using a FTIR spectrophotometer (Nicolet Nexus 670, USA).

2.2.2 Contact angle determination

The contact angle is the angle, conventionally measured through the liquid, where a liquid/vapor interface meets a solid surface. The coal powder sample was first compressed into a cylinder with a diameter of 20 mm and a height of 2 mm. A drop of water was then deposited on the coal cylinder. The dripping process was captured continuously using a high resolution camera in a contact angle measuring meter (Powereach JC2000C, Zhongchen Co., Shanghai). The photo at the moment when the water dropped onto the coal piece was selected and the contact angle between the coal and water was measured.

2.2.3 PCLS preparation

The raw coal and petroleum coke samples were ground with a laboratory-size ball mill, and then passed through a 100 mesh sieve to obtain experimental samples. The PSD of the sample was determined using a Mastersizer 2000 (Malvern Ltd., Britain).

PCLS preparation was standardized and used for all samples tested. At the outset, lignite, petroleum coke, distilled water, and additives were calculated and weighed in predetermined ratios. Furthermore, the weighed water

Table 1 Proximate and ultimate analysis of petroleum coke and XM lignite

Samples	Proximate analysis, %				$Q_{b,d}$, MJ/kg	Ultimate analysis, %				
	M_t	A_d	V_d	FC_d		C_d	H_d	N_d	$S_{t,d}$	O_d
Petroleum coke	0.47	0.43	10.6	89.0	36.0	88.6	3.59	1.11	5.54	0.74
XM lignite	18.4	10.1	30.1	52.8	18.6	65.4	5.44	0.91	1.85	16.3

M_t is the total moisture; all the following are calculated on a dry basis, A_d, V_d, and FC_d refer to ash, volatiles, and fixed carbon, respectively; $S_{t,d}$ refers to total sulfur and $Q_{b,d}$ refers to the bomb calorific value

Fig. 1 Chemical structure of additive (NDF)

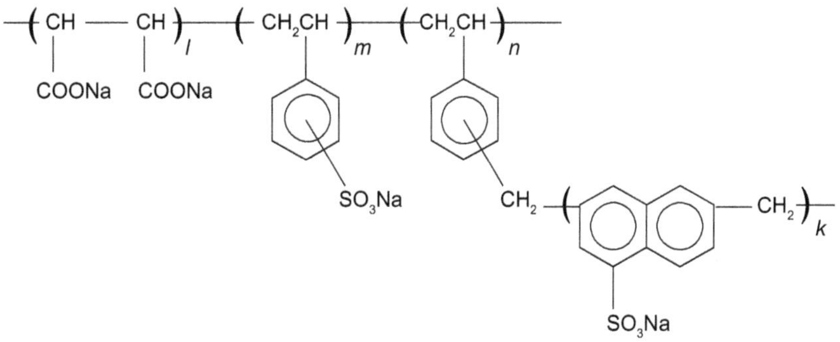

was poured into a 0.5 L stainless steel beaker in which the additives were completely dissolved by stirring with a mechanical mixer at 200 rpm. Subsequently, lignite and petroleum coke blends were slowly poured into the beaker, and the mixture was continuously stirred for 15 min at 1,000 rpm. It was necessary to allow the slurry to stand for 5 min to release the entrapped air before all measurements.

2.2.4 Determination of PCLS properties

The rheological behavior of the PCLSs was measured by a rotational viscometer (HAAKE VT 550, Thermo, USA). During the experiment, the temperature was maintained at 20 ± 0.5 °C using a water bath. The apparent viscosity η could be calculated by measuring the shear stress at a particular shear rate. The shear rate was increased from 10 to 100 s^{-1}, and then held constant at that level for 5 min. The viscosity data were recorded every 30 s during the 5 min period. The characteristic viscosity (η_c), defined as the apparent viscosity of the slurry sample at a shear rate of 100 s^{-1}, was used to evaluate the slurryability. In this study, η_c was calculated as an average of the 10 viscosity values recorded.

The solid concentration of PCLS, ω, was determined from the weight difference before and after drying in an oven at 105 °C for 2 h.

The effects of increasing temperature on the PCLS were considered as the apparent dependence of viscosity on temperature. We recorded viscosity data at a fixed shear rate of 100 s^{-1} as we raised the temperature from 20 to 50 °C.

The water separation ratio (WSR) is defined as the mass ratio of separated supernatant water to the total water in the test slurry after the slurry was left standing for 7 days. For comparison, the η_c of test slurry was fixed at 1,000 mPa s. In the present study, the static stability of PCLS is represented by WSR. The lower the WSR, the higher is the static stability of PCLS.

3 Results and discussion

3.1 Particle-size distribution

PSD is one of the important factors in preparing high-solid-concentration slurries. A PSD for the preparation of high-solid-concentration slurries is one that allows the particle system to attain a maximum packing fraction. The PSD curves for lignite and coke samples (Fig. 2) demonstrated that the petroleum coke sample contained significant amounts of finer particles than the XM lignite sample. The volume average particle diameter was 24 μm for petroleum coke and 61 μm for lignite. In addition, a slight bimodal distribution of lignite could also be observed. The difference in the mean sizes of coke and coal can be effectively used to obtain a lower viscosity, because the finer particles can enter the voids between the larger particles and act as lubricants. This effect increases the relative mobility of the

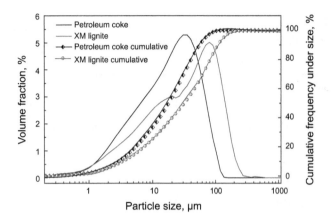

Fig. 2 Particle-size distribution of test samples

particles and decreases the viscosity of the slurry (Roh et al. 1995a, b).

3.2 FTIR analysis

The FTIR analysis results of the petroleum coke and lignite are shown in Fig. 3. There were two evident absorption bands, between 3,600 and 3,200 cm^{-1} and 1,800 and 900 cm^{-1} respectively. The absorption band between 3,600 and 3,200 cm^{-1} was attributed to –OH stretching vibrations of hydroxyl group. Lignite showed strong absorption intensity at this band, indicating it had more content of –OH group than petroleum coke. The strong peak at 1,618 cm^{-1} was associated with –COO– stretching vibrations of carbonyl groups, and the peak at 1,034 cm^{-1} was attributed to ether bonds. These peaks of lignite were also higher than that of petroleum coke, indicating lignite had higher content of oxygen functional groups.

3.3 Contact angle

Contact angle is one of the important indicators to measure the wettability of a surface or material. A low value of

contact angle indicates that the liquid spreads, or wets well, while a high contact angle indicates poor wetting. The two materials had very different contact angles with water. The contact angle between petroleum coke and water was 91°, while that between lignite and water was only 59°, indicating lignite was more hydrophilic than petroleum coke. This result was consistent with FTIR analysis. Due to the high content of oxygen functional groups, the surface of lignite showed strong hydrophilicity, and thus its contact angle with water was smaller. On the contrary, petroleum coke had a hydrophobic surface, which is difficult to wet.

3.4 Slurryability of PCLS

Slurryability is generally measured using the fixed-viscosity solid concentration (ω_0), which is defined as the ω of the slurry at a given η_c (in this case 1,000 mPa s). The higher the ω_0 of a PCLS, the better is the slurryability.

The relationship between η_c and ω for PCLSs with various α values (Fig. 4) revealed that for all the cases of α, η_c significantly increased with ω. This observation can be explained in two possible ways. First, the force of friction between the particles becomes more significant as the number of solid particles in the slurry system increases. Second, the proportion of free water acting as a lubricant in the slurry system decreases, leading to higher viscosity.

Figure 4 also demonstrates that α greatly affects the ω_0 of PCLSs. LWS ($\alpha = 0$) had a very low ω_0 of 46.7 %, whereas PCWS ($\alpha = 100$ %) had a high ω_0 of 71.3 %. The values for ω_0 were 50.0, 51.3, 53.2, 55.7, 57.4, 59.8, and 62.2 % for PCLSs with α values of 20, 30, 40, 50, 60, 70, and 80 %, respectively. On one hand, this may be attributed to the well-developed pore structures and high surface areas in lignite. Water is adsorbed on pore surfaces and forms as pore water. Pore water cannot reduce the mutual

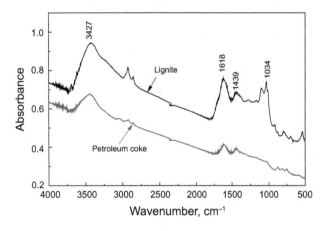

Fig. 3 FTIR spectra of test samples

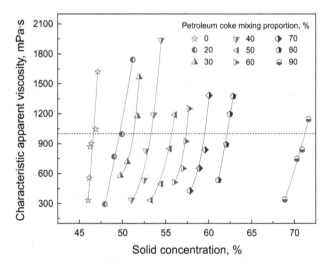

Fig. 4 η_c–ω dependence of PLWS with different α values

friction between the solid particles like free water. As a result, lignite is difficult to use in the preparation of highly concentrated slurries. On the other hand, the content of oxygen is up to 16.3 % for lignite, compared to 0.74 % for petroleum coke. Lignite has a large number of hydrophilic functional groups such as –COOH and –OH, resulting in absorption or trapping of large quantities of water by lignite particles. The high amount of inherent moisture cannot flow freely and is therefore unusable as a lubricant. As a result, the solid concentration of LWS ($\alpha = 0$) is low. However, petroleum coke displays perfect slurryability due to the absence of inherent moisture content. The majority of water can flow freely as lubricant. Thus, the higher the α value, the more mixed petroleum coke there is, which improves the slurryability of PCLSs.

3.5 Rheological characteristics

Rheological behavior is an important criterion of PCLSs that affects the pumping, atomizing, and combustion performance of slurry fuels. Therefore, good rheological characteristics are of great importance for the industrial application of PCLSs. A PCLS is a two-phase solid–liquid fluid, and the rheological characteristics of PCLSs can be described by the power-law model (Roh et al. 1995a, b):

$$\tau = K \cdot \gamma^n$$

where τ is the shear stress, Pa; K is the consistency coefficient, Pa s^n; γ is the shear rate, s^{-1}; and n is the rheological index (dimensionless). For a Newtonian fluid, $n = 1$; for a dilatant fluid, $n > 1$; for a pseudo-plastic fluid, $n < 1$.

3.5.1 Rheological characteristics of LWS ($\alpha = 0$)

The rheological behavior curves for LWS (Fig. 5) show that shear-thinning behavior ($n < 1$) can be observed for LWS, implying that LWS can be regarded as a pseudo-plastic fluid. It has been confirmed that a reticular three-dimensional network structure is formed through hydrogen bonds and dipolar polar groups that bind large amounts of water on the surfaces of lignite particles. In addition, the surfaces of solid particles combined with additives form hydration shells. Thus, LWS has high viscosity and good stability under static conditions. Upon being sheared, the reticular structure of the LWS is destroyed and then part of the water is released as a lubricant. Thus, the viscosity decreases as the shear rate increases owing to the lowering of the friction between the particles, and LWS shows features of a pseudo-plastic fluid (Yang et al. 2008; Zhu et al. 2003).

As shown in Table 2, the solid concentration had a great influence on the rheological index n of LWS. With

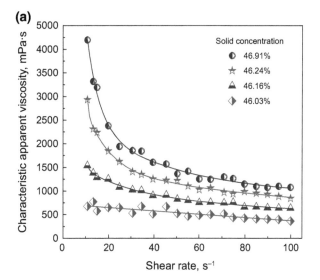

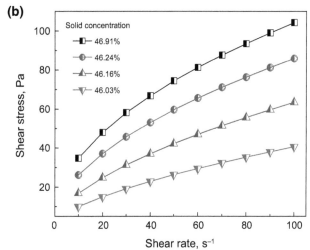

Fig. 5 Rheological properties of LWS. **a** Apparent viscosity versus shear rate ($\alpha = 0$). **b** Shear stress versus shear rate ($\alpha = 0$)

Table 2 Parameters of rheological model for LWS

ω, wt%	η_c, mPa s	K, Pa s^n	n	R^2
46.03	332	2.20	0.633	0.896
46.24	557	4.07	0.596	0.989
46.16	904	7.38	0.533	0.983
46.91	1,047	10.78	0.493	0.956

increasing concentration, the value of n gradually decreased, indicating shearing had a greater influence on the viscosity at higher solid concentration. It is because that greater changes in interior structure of higher solid concentration slurry occur when the slurry is subjected to shear, mainly in the following two aspects: on the one hand, when the solid concentration increases, the amounts of solid particles significantly increase, then more clusters of "water in coal" are formed in the slurry when the coal particles link and coagulate with each other during the preparation process, but the clusters could be deformed and broken up easily when subjected to shear, and then the enclosed water is released (Wang et al. 2011). On the other hand, high speed shearing induces directional migration of particles, and the fine particles fill the interspace of the coarse particles. Consequently, a rigid microstructure assembled by particles is produced, i.e., the maximum packing fraction Φ_m increases with the shearing force. Based on the research of Wildemuth and Williams (1984), the viscosity of slurry declines with increasing Φ_m.

3.5.2 Rheological characteristics of PCWS (α = 100 %)

The rheological behavior curves for PCWS (Fig. 6) show that shear-thickening behavior $(n > 1)$ is observed,

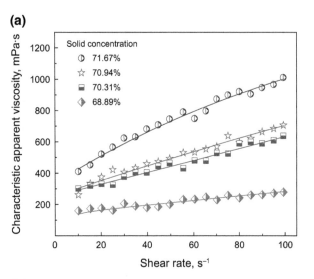

(a)

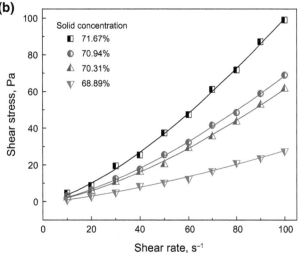

(b)

Fig. 6 Rheological properties of PCWS (α = 100 %). **a** Apparent viscosity versus shear rate (α = 100 %). **b** Shear stress versus shear rate (α = 100 %)

Fig. 7 Rheological properties of PCLSs. **a** Apparent viscosity versus ▶ shear rate (α = 20 %). **b** Shear stress versus shear rate (α = 20 %). **c** Apparent viscosity versus shear rate (α = 30 %). **d** Shear stress versus shear rate (α = 30 %). **e** Apparent viscosity versus shear rate (α = 40 %). **f** Shear stress versus shear rate (α = 40 %). **g** Apparent viscosity versus shear rate (α = 50 %). **h** Shear stress versus shear rate (α = 50 %). **i** Apparent viscosity versus shear rate (α = 60 %). **j** Shear stress versus shear rate (α = 60 %). **k** Apparent viscosity versus shear rate (α = 70 %). **l** Shear stress versus shear rate (α = 70 %). **m** Apparent viscosity versus shear rate (α = 80 %). **n** Shear stress versus shear rate (α = 80 %)

indicating that PCWS is a dilatant fluid. This is because the inter-particle forces keep the petroleum coke particles with hydration shells in an ordered structure under static conditions. Thus, the PCWS shows good fluidity. At a higher shear rate, the shear forces attain larger magnitudes than the inter-particle forces. Consequently, the hydration shell ruptures and petroleum coke particles easily aggregate because of their mutual attraction, causing an increase in viscosity and confirming that PCWS is a dilatant fluid.

Table 3 shows the rheological parameters with different solid concentrations for PCWS. The rheological index n changed slightly, but the consistency coefficient K increased with concentration. At a low concentration, the apparent viscosity η was relatively low and changed slightly when the shear rate γ was increased. At a high concentration, the viscosity of PCWS increased strongly with the shear rate.

3.5.3 Effect of α on rheological characteristics of PCLSs

The rheological characteristics are quite different when lignite and petroleum coke are used to make slurry. Mixing petroleum coke with lignite for preparing PCLS is an effective way to use these two fuels. The rheological behavior curves for PCLSs at different values of α (Fig. 7) and the rheological model parameters of PCLSs (Table 4) show that the consistency coefficient K depended on ω and α, and η_c was strongly related to K, and that α highly influenced the rheological characteristics of PCLSs. Increasing the value of α clearly led to an increase in the value of n, indicating that on adding high quantities of lignite, the pseudo-plastic behavior of PCLSs became more obvious. The PCLSs exhibited pseudo-plastic behavior even for α in the range of 60–70 %. When more petroleum

Table 3 Parameters of rheological model for PCWS (α = 100 %)

ω, wt%	η_c, mPa s	K, Pa s^n	n	R^2
68.89	342	0.04	1.404	0.992
70.31	748	0.07	1.469	0.995
70.94	839	0.07	1.489	0.994
71.67	1,146	0.13	1.445	0.996

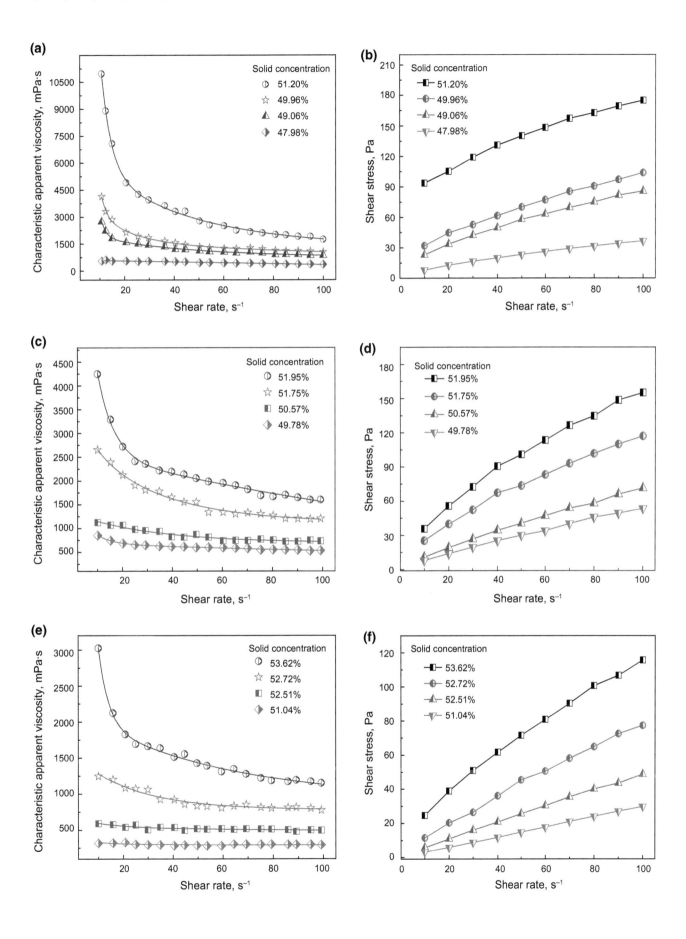

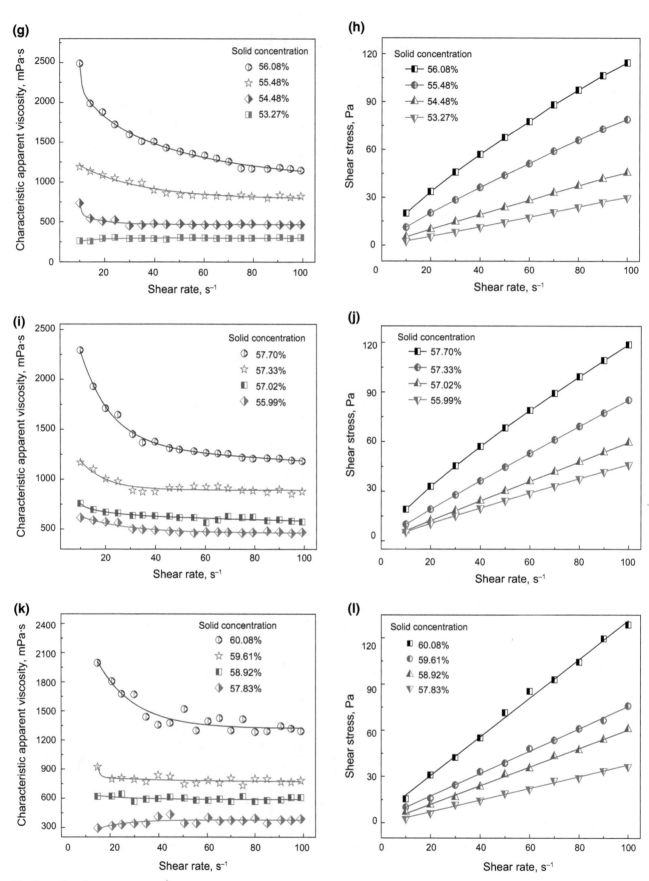

Fig. 7 continued

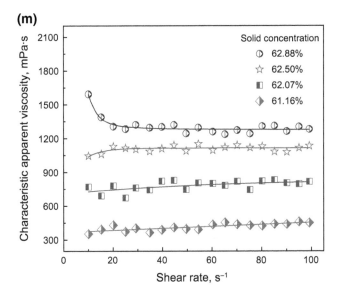

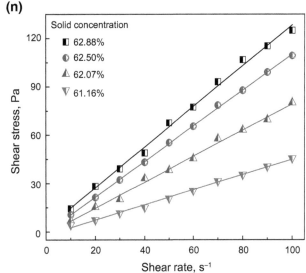

Fig. 7 continued

Table 4 Parameters of the rheological model for PCLSs

α, wt%	ω, wt%	η_c, mPa s	K, Pa s^n	n	R^2
20	47.98	293	1.78	0.657	0.986
	49.06	768	5.60	0.592	0.986
	49.96	994	7.90	0.557	0.984
	51.20	1,741	39.69	0.322	0.985
30	49.78	579	1.14	0.837	0.998
	50.57	716	1.64	0.821	0.992
	51.75	1,177	5.34	0.672	0.994
	51.95	1,567	7.95	0.650	0.991
40	51.04	337	0.28	1.013	0.997
	52.51	534	0.67	0.933	0.996
	52.72	827	1.67	0.834	0.989
	53.62	1,191	5.24	0.669	0.990
50	53.27	333	0.23	1.055	0.998
	54.48	495	0.58	0.951	0.996
	55.48	847	1.57	0.852	0.992
	56.08	1,188	3.43	0.763	0.988
60	55.99	513	0.67	0.923	0.991
	57.02	651	0.64	0.987	0.994
	57.33	923	1.16	0.934	0.991
	57.70	1,251	3.00	0.800	0.989
70	57.83	427	0.37	1.003	0.973
	58.92	654	0.57	1.009	0.995
	59.61	838	0.85	0.978	0.993
	60.08	1,383	2.59	0.850	0.989
80	61.16	534	0.24	1.138	0.995
	62.07	892	0.58	1.072	0.992
	62.50	1,198	0.99	1.026	0.989
	62.88	1,375	1.37	0.987	0.991

coke was added, the PCLSs (when $\alpha > 70\ \%$) exhibited the rheological characteristics of an approximate Newtonian fluid ($n = 1$).

With increasing solid concentration, the value of n gradually decreased, indicating the pseudo-plastic behavior of slurry was enhanced (such as $\alpha = 30\ \%$), or the flow behavior of slurry was changed from dilatant to pseudo-plastic (such as $\alpha = 70\ \%$). It is because that as solid concentration is increased, the changes in interior structure of the slurry become more prominent when the slurry is sheared at a high speed. The detailed explanations are shown in Sect. 3.5.1.

When $\alpha \leq 70\ \%$, the pseudo-plastic behavior of PCLS matched well with the requirements of industrial applications. The sedimentation of solid particles was effectively restrained because of high viscous force and friction under static storage of PCLSs. As a result, the stability of slurry fuel was enhanced. However, the viscosity of slurry declined a lot during the pumping and atomizing processes, which is beneficial for cost reduction.

3.6 Effect of temperature

Temperature has a considerable influence on the properties of the slurry. In general, the slurry is preheated prior to combustion or gasification to obtain optimal atomization effects, and the viscosity of the slurry is expected to decrease as the temperature rises. The relationship between η and temperature (T) at varying values of α (30, 50, and 70 %) is presented in Fig. 8, from which it can be observed that the value of η consistently decreased as the temperature increased from 20 to 50 °C, with a concomitant drop

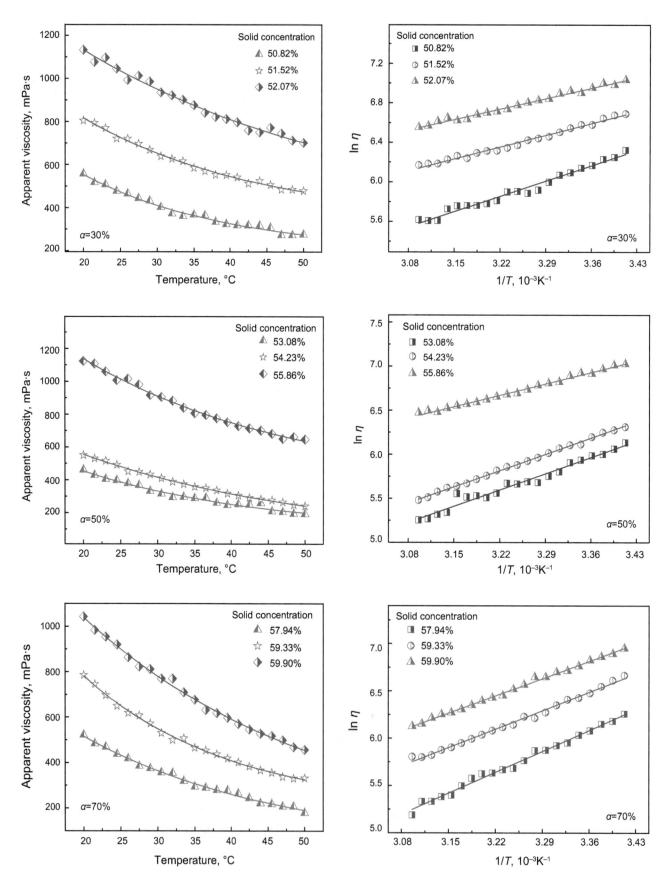

Fig. 8 Effect of temperature on apparent viscosity

in viscosity of up to 60 %. For example, the value of η sharply decreased from 1,043 mPa s to 457 mPa s at $\alpha = 70$ % and $\omega = 59.9$ %. This phenomenon can be attributed to a combination of several factors (Chen et al. 2008; Roh et al. 1995a, b). Primarily, the liquid phase in the PCLS system, essentially free water, possesses good viscosity-reducing properties with increasing temperature, which is the main cause of the decrease in the viscosity of PCLS. Second, as the temperature is increased, the effects of additives that make coal particles more easily dispersed are enhanced. In addition, an increase in the temperature of the system leads to an increase in the kinetic energy of the particles, resulting in their further dispersion and consequent reduction in the value of η.

Earlier studies have described the general consensus about the relationship between η and T in terms of a simple Arrhenius expression for the range of temperatures investigated (Mishra et al. 2002):

$$\eta = A \cdot \exp\left(E/RT\right)$$

Or

$$\ln \eta = E/RT + \ln A,$$

where η is the viscosity; E is the fluid-flow activation energy; T is the temperature in Kelvin; R is the universal gas constant; and A is a fitting parameter.

As indicated in Fig. 8, $\ln(\eta)$ was approximately linear with $1/T$, which implies that the influence of temperature on the viscosity of PCLS is in good agreement with the Arrhenius equation. Table 5 presents the parameters of the Arrhenius correlation for PCLSs with varying ω values. Notably, the value of A rapidly increases with concentration at a constant value of α. This suggests that the value of

Table 5 Parameters for Arrhenius correlation of PCLSs

α, wt%	ω, wt%	E, kJ mol^{-1}	A, mPa s	R^2
30	50.82	18.99	0.223	0.980
	51.52	14.58	2.025	0.988
	52.07	12.62	6.346	0.986
50	53.08	22.21	0.049	0.977
	54.23	21.85	0.071	0.997
	55.86	15.51	1.939	0.992
70	57.94	26.13	0.011	0.994
	59.33	23.77	0.045	0.993
	59.90	21.61	0.147	0.996

A can indicate the concentration of the slurry to some extent. The value of E varies little, suggesting that the fluid-flow activation energy is almost uninfluenced by concentration. The fluid-flow activation energy E refers to the energy needed for particles to overcome the interaction of surrounding molecules and flow into free space, and more importantly, it shows how susceptible viscosity is to changes in temperature. The higher the value of E, the more sensitive viscosity is to variations in temperature. Moreover, the value of E increases with the value of α, which implies that the sensitivity of viscosity to temperature increases with the proportion of petroleum coke.

3.7 Stability properties of PCLS

Stability is an important feature in the quality of PCLS, which determines its efficacy and employability in large-scale industrial applications. In this study, all the slurry samples were allowed to stand, and the water separation and sedimentation were evaluated to assess the stability. It was observed that α highly influenced the sedimentation of PCLSs. Thus, for α in the range of 0–70 %, soft sedimentation was observed at the bottom of the slurry within 72 h, which could be reverted to the initial slurry by agitation even after 7 days. Nevertheless, rigid sedimentation, which could not be reverted to a uniform suspension by agitation, was engendered in PCLSs for $\alpha \geq 70$ %.

We measured WSR (Table 6) to evaluate the static stability of each PCLS. The WSR was below 6 % for all α values. Furthermore, the stability tended to be better for low α PCLSs: the lower the value of α, the lower the WSR (the better the stability). Significantly, the WSR rapidly increased for $\alpha \geq 70$ %, which could be explained as follows. The higher the lignite content, the better the stability of the reticular three-dimensional structure. In contrast, the higher the content of petroleum coke added to PCLS, the more obvious the extent of sedimentation and water separation. A high content of petroleum coke reduced the stability, which can be attributed to the highly hydrophobic surface of petroleum coke that can easily lead to substantial aggregation of particles.

3.8 The optimal value of α

When α increases, the PCLS exhibits better slurryability, but poorer rheological properties and stability. By considering all the factors, the optimal value of α for industrial

Table 6 WSR of PCLSs maintained for 7 days

α, wt%	0	20	30	40	50	60	70	80	100
WSR, %	0	0	0.469	0.629	1.423	1.641	1.971	4.049	5.733

applications is between 60 % and 70 %. Thus, a high-quality slurry fuel with high slurryability ($\omega = 57$–60 %), good stability (WSR < 2 %), and superior pseudo-plastic behaviors ($n \approx 0.9$) can be produced.

(WSR < 2 %), and superior pseudo-plastic behaviors ($n \approx 0.9$) can be produced.

Acknowledgments The authors acknowledge the financial support provided by the National Basic Research Program of China (Grant No. 2010CB227001).

4 Conclusions

Petroleum coke and lignite are promising fuels that have different slurrying characteristics. The strongly hydrophobic surface of petroleum coke can enhance the solid concentration of the slurry, but leads to poor stability. In contrast, the high content of inner moisture and the presence of oxygen functional groups in lignite reduce its usefulness in preparing high-concentration slurries, but impart lignite with a three-dimensional network structure that provides both fluidity and stability. In our study, we verified that mixing petroleum coke with lignite in PCLS combines their complementary strengths to make PCLS useful for combustion and gasification. The main conclusions are as follows:

(1) The fixed-viscosity solid concentration (ω_0) of PCLS positively correlates with α. The ω_0 of LWS ($\alpha = 0$) is 46.7 % compared to 71.3 % for PCWS ($\alpha = 100$ %). PCLSs have α values between those of LWS and PCWS. The characteristic apparent viscosity η_c increases with ω. Mixing fine petroleum coke with coarse lignite is conducive to improving the slurryability of PCLS.

(2) The rheological behavior of PCLS perfectly fits the power-law model. LWS performs as a pseudo-plastic fluid. As α increases, the slurry behaves as an approximate Newtonian fluid, and then turns to a dilatant fluid, which exhibits shear-thickening behavior. As ω increases, the consistency coefficient K increases, while the rheological index n decreases.

(3) The η_c of PCLS decreases as temperature increases. The viscosity drops by up to 60 % as temperature increases from 20 to 50 °C. The fluid-flow activation energy E increases with α, indicating that viscosity becomes more sensitive to decrease as temperature increases.

(4) LWS ($\alpha = 0$) provides good stability without water separation and rigid sedimentation. As more petroleum coke is added, the rigid sedimentation and WSR increase, lowering the stability of PCLS.

(5) After considering all of the parameters that we studied, including slurryability, fluidity, and stability of PCLS, the optimal value of α for industrial applications is 60–70 %. At this concentration of petroleum coke, a high-quality slurry fuel with high slurryability ($\omega = 57$–60 %), good stability

References

Chen LY, Duan YF, Liu M, et al. True rheological behavior of coal-water slurry. J Power Eng. 2008;28(5):753–8 (in Chinese).

Gao FY, Liu JZ, Wang CC, et al. Effects of the physical and chemical properties of petroleum coke on its slurryability. Pet Sci. 2012;9(2):251–6.

Gao FY, Liu JZ, Wang CC, et al. Slurryability of petroleum coke and rheological characteristics and stability of PCWS. CIESC J. 2010;61(11):2912–8 (in Chinese).

Goudoulas TB, Kastrinakis EG, Nychas SG. Preparation and rheological characterization of lignite-water slurries. Energy Fuels. 2010;24:496–502.

He QH, Wang R, Wang WW, et al. Effect of particle size distribution of petroleum coke on the properties of petroleum coke-oil slurry. Fuel. 2011;90(9):2896–901.

Milenkova KS, Borrego AG, Alvarez D, et al. Devolatilisation behaviour of petroleum coke under pulverised fuel combustion conditions. Fuel. 2003;82(15–17):1883–91.

Mishra SK, Senapati PK, Panda D. Rheological behavior of coal-water slurry. Energy Sources. 2002;24(2):159–67.

Roh NS, Shin DH, Kim DC, et al. Rheological behavior of coal-water mixtures. 1. Effect of coal type, loading and particle-size. Fuel. 1995a;74(8):1220–5.

Roh NS, Shin DH, Kim DC, et al. Rheological behavior of coal-water mixtures. 2. Effect of surfactants and temperature. Fuel. 1995b;74(9):1313–8.

Vitolo S, Belli R, Mazzanti M, et al. Rheology of coal-water mixtures containing petroleum coke. Fuel. 1996;75(3):259–61.

Wang JS, Anthony EJ, Abanades JC. Clean and efficient use of petroleum coke for combustion and power generation. Fuel. 2004;83(10):1341–8.

Wang RK, Liu JZ, Yu YJ, et al. The slurrying properties of coal water slurries containing raw sewage sludge. Energy Fuels. 2011;25(2): 747–52.

Wang ZQ, Wang HF, Guo QJ. Effect of ultrasonic treatment on the properties of petroleum coke oil slurry. Energy Fuels. 2006;20(5):1959–64.

Wildemuth CR, Williams MC. Viscosity of suspensions modeled with a shear-dependent maximum packing fraction. Rheol Acta. 1984;23(6):627–35.

Wu YQ, Wu SY, Gu J, et al. Differences in physical properties and CO_2 gasification reactivity between coal char and petroleum coke. Process Saf Environ Prot. 2009;87(5):323–30.

Xu RF, He QH, Cai J, et al. Effects of chemicals and blending petroleum coke on the properties of low-rank Indonesian coal water mixtures. Fuel Process Technol. 2008;89(3):249–53.

Yang BL, Gong KF, Zou JJ, et al. Slurryability of lignite and petroleum coke mixture. J Fuel Chem Technol. 2008;36(4): 391–6 (in Chinese).

Yu YJ, Liu JZ, Wang RK, et al. Effect of hydrothermal dewatering on the slurryability of brown coals. Energy Convers Manag. 2012;57:8–12.

Zhan XL, Zhou ZJ, Kang WZ, et al. Promoted slurryability of petroleum coke-water slurry by using black liquor as an additive. Fuel Process Technol. 2010;91(10):1256–60.

Zhan XL, Jia J, Zhou ZJ, et al. Influence of blending methods on the co-gasification reactivity of petroleum coke and lignite. Energy Convers Manag. 2011;52(4):1810–4.

Zhou ZJ, Hu QJ, Liu X, et al. Effect of iron species and calcium hydroxide on high-sulfur petroleum coke CO_2 gasification. Energy Fuels. 2012;26(3):1489–95.

Zhu SQ, Zou LZ, Huang B, et al. Study on the interaction characteristics between different CWS dispersants and coals I. Effect of the interaction of complex coal particles on CWS rheological behavior. J Fuel Chem Technol. 2003;31(6):519–24 (in Chinese).

7

The coupling of dynamics and permeability in the hydrocarbon accumulation period controls the oil-bearing potential of low permeability reservoirs: a case study of the low permeability turbidite reservoirs in the middle part of the third member of Shahejie Formation in Dongying Sag

Tian Yang[1,2,3] · Ying-Chang Cao[1,2] · Yan-Zhong Wang[1,2] · Henrik Friis[3] ·
Beyene Girma Haile[4] · Ke-Lai Xi[1,2,4] · Hui-Na Zhang[1,2]

Abstract The relationships between permeability and dynamics in hydrocarbon accumulation determine oil-bearing potential (the potential oil charge) of low permeability reservoirs. The evolution of porosity and permeability of low permeability turbidite reservoirs of the middle part of the third member of the Shahejie Formation in the Dongying Sag has been investigated by detailed core descriptions, thin section analyses, fluid inclusion analyses, carbon and oxygen isotope analyses, mercury injection, porosity and permeability testing, and basin modeling. The cutoff values for the permeability of the reservoirs in the accumulation period were calculated after detailing the accumulation dynamics and reservoir pore structures, then the distribution pattern of the oil-bearing potential of reservoirs controlled by the matching relationship between dynamics and permeability during the accumulation period were summarized. On the basis of the observed diagenetic features and with regard to the paragenetic sequences, the reservoirs can be subdivided into four types of diagenetic facies. The reservoirs experienced two periods of hydrocarbon accumulation. In the early accumulation period, the

reservoirs except for diagenetic facies A had middle to high permeability ranging from 10×10^{-3} μm^2 to 4207×10^{-3} μm^2. In the later accumulation period, the reservoirs except for diagenetic facies C had low permeability ranging from 0.015×10^{-3} μm^2 to 62×10^{-3} μm^2. In the early accumulation period, the fluid pressure increased by the hydrocarbon generation was 1.4–11.3 MPa with an average value of 5.1 MPa, and a surplus pressure of 1.8–12.6 MPa with an average value of 6.3 MPa. In the later accumulation period, the fluid pressure increased by the hydrocarbon generation process was 0.7–12.7 MPa with an average value of 5.36 MPa and a surplus pressure of 1.3–16.2 MPa with an average value of 6.5 MPa. Even though different types of reservoirs exist, all can form hydrocarbon accumulations in the early accumulation period. Such types of reservoirs can form hydrocarbon accumulation with high accumulation dynamics; however, reservoirs with diagenetic facies A and diagenetic facies B do not develop accumulation conditions with low accumulation dynamics in the late accumulation period for very low permeability. At more than 3000 m burial depth, a larger proportion of turbidite reservoirs are oil charged due to the proximity to the source rock. Also at these depths, lenticular sand bodies can accumulate hydrocarbons. At shallower depths, only the reservoirs with oil-source fault development can accumulate hydrocarbons. For flat surfaces, hydrocarbons have always been accumulated in the reservoirs around the oil-source faults and areas near the center of subsags with high accumulation dynamics.

✉ Ying-Chang Cao
   cyc8391680@163.com

[1] School of Geosciences, China University of Petroleum, Qingdao 266580, Shandong, China

[2] Laboratory for Marine Mineral Resources, Qingdao National Laboratory for Marine Science and Technology, Qingdao 266071, China

[3] Department of Geoscience, Aarhus University, Høegh-Guldbergs Gade 2, 8000 Aarhus C, Denmark

[4] Department of Geosciences, University of Oslo, P.O. Box 1047, Blindern, 0316 Oslo, Norway

Edited by Jie Hao

Keywords Reservoir porosity and permeability evolution · Accumulation dynamics · Cutoff-values of permeability in the accumulation period · Oil-bearing potential · Low permeability reservoir · The third member of the Shahejie Formation · Dongying Sag

1 Introduction

With the increasing interest in oil and gas exploration and development, low permeability clastic rock reservoirs are becoming key exploration target areas (Yang et al. 2010; Cao et al. 2012). The low permeability clastic rock reservoirs have gone through complex diagenetic events (Yang et al. 2010; Wang et al. 2011). The distribution of sandstone porosity is not consistent with the hydrocarbon accumulation. The porosity of sandstone during the accumulation period is the key factor to determine the oiliness of the reservoirs (Cao et al. 2012; Liu et al. 2014a; Wang et al. 2014a). Some researchers have attempted to extract data from the porosity of low permeability clastic rock reservoirs during the accumulation period (Cao et al. 2011, 2012, 2013; Wang et al. 2013a; Liu et al. 2014a). However, they did not calculate the cutoff values for porosity of the reservoir under the control of accumulation dynamics during the accumulation period (Pan et al. 2011; Wang et al. 2014a; Liu et al. 2014a). The distribution of the oil-bearing potential of reservoirs is still poorly understood. The relationships between porosity and the oil-bearing potential of turbidite reservoirs of the middle part of the third member of Shahejie Formation (Es_3^z) in Dongying Sag are complex, even though the reservoirs have similar accumulation conditions. The high or low porosity and permeability sandstone reservoirs either contain oil or not. Liu et al. (2014a, b) analyzed the relationship between porosity and the cutoff-values for porosity in the early accumulation period of Es_3^z turbidite reservoirs in Niuzhuang subsag with the guide of porosity estimation and effect-oriented simulation. They concluded that the porosity of reservoirs in the early accumulation period was higher than the cutoff-values for porosity of the reservoirs. So the reservoirs could be charged with oil. The permeability is the main controlling factor for percolation and the development of low permeability reservoirs (Meng et al. 2013). There were several stages of accumulation for the Es_3^z turbidite reservoirs in the Dongying Sag and the later accumulation period was the most important (Cai 2009). The permeability and the cutoff-values at the later accumulation period are the most important for the distribution of the oil-bearing potential of reservoirs today.

On the basis of previous studies, taking the Es_3^z turbidite reservoirs as an example, the permeability of the reservoirs in the accumulation period was estimated. The permeability estimation method was based on the paragenetic sequence of diagenetic minerals and the reservoir pore-throat geometry. The cutoff-values for permeability of reservoirs in the accumulation period are calculated after the estimation of accumulation dynamics and reservoir pore-throat geometries, and finally the distribution pattern of the oil-bearing potential of the reservoirs is determined. This can provide theoretical guidance for the exploration and development of low permeability turbidite reservoirs.

2 Geological background

The Dongying Sag is a sub-tectonic unit lying in the southeastern part of the Jiyang Depression of the Bohai Bay Basin, East China. It is a Mesozoic-Cenozoic half graben rift-downwarped basin with lacustrine facies directly deposited on Paleozoic bedrocks (Cao et al. 2014; Wang et al. 2014b). The Dongying Sag is bounded to the east by the Qingtuozi Salient, to the south by the Luxi Uplift and Guangrao Salient, to the west by the Linfanjia and Gaoqing salients, and to the north by the Chenjiazhuang-Binxian Salient. The NE-trending sag covers an area of 5850 km^2 (Fig. 1). It is a half graben with a faulted northern margin and a gentle southern margin. Horizontally, this sag is further subdivided into several secondary structural units, such as the northern steep slope zone, middle uplift belt, and the Lijin, Minfeng and Niuzhuang trough zones, Boxing subsag, and the southern gentle zone (Zhang et al. 2014). The sag is filled with Cenozoic sediments, which are formations from the Paleogene, Neogene, and Quaternary periods. The formations from the Paleogene period are the Kongdian (Ek), Shahejie (Es), and Dongying (Ed); the formations from the Neogene period are the Guantao (Ng) and Minghuazhen (Nm); and the formation from the Quaternary period is the Pingyuan (Qp). Detailed descriptions of the Paleogene stratigraphy have been provided by several authors (Zhang et al. 2004, 2010; Guo et al. 2012) (Fig. 2).

During the deposition of the third member of the Shahejie Formation, tectonic movement was strong, and the basin subsided rapidly reaching its maximum depth. As a result, large amounts of detrital materials were transported into the basin and formed plentiful source rocks and turbidites in deep-water environments in the depressed zone and uplifted zone (Wang et al. 2013b; Yang et al. 2015) (Fig. 3). The thickness of single sand layers of turbidite reservoirs is 0.1–0.5 m; the accumulation thickness is 10–158 m. Turbidity current deposits with Bouma sequences and debris flow deposits with massive bedding are most common. The east slope of the Niuzhuang subsag, Liangjialou, and the front of the Dongying delta are places where a large volume of turbidites are distributed (Yang et al. 2015). Most turbidite reservoirs are low permeability with complex oil-bearing characteristics.

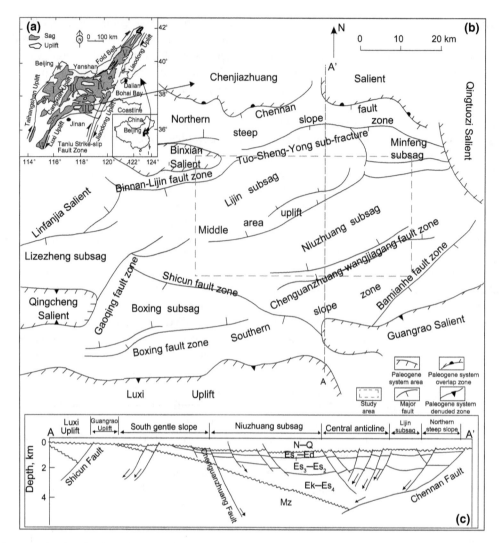

Fig. 1 **a** Location map showing the six major sub basins of the Bohai Bay Basin. **b** Structural map of the Dongying Sag. The area in the *green line box* is the study area (After Liu et al. 2014a). **c** N–S cross section (A′–A) of the Dongying Sag showing the various tectonic-structural zones and key stratigraphic intervals

3 Materials and methods

Over 1500 m of representative cores of turbidite in the target formation have been described. 119 typical samples were taken from the core. Thin section examination and porosity and permeability testing of all 119 samples were undertaken. Mercury injection testing of 90 samples, scanning electron microscopy (SEM) examination of 15 samples, cathode luminescence testing of 17 samples, fluorescence thin section observation of 17 samples, and fluid inclusion testing of 53 samples were undertaken. The core samples were provided by the Geological Scientific Research Institute of the Sinopec Shengli Oilfield Company. Porosity, permeability, and mercury injections were measured at the Exploration and Development Research Institute of the Sinopec Zhongyuan Oilfield Company as were the SEM examinations. Porosity and permeability

were tested by a 3020-62 helium porosity analyzer and GDS-9F gas permeability analyzer at common temperature and humidity. Mercury injection was tested by a 9505 mercury injection analyzer at 22 °C and 60 % humidity. Samples were examined by a JSM-5500LVSEM combined with QUANTAX400 energy dispersive X-ray microanalyser (EDX). The thin sections and fluorescence thin sections were prepared by the CNPC Key Laboratory of Oil and Gas reservoirs at the China University of Petroleum and were examined using an Axioscope A1 APOL digital polarizing microscope produced by the German company Zeiss. The cathodoluminescence was studied using an Imager D2 m cathode luminescence microscope also produced by Zeiss. The fluid inclusions were analyzed using a THMSG600 conventional inclusion temperature measurement system produced by the British Company Linkam. Sandstone composition analysis data of 2314 samples and

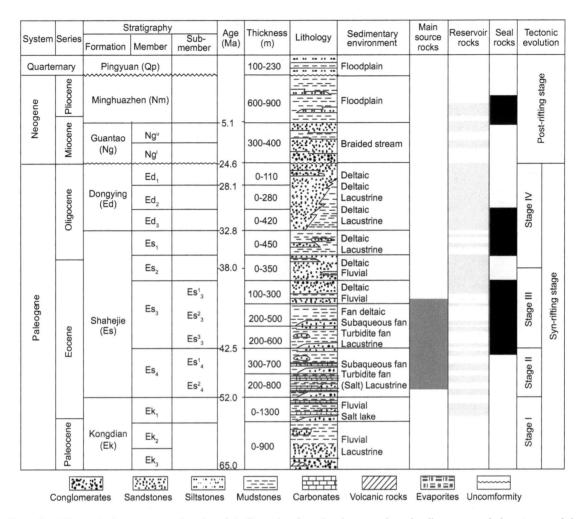

Fig. 2 Generalized Cenozoic Quaternary stratigraphy of the Dongying Sag, showing tectonic and sedimentary evolution stages and the major petroleum system elements (After Yuan et al. 2015)

porosity and permeability testing of 7433 samples of the research area have been collected from the Geological Scientific Research Institute of the Sinopec Shengli Oilfield Company.

4 Characteristics and porosity–permeability evolution of low permeability turbidite reservoirs

4.1 Characteristics of low permeability turbidite reservoirs

4.1.1 Petrography

Es_3^z turbidite sandstones from the Dongying Sag predominantly belong to lithic arkose families based on the sandstones classification scheme of Folk (1974) (Fig. 4). The reservoirs are mainly composed of fine to medium grained sandstones. Based on the amount of framework grains, the quartz content is 29 %–69.2 % with an average of 43.5 %; the feldspar content is 14.3 %–47 % with an average of 33.7 %; the content of rock fragments is 2 %–44.2 % with an average of 22.8 %. The mud content is 0.5 %–48 % with an average of 11.0 %, and the cement content is 0.5 %–34.6 % with an average of 8.2 %. The compositional maturity is 0.41–2.25 with an average of 0.8, and detrital grains are moderately sorted, with sub-angular or sub-rounded shapes.

4.1.2 Reservoir features

(1) Porosity–permeability

Based on the porosity–permeability data, the study area is characterized by low permeability with an average porosity and permeability value of 17.1 % and 38.1 × 10^{-3} μm^2, respectively. It contains 31 % low porosity reservoirs, 69 % medium to high porosity reservoirs, 88 % low

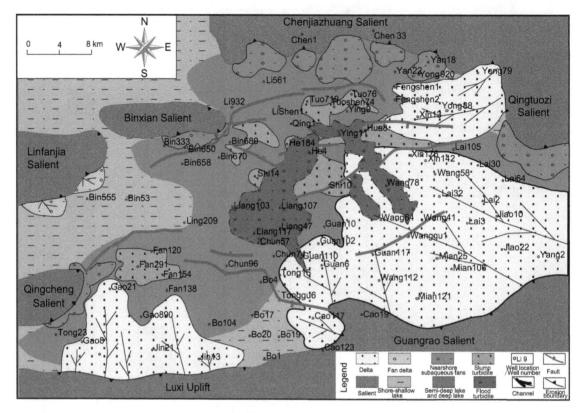

Fig. 3 Sedimentary facies distribution of Es_3^z in Dongying Sag

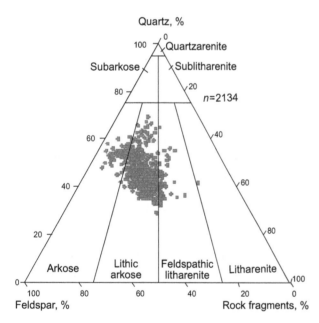

Fig. 4 Triangular plot of sandstones of the low permeability Es_3^z turbidite reservoirs

permeability reservoirs, and 12 % medium to high permeability reservoirs. Low permeability reservoirs with middle-high porosity are most common with 59 % of the total reservoirs (Fig. 5).

(2) Reservoir space

The reservoir space consists of primary pores, mixed pores, and secondary pores and gaps. Primary pores include the remaining intergranular pores after compaction and cementation and micropores in clay mineral matrices making up the main pore type (Fig. 6e, f, g). Expansion of pores by dissolution is the main kind of mixed pores (Fig. 6h). There are various kinds of secondary pores and gaps containing dissolution pores in particles and cements (Fig. 6k, l), moldic pores (Fig. 6i), intergranular micropores of kaolinite (Fig. 6m, n, o and p), microfractures and diagenetic contraction fractures. As one kind of gravity flow deposits, turbidite is characterized by a large amount of matrix which contains significant amounts of primary micropores. During the process of diagenetic evolution, additional intergranular micropores are developed due to the transformation from feldspar to kaolinite (Bjørlykke 2014; Giles and de Boer 1990) (Fig. 6m, n, o). The large proportion of micropores results in much lower permeability of reservoirs than that of other reservoirs with the same porosity (Yuan et al. 2013, 2015; Cao et al. 2014). So middle and high porosity low permeability reservoirs are common.

(3) The characteristics of pore throat structure

Using mercury injection data, we classify pore-throat structures according to the parameters of displacement pressure (P_d)

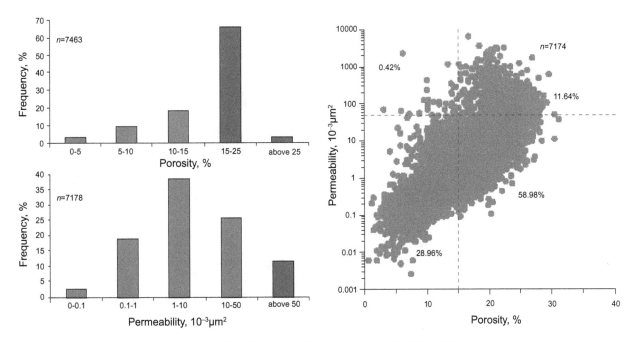

Fig. 5 Plots illustrating the porosity and permeability distribution of the low permeability Es$_3^z$ turbidite reservoirs

and median capillary pressure (P_{50}) (Wang et al. 2014a). First, reservoirs are classified into six types according to displacement pressure (P_d) IA ($P_d \leq 0.05$ MPa), IB (0.05–0.1 MPa P_d), IIA (0.1–0.5 MPa P_d), IIB (0.5–2 MPa P_d), IIIA (2–5 MPa P_d), and IIIB ($P_d > 5$ MPa). Second, each type is further divided into six units according to median capillary pressure (P_{50}) $P_{50} \leq 0.3$ MPa, 0.3–1.5 MPa P_{50}, 1.5–5 MPa P_{50}, 5–20 MPa P_{50}, 20–40 MPa P_{50}, $P_{50} > 40$ MPa. If the P_{50} datum of a sample is not in accordance with the overall characteristics of a unit, then the sample is assigned to the lower unit (Wang et al. 2014a). We divide the Es$_3^z$ turbidite reservoirs in the Dongying Sag into three broad types and six types. Then we correlate K/Φ with K for each type of reservoir (Fig. 7). So, we can determine the ranges of permeability and the ratio of permeability to porosity corresponding to various types of reservoirs (Table 1). Reservoirs with different kinds of pore throat structures have the same power function relationship between K/Φ and K. This reflects that the permeability of low permeability reservoirs is controlled by pore throat structures. However, different kinds of reservoirs have different ranges of permeability (Fig. 7). Good pore throat structures are characterized by lower P_d and P_{50}, as well as higher K/Φ and K values; poor pore throat structures are characterized by higher P_d and P_{50} and lower K/Φ and K values.

4.1.3 Diagenesis features

(1) Diagenetic events

The major diagenetic events in the research area include compaction, cementation, replacement, and dissolution.

Fig. 6 Typical diagenesis characteristics and reservoir pore types of ▶ the low permeability Es$_3^z$ turbidite reservoirs. **a** Wangxie 543, 3177.3 m (–), calcite; **b** He 140, 2976.6 m (CL), calcite; **c** Shi 101, 3259.5 m (–), quartz overgrowth; **d** He 135, 3030.87 m (CL), quartz overgrowth; **e** Niu 42, 3258.6 m (–), grain point contact; **f** He 155, 2987.04 m (–), primary pore; **g** Shi 101, 3258.6 m (SEM), primary pore; **h** Hao 7, 2961.1 m (–), dissolution expanding pore; **i** Wangxie 543, 3184.5 m (–), moldic pore; **j** Wangxie 543, 3180.6 m (SEM), feldspar dissolution pore; **k** Dongke 1, 3333.65 m (–), ankerite dissolution pore; **l** Dongke 1, 3333.65 m (SEM), ankerite dissolution pore; **m** Nan 1, 3403.35 m (–), kaolinite replaces feldspar; **n** He 155, 2987.04 m (SEM), kaolinite replaces feldspar; **o** Hao 5, 3142.01 m (SEM), kaolinite filling pore; **p** Wangxie 543, 3180.6 m (SEM), kaolinite part illitization. *Q* quartz; *F* feldspar; *R* rock fragments; *M* matrix; *Qa* quartz overgrowth; *Ka* kaolinite; *Il* illite; *Cc* carbonate cement; *FD* feldspar dissolution; *CD* carbonate dissolution; *PP* primary pore; (–) plane-polarized light; *CL* cathodoluminescence; *SEM* scanning electron microscope

Grains are arranged mainly by point contacts and point-line contacts, reflecting moderate compaction (Fig. 6e). The reservoirs are mainly carbonate cemented. The first groups of carbonate cements are calcite and ferran calcite. Calcite and ferran calcite always occur in the form of basal cementation (Fig. 6a) or porous cementation (Fig. 6b). The second groups of carbonate cements are dolomite, ankerite, and siderite. As revealed from our observations, dolomite, ankerite, and siderite always develop euhedral crystals (Fig. 6k). Quartz overgrowth is the main kind of siliceous cementation (Fig. 6c, d). Two phases of quartz overgrowths can be identified by cathodoluminescence microscopy. The first phase of quartz overgrowth is dark black and the second phase is brown as also described by Lander et al. (2008) and

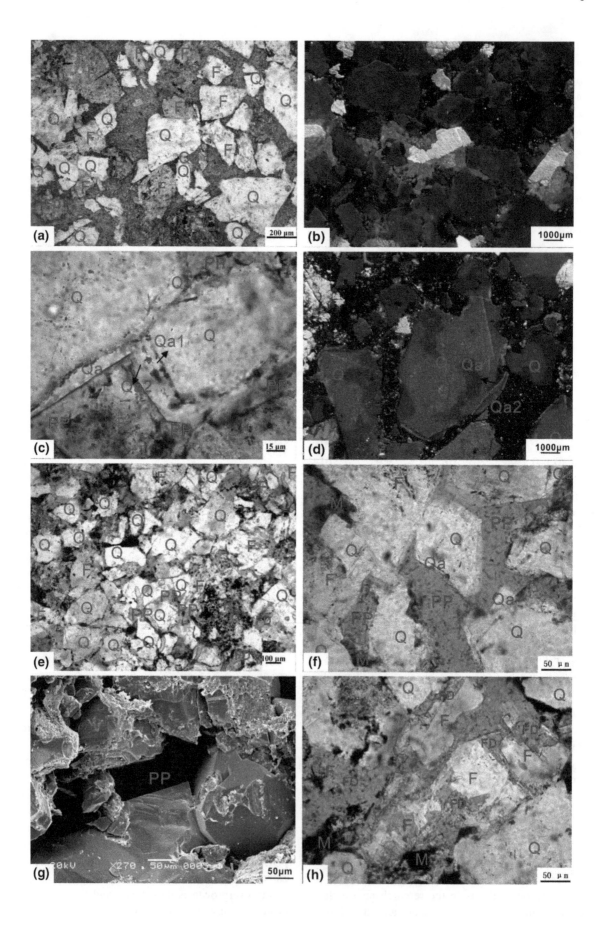

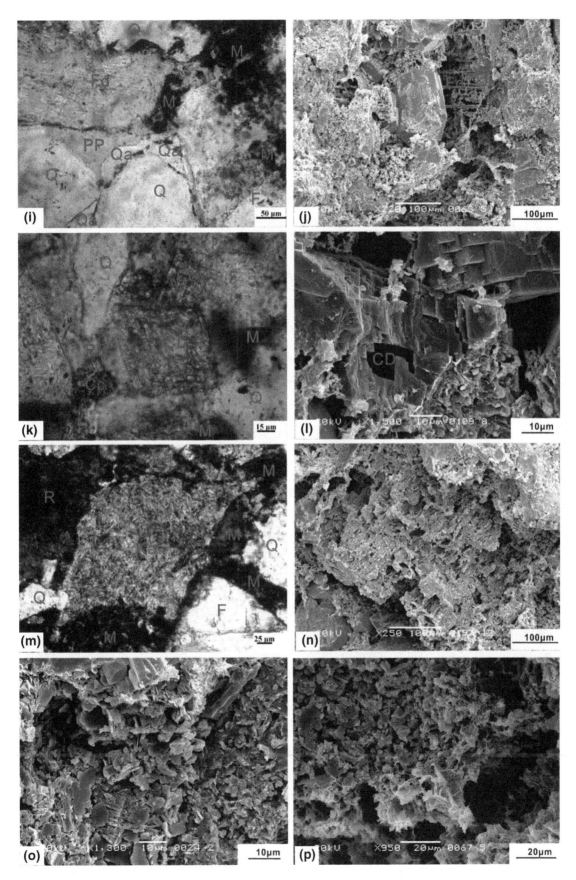

Fig. 6 continued

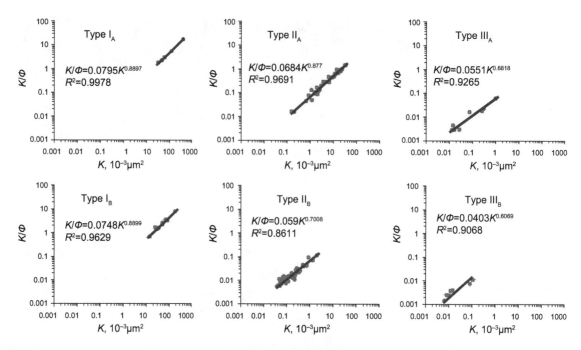

Fig. 7 Pore-throat structure types and their porosity–permeability relationships of the low permeability Es_3^z turbidite reservoirs

Table 1 Ranges of K and K/Φ of different pore structures of the low permeability Es_3^z turbidite reservoirs

Type of pore-throat structure	K, 10^{-3} μm^2	K/Φ	P_d, MPa	P_{50}, MPa
IA	>30.6	>1.52	0.02–0.05	0.26–0.61
IB	13.9–183.34	0.68–7.85	0.06–1	0.16–1.26
IIA	0.15–34.3	0.016–1.54	0.15–0.5	0.48–4.41
IIB	0.037–1.95	0.0056–0.12	0.15–2	2.84–22.35
IIIA	0.013–0.96	0.0027–0.058	0.8–4	17.36–74.12
IIIB	<0.11	<0.011	3–8	47.8–73.53

Tournier et al. (2010). Kaolinite is the most important kind of clay mineral (Fig. 6m, n, o). Kaolinite mainly occurs as euhedral booklets and vermicular aggregates with abundant intercrystalline microporosity. The margin of kaolinite is fibrous as a result of illitization (Fig. 6p). The dissolution of feldspar (Fig. 6h, i, j), lithic fragments, carbonate cements, and other minerals which are unstable in the acid environment can form honeycomb-shaped dissolution expanding pores with curved outlines (Fig. 6k, l). Besides this, quartz and quartz overgrowths have been slightly dissolved. Replacement between carbonate cements (Fig. 6d), between carbonate cements and detrital particles (Fig. 6b), between kaolinite and feldspar (Fig. 6c) all occurred. Replacement between carbonate cements mainly results in dolomite replacing calcite, ferroan calcite replacing calcite, ankerite replacing calcite, and ankerite replacing ferroan calcite.

(2) Paragenesis of diagenetic minerals

On the basis of previous studies (Jiang et al. 2003), the analysis of the fluorescence color of hydrocarbon inclusions and thermometry analysis of aqueous inclusions which were captured at the same time as hydrocarbon inclusions can identify two periods of hydrocarbon accumulation. The first period of hydrocarbon accumulation is from 27.5 to 24.6 Ma, and the second period is from 13.8 Ma until now. From observations using cathodoluminescence and polarizing microscopy, two phases of quartz overgrowths can be recognized. There are some hydrocarbon inclusions and oil absorption on clay minerals located in the boundaries between quartz grains and overgrowth rims (Fig. 8i, k) as also described by Girard et al. (2002) and Higgs et al. (2007). The color of those organic materials is orange to yellow in fluorescence microscopy which reflects the low maturity of hydrocarbon (Liu et al. 2014c; Chen 2014). It can be inferred that the first phase of quartz overgrowths formed after the early period hydrocarbon filling. The homogenization temperature of the aqueous inclusions in the first phase of quartz overgrowths ranges from 98 to 118 °C with an average of 106 °C (Fig. 9). The color of hydrocarbon inclusions in the second phase of quartz overgrowths is blue and white under the fluorescence microscope which reflects a high hydrocarbon

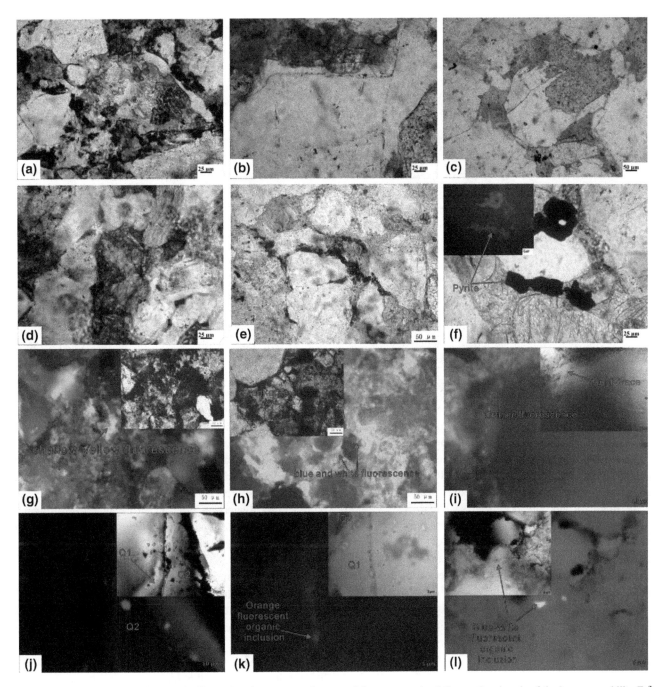

Fig. 8 Optical microscope micrographs illustrating the texture and nature of the paragenesis of diagenetic minerals of the low permeability Es_3^z turbidite reservoirs. **a** Niu 24, 3175.61 m (−), feldspar dissolution pore filled by ankerite; **b** Niu 30, 2871.85 m (−), ankerite replaced quartz overgrowth; **c** Niu 83, 3199.83 m (−), feldspar dissolution pore filled by kaolinite; **d** Niu 30, 2891.62 m (−), ankerite replaced quartz ferroan calcite; **e** Liang 49, 2836.13 m (−), siderite growth around a quartz particle; **f** Niu 128, 3059.55 m (−), pyrite replaced carbonate cements; **g** Niu 43, 3266.80 m (FL), first period oil filling after feldspar dissolution; **h** Liang 49, 2838.13 m (FL), blue in cleavage crack and margin of ankerite; **i** Shi 101, 3263.9 m (FL), orange fluorescence in quartz overgrowth dust trace; **j** Niu 42, 3261.9 m (FL), blue-white fluorescent organic inclusion in Q2; **k** Niu 42, 3261.9 m (FL), orange fluorescent organic inclusion in Q1; **l** Nan 1, 3401.75 m (FL), blue-white fluorescent organic inclusion in ankerite. − plane-polarized light; *FL* fluorescence; *Q1* Quartz overgrowth in the first phase; *Q2* Quartz overgrowth in the second phase

maturity (Fig. 8j) (Chen 2014). It can be concluded that the quartz overgrowths formed after the late period hydrocarbon fill. The homogenization temperature of the aqueous inclusions in the second phase of quartz overgrowths ranges from 120 to 146 °C with an average of 134 °C (Fig. 9). Temperatures calculated from the O isotope ratios in early carbonate cements (dolomite and calcite) range from 66 to 102 °C (Guo et al. 2014), and temperatures

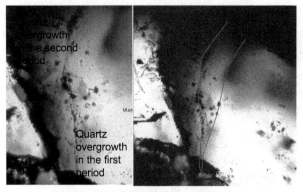
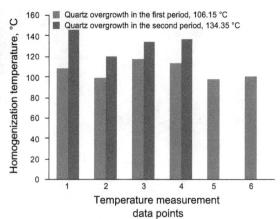

Niu 42, 3261.9, fluid inclusion distribution
in the two periods of quartz overgrowth

Fig. 9 Fluid inclusion homogenization temperatures of the two phases of quartz overgrowths of the low permeability Es_3^z turbidite reservoirs

calculated from the isotope ratios in late carbonate cements (ferroan calcite and ankerite) range from 110 to 147 °C (Zhang 2012). There are some blue and white color hydrocarbon inclusions in the ankerite under fluorescence microscopy (Fig. 8l), and cleavage cracks and the edges of ankerite grains are impregnated by hydrocarbon with blue-white fluorescence (Fig. 8h) (Wilkinson et al. 2006). We can infer that the ankerite formed at the same time as hydrocarbon charging.

The siderites and some micritic carbonate have grown around the quartz particles without quartz overgrowths (Fig. 8e), showing that siderite cements formed earlier than the quartz overgrowths. The feldspar dissolution pores were filled by ankerite (Fig. 8a), so feldspar dissolution occurred earlier than ankerite cementation. Ankerite cementation occurred later than quartz overgrowth reflected by the replacement relation between ankerite and quartz overgrowth (Fig. 8b). Ankerite replaced ferroan calcite (Fig. 8d), so ankerite cementation occurred later than ferroan calcite. The feldspar dissolution pores were filled by kaolinite (Fig. 8c), so feldspar dissolution took place earlier than kaolinite cementation. Pyrite replaces carbonate cements (Fig. 8f), so pyrite formed later than carbonate cements.

After the analysis of timing and order of hydrocarbon filling and formation of various authigenic minerals, the paragenesis of authigenic minerals was determined. Siderite/micritic carbonate → first dissolution of feldspar → the beginning of the first hydrocarbon filling → first quartz overgrowth/authigenic kaolinite precipitation → the first group of carbonate cementation → the end of the first hydrocarbon filling → dissolution of quartz/feldspar overgrowth → second dissolution of feldspar and carbonate cementation → the beginning of the second hydrocarbon filling → second quartz overgrowth/authigenic kaolinite precipitation → the second group of carbonate cementation/

pyrite cementation. Compaction existed throughout the entire burial and evolutional processes.

According to the burial history and organic evolution history analysis for the reservoirs in the research area, combined with the diagenetic environment implied by authigenic minerals, the reservoir experienced a diagenetic environment evolution from slightly alkaline → acid → alkaline → slightly acidic now. The early slightly alkaline diagenetic environment was controlled by the original sedimentary water from 42 to 38 Ma (Qi et al. 2006). With the increase of burial depth, a larger amount of organic acid was produced from the evolution of organic matter in high-quality source rocks in Es_4^x and Es_4^s (Surdam et al. 1989). The diagenetic pore-water became acidic, which lasted from 38 to 28 Ma, and the temperature of reservoirs was from 80 to 120 °C. With further increase in burial depth, organic acid decarboxylation and the alkaline fluid from the gypsum in Es_4^x dominated the diagenetic environment from 28 to 16.4 Ma (Wang 2010). The strata were uplifted by the Dongying Movement, and organic acid was generated again. The diagenetic pore water became acid again from 16.4 to 5 Ma. From 5 Ma to now, organic acid was generated from source rock in Es_3^z. As a result of this process, the diagenetic pore water is considered to have remained acidic.

4.2 Porosity–permeability evolution of Es_3^z low-permeability turbidity reservoirs

Based on the diagenetic features and paragenetic sequences, the porosity and permeability estimation method for the geological history of the reservoirs has been used (Wang et al. 2013a; Cao 2010). According to this method, we can determine the porosity and permeability of the reservoirs in the accumulation period. First, we take the thin sections of reservoir samples as the study object. After the analysis of the paragenetic sequence and diagenetic fluid evolution combined with the study of burial history, we determine the geological time and burial depth of diagenetic events. Second, we fit the function of plane porosity and visual reservoir porosity from the analysis of thin sections, and then we can calculate the contributions of different dissolution pores and authigenic minerals to porosity increase or decrease. After the calculation of initial porosity, the evolution of porosity can be estimated with the principle of inversion and back-stripping constraint of the diagenetic paragenetic sequences. Third, the evolution history of actual porosity with geological time or burial depth with different diagenetic characteristics can be established quantitatively combined with the chart of mechanical and thermal compaction correction. Fourth, on the basis of characteristics of pore throat structure, according to the back-stripping constraint result of plane porosity and the principle of equivalent expanding, the pore throat structures of reservoirs can be estimated at the geological time of the main diagenetic events. Finally, according to the relationship between pore throat structure and porosity, the evolution of permeability in geological time can be estimated with the relationships of porosity and permeability in different kinds of pore throat structures. Taking the turbidite reservoir at the Niu107 well at 3025.5 m as example (Fig. 10), the estimated permeability

of 0.31×10^{-3} μm^2 is close to the actual measured permeability of 0.307×10^{-3} μm^2.

On the basis of diagenetic paragenetic sequences and the type and strength of diagenetic events, the reservoir can be divided into four types of diagenetic facies. These are strong compaction—weak dissolution of feldspar—weak cementation of carbonate: Diagenetic facies (A); weak compaction—weak dissolution of feldspar—strong cementation of carbonate: Diagenetic facies (B); weak compaction—strong dissolution of feldspar—weak cementation of carbonate: Diagenetic facies (C); and medium compaction—medium dissolution of feldspar—medium cementation of carbonate: Diagenetic facies (D). Thin sandstones mainly develop diagenetic facies A and diagenetic facies B. Thick sandstones develop diagenetic facies A and B in the reservoirs adjacent to mudstones, and diagenetic facies C and D in the middle of sandstones (McMahon et al. 1992). Typical samples of different kinds of diagenetic facies were selected and their evolution of porosity–permeability were estimated (Fig. 11). The results show that in the early accumulation period, all reservoirs except for reservoirs with diagenetic facies A have middle-high permeability ranging from 10×10^{-3} μm^2 to 4207×10^{-3} μm^2. In the later accumulation period, all reservoirs except for reservoirs with diagenetic facies C have low permeability ranging from 0.015×10^{-3} μm^2 to 62×10^{-3} μm^2.

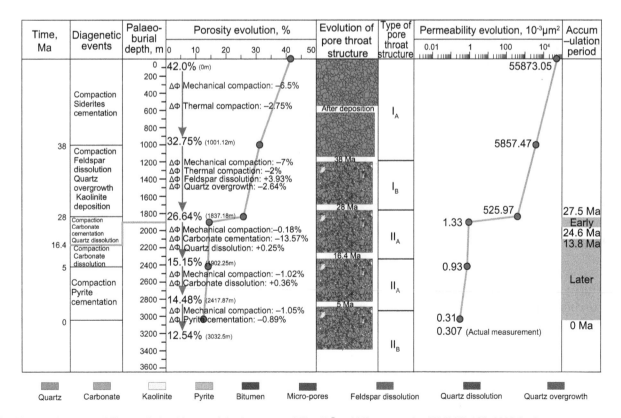

Fig. 10 Porosity-permeability evolution history of the low permeability Es_3^2 turbidite reservoirs (Well Niu107, 3032.5 m)

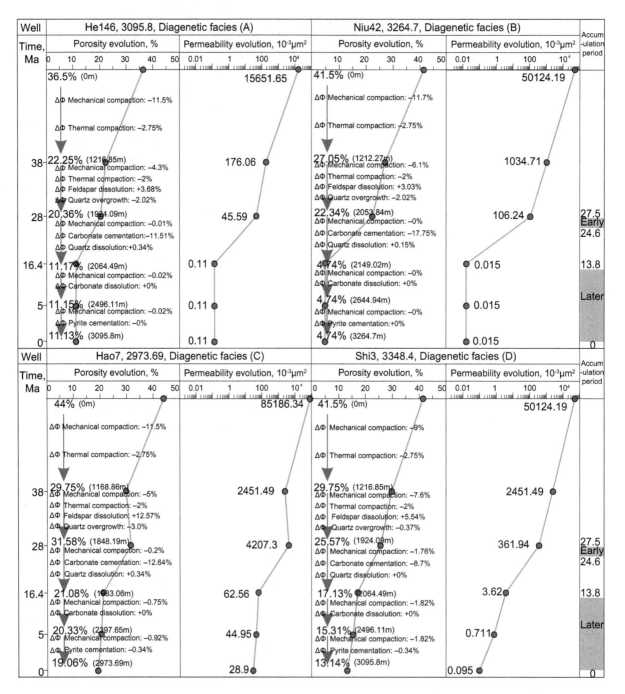

Fig. 11 Porosity-permeability evolution history of different diagenetic facies low permeability Es$_3^z$ turbidite reservoirs

5 Cutoff-values for porosity and permeability of turbidite reservoirs in the accumulation period

Capillary pressure (Pc) is the most important resistance force in hydrophilic reservoir rocks. Only when the dynamic force surpasses the resistance force, can petroleum seep into rocks and form petroleum reservoirs (Hao et al. 2010). We calculated the cutoff-values for porosity and permeability in the accumulation period under the constraint of accumulation dynamics and pore throat structure (Wang et al. 2014a). The method procedure includes: (1) establishing a functional relationship between oil–water interfacial tension and formation temperature; (2) calculating lower limiting values of maximum connected pore-throat radius according to formation temperature and dynamic forces of each reservoir interval; (3) correlating permeability with maximum connected pore-throat radius and then obtaining cutoff-values for permeability in the accumulation period; and (4) calculating cutoff-values for porosity on the basis of cutoff-values for permeability according to specific correlations suitable for the type of pore-throat structure (Wang et al. 2014a).

According to the test data of oil–water interfacial tension (δ) for different formation temperature (T) in the Es$_3$ and Es$_4$ reservoirs in the Dongying Sag, the functional relationship can be written as (Wang et al. 2014a):

$$\delta = 40.5 \times T^{-0.149}, R^2 = 0.65 \tag{1}$$

This equation could be used to calculate the oil–water interfacial tension at any given formation temperature. For example, for a formation temperature of 125 °C which is close to the actual formation temperature of Es$_3^z$ in the research area, the calculated oil–water interfacial tension is 19.7 mN/m. For a fixed critical accumulation dynamic value P_f, we can get cutoffs of maximum connected pore throat radius using equation $r_0 = 2\delta\cos\theta/P_f$ when the wetting contact angle of oil–water is 0° and interfacial tension at 125 °C is 19.7 mN/m (Table 2).

Establishing a correlation between permeability and maximum connected pore-throat radius using mercury injection data (Fig. 12), we find that there is a good exponential relationship between permeability and the maximum connected pore-throat radius as:

$$K = 0.3927 \times r_0^{1.7992}, R^2 = 0.8275, \tag{2}$$

where K is the permeability, 10^{-3} μm^2; r_0 is the maximum connected pore-throat radius, μm.

Substituting the limiting value of the maximum connected pore-throat radius under different critical accumulation dynamics into Eq. (2), a series of cutoff-values for permeability in the accumulation period can be obtained at 125 °C (Table 2).

On the basis of the classification of pore-throat structures, according to the functional relationships between K and K/Φ of different pore-throat structures as well as their variation ranges (Fig. 4, Table 1), we calculated cutoff-values for porosity according to variation ranges of permeability in Table 1 and regarded those values as cutoff-values for porosity in the accumulation period for the corresponding type of pore-throat structures under different critical accumulation dynamics. With the same method, we can calculate the cutoff-values for porosity and permeability in the accumulation period for the corresponding type of pore-throat structures under different critical accumulation dynamics at different formation temperatures (Fig. 13).

6 Control on the oil-bearing potential of a reservoir by the relationship between permeability and dynamics in the accumulation period

6.1 Accumulation dynamics estimation

The turbidite reservoirs are located in overpressured formations of the Dongying Sag. Overpressure is the main dynamic controlling hydrocarbon accumulation (Zhuo et al. 2006; Sui et al. 2008; Gao et al. 2010). Disequilibrium stresses under a high subsidence rate or rapid burial and hydrocarbon generation are the two possible overpressure generating mechanisms in sedimentary basins (Bao et al. 2007; Bloch et al. 2002; Taylor et al. 2010). By means of fluid inclusion PVT simulation, the minimum fluid pressure in the hydrocarbon accumulation period can be obtained. According to basin modeling techniques, fluid pressure resulting from disequilibrium compaction can be determined (the balance pressure between sandstones and mudstones). The differences between those two pressures are the increased minimum fluid pressure of hydrocarbon generation. For an isolated lenticular sand body without faults, fluid pressures generated by disequilibrium compaction would transfer from mudstones to sandstones to reach a balance of fluid pressure (Cai et al. 2009). So the fluid pressure generated by hydrocarbon generation is the main accumulation dynamic. For a sand body with faults developed, the surplus pressure which is the difference between fluid pressure and hydrostatic pressure will result in fluid migration through the faults which is the main accumulation dynamic (Zhuo et al. 2006; Cai et al. 2009). According to the estimations of the accumulation dynamics of reservoirs in the research area (Table 3), in the early accumulation period the fluid pressure increase by hydrocarbon generation is 1.4–11.3 MPa with an average of 5.14 MPa, and the surplus pressure is 1.8–12.6 MPa with an average of 6.3 MPa. In the late accumulation period the fluid pressure increased by hydrocarbon generation is 0.7–12.7 MPa with an average of 5.4 MPa, and the surplus pressure is 1.3–16.2 MPa with an average of 6.6 MPa. The accumulation dynamics in the later accumulation period are stronger than those in the early accumulation period.

6.2 Coupling of dynamics and permeability in the hydrocarbon accumulation period

The estimation of the permeability of reservoirs with different diagenetic facies indicated that the permeability of the

Table 2 Cutoff-values for porosity and permeability of the low permeability Es_3^z turbidite reservoirs under the constraint of the accumulation dynamics and pore throat structure and at 125 °C formation temperature

Accumulation dynamics P_f, MPa	Maximum connected pore-throat radius r_0, μm	K_{cutoff}, 10^{-3} μm^2	Φ_{cutoff}, %					
			Φ_{IA}	Φ_{IB}	Φ_{IIA}	Φ_{IIB}	Φ_{IIIA}	Φ_{IIIB}
0.01	48.45	422.82	24.04	–	–	–	–	–
0.02	24.22	121.49	21.62	22.68	–	–	–	–
0.024	20.19	87.51	21.02	21.87	–	–	–	–
0.026	18.63	75.78	20.77	21.53	–	–	–	–
0.03	16.15	58.58	20.32	20.93	–	–	–	–
0.04	12.11	34.91	19.44	19.77	22.63	–	–	–
0.05	9.69	23.37	–	18.91	21.54	–	–	–
0.055	8.81	19.68	–	18.56	21.09	–	–	–
0.06	8.07	16.83	–	18.24	20.69	–	–	–
0.065	7.45	14.57	–	17.96	20.33	–	–	–
0.07	6.92	12.75	–	–	20.0	–	–	–
0.075	6.46	11.27	–	–	19.69	–	–	–
0.08	6.06	10.03	–	–	19.41	–	–	–
0.09	5.38	8.12	–	–	18.91	–	–	–
0.1	4.84	6.71	–	–	18.48	–	–	–
0.2	2.42	1.93	–	–	15.85	20.63	–	–
0.3	1.62	0.93	–	–	14.49	16.59	17.74	–
0.32	1.51	0.83	–	–	14.29	16.02	17.09	–
0.4	1.21	0.55	–	–	13.60	14.21	15.04	–
0.49	1	0.39	–	–	13.0	12.74	13.39	–
0.5	0.97	0.37	–	–	12.94	12.60	13.24	–
0.7	0.69	0.2	–	–	12.01	10.51	10.92	–
0.9	0.54	0.13	–	–	–	9.18	9.46	–
1	0.48	0.11	–	–	–	8.68	8.90	10.29
1.2	0.4	0.077	–	–	–	7.86	8.02	9.05
1.4	0.35	0.058	–	–	–	7.24	7.34	8.11
1.5	0.32	0.051	–	–	–	6.97	7.06	7.73
1.6	0.3	0.046	–	–	–	6.74	6.80	7.38
2	0.24	0.031	–	–	–	–	5.99	6.30
2.2	0.22	0.026	–	–	–	–	5.67	5.89
2.4	0.2	0.022	–	–	–	–	5.39	5.54
2.5	0.19	0.021	–	–	–	–	5.27	5.38
3	0.16	0.015	–	–	–	–	4.75	4.73
3.2	0.15	0.013	–	–	–	–	4.57	4.52
3.4	0.14	0.012	–	–	–	–	–	4.33
3.6	0.14	0.011	–	–	–	–	–	4.16
4	0.12	0.009	–	–	–	–	–	3.86
6	0.08	0.004	–	–	–	–	–	2. 90
8	0.06	0.003	–	–	–	–	–	2.37
10	0.05	0.002	–	–	–	–	–	2.02

reservoir ranged from 10×10^{-3} μm^2 to 4207×10^{-3} μm^2 in the early accumulation period. When there is no fault development in the sand body, the fluid pressure increased by hydrocarbon generation is the main accumulation dynamic. The fluid pressure increased by hydrocarbon generation is 1.4–11.3 MPa with an average value of 5.1 MPa. Using the minimum accumulation dynamics of 1.4 MPa, the maximum cutoff-value for permeability in the accumulation period at a formation temperature of 125 °C was calculated to be 0.058×10^{-3} μm^2. The permeability of

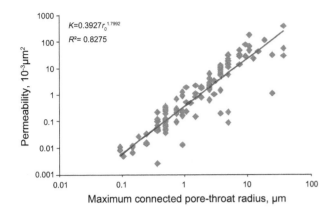

Fig. 12 The relationship between permeability and maximum connected pore-throat radius of the low permeability Es_3^z turbidite reservoirs

the reservoir in the early accumulation period was much higher than this cutoff-value, so all the studied reservoirs could accumulate hydrocarbon. When there is fault development in the sand body, the surplus pressure is the main accumulation dynamic. The surplus pressure is 1.8–12.6 MPa with an average of 6.3 MPa. Using the minimum accumulation dynamics of 1.8 MPa, the maximum cutoff-value for permeability in the accumulation period at a formation temperature of 125 °C was calculated to be 0.037×10^{-3} μm^2. The permeability of the reservoirs in the early accumulation period was much higher than this cutoff-value, so all the studied reservoirs could accumulate hydrocarbon.

In the later accumulation period the permeability had been reduced and was in the range of 0.015×10^{-3} μm^2 to 62×10^{-3} μm^2. The fluid pressure increased by hydrocarbon generation was 0.7 MPa to 12.7 MPa with an average of 5.4 MPa in the late accumulation period. Using the minimum accumulation dynamics of 0.7 MPa, the

maximum cutoff-value for permeability in the accumulation period at a formation temperature of 125 °C was calculated to be 0.203×10^{-3} μm^2. Using the maximum accumulation dynamics of 12.7 MPa, the minimum cutoff-value for permeability in the accumulation period at a formation temperature of 125 °C is calculated to be 0.001×10^{-3} μm^2. So, at the high level of accumulation dynamics hydrocarbon can accumulate in all studied reservoirs. Reservoirs with diagenetic facies A and diagenetic facies B do not develop accumulation conditions at the low level of accumulation dynamics, because the permeability of reservoirs with diagenetic facies A and diagenetic facies B is lower than the maximum cutoff value. The surplus pressure is 1.3–16.2 MPa with an average of 6.6 MPa. Using the minimum accumulation dynamics 1.3 MPa, the maximum cutoff-value for permeability in the accumulation period at a formation temperature of 125 °C was calculated to be 0.066×10^{-3} μm^2. Using the maximum accumulation dynamics 16.2 MPa, the minimum cutoff-value for permeability in the accumulation period at a formation temperature of 125 °C was calculated to be 0.0007×10^{-3} μm^2. So, different kinds of reservoirs can all accumulate hydrocarbon with high accumulation dynamics. Reservoirs with diagenetic facies A and diagenetic facies B do not develop accumulation conditions at the low level of accumulation dynamics.

6.3 Distribution of hydrocarbon resources

The Es_3^z source rocks are not fully mature, because the burial depth of the Es_3^z turbidite reservoirs ranges from 2500 to 3500 m. Oil–source correlation analysis indicated that the oil of the low permeability turbidite reservoirs in the early accumulation period comes from source rocks in Es_3^x and Es_4^s (Cai 2009). The source rocks are overlain by

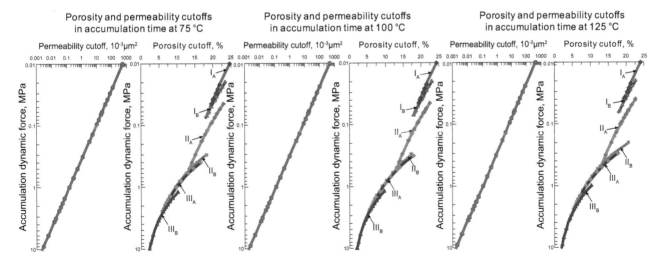

Fig. 13 Cutoff-values for porosity and permeability under different formation temperatures of the low permeability Es_3^z turbidite reservoirs

Table 3 Estimates of accumulation dynamics of the low permeability Es_3^2 turbidite reservoirs (Zhang 2014)

Well	Depth, m	Paleo-fluid pressure, MPa	Paleopressure after disequilibrium compaction, MPa	Geological time, Ma	Palaeo-burial depth, m	Hydrostatic pressure, MPa	Pressure generated by hydrocarbon, MPa	Surplus pressure, MPa	Pressure coefficient	Accumulation period
Xin 154	2936	31.9	31.2	2.3	2800	28	0.7	3.9	1.14	Late
Xin 154	2939	22	19.5	27.5	1950	19.5	2.5	2.5	1.13	Early
Xin 154	2939	26.8	23.7	7.5	2350	23.5	3.1	3.3	1.14	Late
Xin 154	2942.8	29.6	31.2	2.3	2800	28	–	1.6	1.06	Late
Niu 108	3146.5	23.8	22.4	25.3	2200	22	1.4	1.8	1.08	Early
Niu 108	3146.5	33.2	21.6	11	2000	20	11.6	13.2	1.66	Late
Niu 108	3146.5	18.6	21.6	11	2000	20	–	–	0.93	Late
Niu 35	2991.7	22.3	21.4	9.8	2100	21	0.9	1.3	1.06	Late
Niu 107	3272.5	34.6	23.9	9.5	2290	22.9	10.7	11.7	1.51	Late
Shi128	3099	33.3	32.2	2.6	2850	28.5	1.1	4.8	1.17	Late
Shi 128	3099	27.8	22	9.3	2140	21.4	5.8	6.4	1.30	Late
Niu 20	3073	43	30.3	3	2680	26.8	12.7	16.2	1.60	Late
Niu 20	3073	28.3	22.5	9.1	2150	21.5	5.8	6.8	1.32	Late
Niu 24	3159.2	27.9	24.3	25.9	2150	21.5	3.6	6.4	1.30	Early
Niu 24	3159.2	29.1	25.4	6	2450	24.5	3.7	4.6	1.19	Late
Niu 24	3159.2	32.9	31.2	3.6	2710	27.1	1.7	5.8	1.21	Late
Niu 24	3175.6	38.6	27.3	4.7	2630	26.3	11.3	12.3	1.47	Late
Niu 24	3175.6	32.2	25.3	24.8	2400	24	6.9	8.2	1.34	Early
Niu 24	3175.6	36.6	25.3	24.8	2400	24	11.3	12.6	1.53	Early
Niu 24	3175.6	27	23.7	9.8	2350	23.5	3.3	3.5	1.15	Late
Niu 24	3175.6	26.4	23.7	9.8	2350	23.5	2.7	2.9	1.12	Late

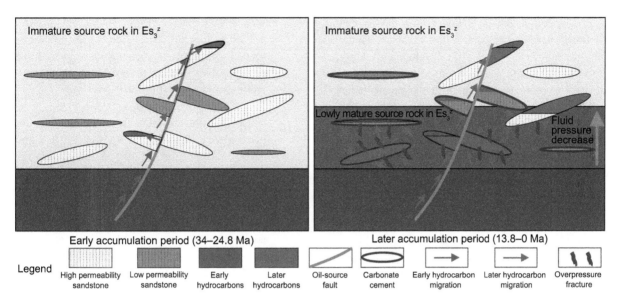

Fig. 14 The hydrocarbon accumulation patterns of low permeability Es_3^z turbidite reservoirs

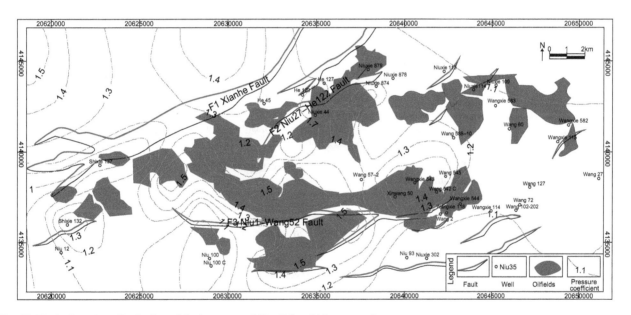

Fig. 15 The hydrocarbon distribution of the low permeability Es_3^z turbidite reservoirs

the reservoir rocks, and the generated oil migrates from the lower part to the upper part (the reservoir). Faults in source rocks controlled the accumulation of reservoirs. The oil of the low permeability turbidite reservoirs in the late accumulation period comes from source rock in Es_3^x, Es_4^s, and Es_3^z (Li et al. 2007). The source rocks are either below the reservoirs or both the source rocks and reservoir rocks are from the same formation.

Surplus pressure is the main accumulation dynamic in the early accumulation period. In the early accumulation period, the permeability of all reservoirs is higher than the cutoff-values for permeability. So, the sand bodies with fault development and connected with source rock in Es_3^x

and Es_4^s accumulate hydrocarbon easily. Duo to heterogeneity caused by diagenetic processes (Liu et al. 2014a), hydrocarbon accumulation mostly occurred in the reservoirs with high permeability under the control of oil-source faults (Fig. 14). The fluid pressure increased by hydrocarbon generation is the main accumulation dynamic for isolated lenticular sand bodies without faults in the later accumulation period. All types of reservoirs with high accumulation dynamics can accumulate hydrocarbon. Reservoirs with diagenetic facies A and diagenetic facies B do not develop accumulation conditions at the low level of accumulation dynamics in the later accumulation period because the permeability of reservoir with diagenetic facies

A and diagenetic facies B is lower than the maximum cutoff value. Surplus pressure is the main accumulation dynamic for sand bodies with fault development. All types of reservoirs with high accumulation dynamics can accumulate hydrocarbon. Reservoirs with diagenetic facies A and diagenetic facies B do not develop accumulation conditions at the low level of accumulation dynamics. Hydrocarbon always accumulated in reservoirs with high accumulation dynamics and oil-source faults development. Source rocks of the lower part of Es_3^z have a high maturity when the burial depth of turbidite reservoir is more than 3000 m (Hao et al. 2006). The oil in the reservoirs came from the source rocks both at the same burial depth as the reservoirs and from a deeper burial depth than the reservoirs. The closer to the source rocks, the higher accumulation dynamics and the higher the hydrocarbon-filling degree. So isolated lenticular sand bodies can accumulate hydrocarbon. As the distances from source rocks to reservoirs increases, the accumulation dynamics for the reservoirs decrease and the hydrocarbon-filling degree decreases as well. The distance limit for an isolated lenticular sand body to accumulate hydrocarbon is about 225 m from the lower part of source rocks (Song et al. 2014) (Fig. 14). When the burial depth of turbidite reservoir is less than 3000 m, the oil in the reservoirs came from the source rocks at deeper burial depth than the reservoirs. The oil-source faults controlled the accumulation of reservoirs. Taking the Niuzhuang subsag as an example, hydrocarbon always accumulated in reservoirs around the oil-source faults and areas near the center of subsag with high accumulation dynamics (Fig. 15).

7 Conclusions

(1) Es_3^z turbidite sandstones in the Dongying Sag are mostly lithic arkoses, and composed of mainly fine to medium sized grains. Low permeability reservoirs with middle to high porosity are most common, and the reservoir space is mainly primary pores. There are three broad types of pore throat structures which are subdivided into six sub-types. The major diagenetic events are mechanical compaction, cementation, replacement, and dissolution. The diagenetic paragenesis is siderite/micritic carbonate → first dissolution of feldspar → the beginning of the first hydrocarbon filling → first quartz overgrowth/authigenic kaolinite precipitation → the first group of carbonate cementation → the end of the first hydrocarbon filling → dissolution of quartz/feldspar overgrowth → second dissolution of feldspar and carbonate cementation → the beginning of the second hydrocarbon filling → second quartz overgrowth/authigenic kaolinite precipitation → the second group of carbonate cementation/pyrite cementation. Compaction existed throughout the entire burial and evolutional processes.

(2) In the early accumulation period, the reservoirs except for diagenetic facies A had middle to high permeability ranging from 10×10^{-3} μm² to 4207.3×10^{-3} μm², all the studied reservoirs can accumulate hydrocarbon. In the later accumulation period the reservoirs except for diagenetic facies C have low permeability ranging from 0.015×10^{-3} μm² to 62×10^{-3} μm², all the studied reservoirs can accumulate hydrocarbon at the high level of accumulation dynamics. Reservoirs with diagenetic facies A and diagenetic facies B do not develop accumulation conditions at the low level of accumulation dynamics.

(3) The hydrocarbon-filling degree is higher when the burial depth of turbidite reservoirs is more than 3000 m. Isolated lenticular sand bodies can accumulate hydrocarbon. When the burial depth of turbidite reservoirs is less than 3000 m, isolated lenticular sand bodies cannot accumulate hydrocarbon. Hydrocarbons always accumulate in reservoirs around the oil-source faults and areas near the center of subsags with high accumulation dynamics.

Acknowledgments   This work is supported by the National Natural Science Foundation of China (Grant No. U1262203), the National Science and Technology Special Grant (No. 2011ZX05006-003), the Fundamental Research Funds for the Central Universities (Grant No. 14CX06070A), and the Chinese Scholarship Council (No. 201506450029). The Shengli Oilfield Company of SINOPEC provided all the related core samples and some geological data. The authors wish to thank editors and reviewers for their thorough and very constructive review that greatly improved the manuscript.

References

Bao XH, Hao F, Fang Y, et al. Evolution of geopressure fields in the Niuzhuang Sag of the Dongying Depression and their effect on petroleum accumulation. Earth Sci J China Univ Geosci. 2007;32(2):241–6 (**in Chinese**).

Bjørlykke K. Relationships between depositional environments, burial history and rock properties. Some principal aspects of diagenetic process in sedimentary basins. Sed Geol. 2014;301:1–14.

Bloch S, Lander RH, Bonnell L. Anomalously high porosity and permeability in deeply buried sandstone reservoirs: origin and predictability. AAPG Bull. 2002;86(2):301–28.

Cai LM. The Dynamic Process of Hydrocarbon Migration and Accumulation in Lenticular Sandstones of the Third Middle Member of the Shahejie Formation in the Niuzhuang Subsag, Dongying Depression. Wuhan: China University of Geosciences; 2009 (in Chinese).

Cai LM, Chen HH, Li CQ, et al. Reconstruction of the paleo-fluid potential field of Es$_3$ in the Dongying Sag of the Jiyang depression with systematic fluid inclusion analysis. Oil Gas Geol. 2009;30(1):19–25 (in Chinese).

Cao YC. Vertical evolution and quantitative characterization of clastic reservoirs in the Shahejie Formation, Dongying Sag. Dongying: Geological Science Research Institute of Shengli Oilfield, Sinopec; 2010 (in Chinese).

Cao YC, Chen L, Wang YZ, et al. Diagenetic evolution of Es$_3$ reservoirs and its influence on reservoir properties in the northern Minfeng sub-sag of the Dongying Sag. J China Univ Pet. 2011;35(5):6–13 (in Chinese).

Cao YC, Yuan GH, Wang YZ, et al. Genetic mechanisms of low permeability reservoirs of the Qingshuihe Formation in the Beisantai area, Junggar Basin. Acta Pet Sinica. 2012;33(5):758–71 (in Chinese).

Cao YC, Ma BB, Wang YZ, et al. Genetic mechanisms and classified evaluation of low permeability reservoirs of Es$_4^s$ in the north zone of Bonan Sag. Nat Gas Geosci. 2013;24(5):865–78 (in Chinese).

Cao YC, Yuan GH, Li XY, et al. Characteristics and origin of abnormally high porosity zones in buried Paleogene clastic reservoirs in the Shengtuo area, Dongying Sag, East China. Pet Sci. 2014;11:346–62.

Chen HH. Microspectrofluorimetric characterization and thermal maturity assessment of individual oil inclusions. Acta Pet Sin. 2014;35(3):584–90 (in Chinese).

Folk RL. Petrology of sedimentary rocks. Austin: Hemphill; 1974. p. 182.

Gao YJ, Wang YS, Yu YL, et al. Coupling of driving force and resistance for the migration of oil and gas at the periods of hydrocarbon accumulation in the southern slope of the Dongying Sag: taking the Jin 8-Bin 188 section as an example. Geoscience. 2010;24(6):1148–56 (in Chinese).

Giles MR, de Boer RB. Origin and significance of redistributional secondary porosity. Mar Pet Geol. 1990;7:378–97.

Girard J, Munz IA, Johansen H, et al. Diagenesis of the Hild Brent sandstones, northern North Sea: isotopic evidence for the prevailing influence of deep basinal water. J Sediment Res. 2002;72(6):746–59.

Guo XW, Liu KY, He S, et al. Petroleum generation and charge history of the northern Dongying Depression, Bohai Bay Basin, China: insight from integrated fluid inclusion analysis and basin modelling. Mar Pet Geol. 2012;32(1):21–35.

Guo J, Zeng JH, Song GQ, et al. Characteristics and origin of carbonate cements in the Shahejie Formation of the Central Uplift Belt in the Dongying Depression. Earth Sci J China Univ Geosci. 2014;39(5):565–76 (in Chinese).

Hao XF, Chen HH, Gao QL, et al. Micro-charging processes of hydrocarbon in the Niuzhuang lentoid sandy reservoirs, Dongying depression. Earth Sci J China Univ Geosci. 2006;31(2):182–90 (in Chinese).

Hao F, Zou HY, Gong ZS. Preferential petroleum migration pathways and prediction of petroleum occurrence in sedimentary basins: a review. Pet Sci. 2010;7:2–9.

Higgs KE, Zwingmann H, Reyes AG, et al. Diagenesis, porosity evolution, and petroleum emplacement in tight gas reservoirs, Taranaki Basin, New Zealand. J Sediment Res. 2007;77:1003–25.

Jiang YL, Liu H, Zhang Y, et al. Analysis of the petroleum accumulation phase in the Dongying Sag. Oil Gas Geol. 2003;24(3):215–8 (in Chinese).

Lander RH, Larese RE, Bonnell LM. Toward more accurate quartz cement models: the importance of euhedral versus noneuhedral growth rates. AAPG Bull. 2008;92(11):1537–63.

Li SM, Qiu GQ, Jiang ZX, et al. Origin of the subtle oils in the Niuzhuang Sag. Earth Sci J China Univ Geosci. 2007;32(2):241–6 (in Chinese).

Liu MJ, Liu Z, Wang B, et al. The controls of sandstone paleo-porosity in the accumulation period to hydrocarbon distribution: a case study from the Es$_3^{Mid}$ in the Niuzhuang Sag, Dongying Depression. Chin J Geol. 2014a;49(1):147–60 (in Chinese).

Liu MJ, Liu Z, Sun XM, et al. Paleoporosity and critical porosity in the accumulation period and their impacts on hydrocarbon accumulation—A case study of the middle Es$_3$ member of the Paleogene Formation in the Niuzhuang Sag, Dongying Depression, Southeastern Bohai Bay Basin, East China. Pet Sci. 2014b;11:495–507.

Liu MJ, Liu Z, Liu JJ, et al. Coupling relationship between sandstone reservoir densification and hydrocarbon accumulation: a case from the Yanchang Formation of the Xifeng and Ansai areas, Ordos Basin. Pet Explor Dev. 2014c;41(2):168–75 (in Chinese).

McMahon PB, Chapelle FH, Falls WF, et al. Role of microbial processes in linking sandstone diagenesis with organic-rich clays. J Sediment Pet. 1992;62(1):1–10.

Meng WB, Lu ZX, Tang Y, et al. Reservoir permeability prediction based on sandstone texture classification. J China Univ Pet. 2013;37(2):1–6 (in Chinese).

Pan GF, Liu Z, Fan S, et al. The study of lower limit of porosity for oil accumulation in the Chang-8 Member, Zhenjing Area, Ordos Basin. Geoscience. 2011;25(2):271–8 (in Chinese).

Qi BW, Lin CM, Qiu GQ, et al. Reservoir diagenesis of the intermediate section in Member 3 of the Paleogene Shahejie Formation in the Niuzhuang Sub-sag, Shandong Province. J Palaeogeogr. 2006;8(4):519–30 (in Chinese).

Song GQ, Hao XF, Liu KQ, et al. Tectonic evolution, sedimentary system and petroleum distribution patterns in a dustpan-shaped rift basin: a case study from Jiyang Depression, Bohai Bay Basin. Oil Gas Geol. 2014;35(3):303–10 (in Chinese).

Sui FG, Hao XF, Liu Q, et al. Formation dynamics and quantitative prediction of hydrocarbons of the superpressure system in the Dongying Sag. Acta Geol Sin (English Edition). 2008;82(1):164–73.

Surdam RC, Crossey LJ, Hagen SE, et al. Organic–inorganic interactions and sandstone diagenesis. AAPG Bull. 1989;73(1):1–12.

Taylor TR, Giles MR, Hathon LA, et al. Sandstone diagenesis and reservoir quality prediction: models, myths, and reality. AAPG Bull. 2010;94(8):1093–132.

Tournier F, Pagel M, Portier E, et al. Relationship between deep diagenetic quartz cementation and sedimentary facies in a Late Ordovician glacial environment (Sbaa Basin, Algeria). J Sediment Res. 2010;80:1068–84.

Wang YZ. Genetic mechanism and evolution model of the secondary pore development zone of the Paleogene in the North Zone, Dongying Depression. Qingdao: China University of Petroleum (Huadong); 2010 (in Chinese).

Wang JM, Liu SF, Li J, et al. Characteristics and causes of Mesozoic reservoirs with extra-low permeability and high water cut in northern Shaanxi. Pet Explor Dev. 2011;38(5):583–88 (in Chinese).

Wang YZ, Cao YC, Xi KL, et al. A method for estimating porosity evolution of clastic reservoirs with geological time: a case study from the upper Es$_4$ submember in the Dongying Depression,

Jiyang Subbasin. Acta Pet Sin. 2013a;34(6):1100–111 **(in Chinese)**.

Wang JD, Li SZ, Santosh M, et al. Lacustrine turbidites in the Eocene Shahejie Formation, Dongying Sag, Bohai Bay Basin, North China Craton. Geol J. 2013b;48(5):561–78.

Wang YZ, Cao YC, Song GQ, et al. Analysis of petrophysical cutoffs of reservoir intervals with production capacity and with accumulation capacity in clastic reservoirs. Pet Sci. 2014a;11:211–21.

Wang YZ, Cao YC, Ma BB, et al. Mechanism of diagenetic trap formation in nearshore subaqueous fans on steep rift lacustrine basin slopes—A case study from the Shahejie Formation on the north slope of the Minfeng Subsag, Bohai Basin, China. Pet Sci. 2014b;11:481–94.

Wilkinson M, Haszeldine RS, Fallick AE. Hydrocarbon filling and leakage history of a deep geopressured sandstone, Fulmar Formation, United Kingdom North Sea. AAPG Bull. 2006;90(12):1945–61.

Yang ZM, Yu RZ, Su ZX, et al. Numerical simulation of the nonlinear flow in ultra-low permeability reservoirs. Pet Explor Dev. 2010;37(1):94–8 **(in Chinese)**.

Yang T, Cao YC, Wang YZ, et al. Status and trends in research on deep-water gravity flow deposits. Acta Geol Sin (English Edition). 2015;89(2):801–22.

Yuan GH, Cao YC, Xi KL, et al. Feldspar dissolution and its impact on physical properties of Paleogene clastic reservoirs in the northern slope zone of the Dongying Sag. Acta Pet Sin. 2013;35(4):853–66 **(in Chinese)**.

Yuan GH, Gluyas J, Cao YC, et al. Diagenesis and reservoir quality evolution of the Eocene sandstones in the northern Dongying Sag, Bohai Bay Basin, East China. Mar Pet Geol. 2015;62:77–89.

Zhang SW, Wang YS, Shi DS, et al. Fault-fracture mesh petroleum plays in the Jiyang Superdepression of the Bohai Bay Basin, eastern China. Mar Pet Geol. 2004;21(6):651–68.

Zhang JL, Li DY, Jiang ZQ. Diagenesis and reservoir quality of the fourth member sandstones of Shahejie Formation in Huimin Depression, Eastern China. J Cent South Univ Technol. 2010;17(1):169–79.

Zhang Q, Zhu XM, Steel RJ, et al. Variation and mechanisms of clastic reservoir quality in the Paleogene Shahejie Formation of the Dongying Sag, Bohai Bay Basin, China. Pet Sci. 2014;11:200–10.

Zhang XF. Synergistic diagenesis of argillaceous source rocks and sandstones of the Shahejie Formation in the southern Dongying Depression and their petroleum geological significance. Ph.D. Thesis. Nanjing: Nanjing University; 2012 **(in Chinese)**.

Zhang W. The heterogeneity of the turbidite in the Niuzhuang area and its control action on the hydrocarbon accumulation. Dongying: Post-doctoral Scientific Research Workstation of Shengli Oilfield, Sinopec; 2014 **(in Chinese)**.

Zhuo QG, Jiang YL, Sui FG. Research on patterns of reservoiring dynamics in sand lens reservoirs in the Dongying Sag. Oil Gas Geol. 2006;27(5):620–9 **(in Chinese)**.

8

The influence of DC stray current on pipeline corrosion

Gan Cui[1] · Zi-Li Li[1] · Chao Yang[1] · Meng Wang[1]

Abstract DC stray current can cause severe corrosion on buried pipelines. In this study, firstly, we deduced the equation of DC stray current interference on pipelines. Next, the cathode boundary condition was discretized with pipe elements, and corresponding experiments were designed to validate the mathematical model. Finally, the numerical simulation program BEASY was used to study the corrosion effect of DC stray current that an auxiliary anode bed generated in an impressed current cathodic protection system. The effects of crossing angle, crossing distance, distance of the two pipelines, anode output current, depth, and soil resistivity were investigated. Our results indicate that pipeline crossing substantially affects the corrosion potential of both protected and unprotected pipelines. Pipeline crossing angles, crossing distances, and anode depths, our results suggest, have no significant influence. Decreasing anode output current or soil resistivity reduces pipeline corrosion gradually. A reduction of corrosion also occurs when the distance between two parallel pipelines increases.

Keywords DC stray current · BEASY · Numerical simulation · DC interference corrosion

✉ Zi-Li Li
cygcx@163.com

Gan Cui
chennacuigan@163.com

[1] College of Pipeline and Civil Engineering, China University of Petroleum (East China), Qingdao 266580, Shandong, China

Edited by Yan-Hua Sun

1 Introduction

Stray current refers to the current that flows elsewhere rather than along the intended current path. It is an important cause of corrosion and leakage of underground metal pipelines (Li et al. 2010; Guo et al. 2015). Stray current corrosion is essentially electrochemical corrosion (Bertolini et al. 2007). Because of the high electrical conductivity of buried steel pipelines, potential differences with the less conductive environment are formed when stray current flows through the pipe (Brichau et al. 1996) effectively creating a corrosion cell. The corrosion caused by stray current is more serious than soil corrosion under normal conditions (the potential difference of soil corrosion is only about 0.35 V without stray current, but the pipe-to-soil potential can be as high as 8–9 V when a stray current exists) (Brichau et al. 1996). The stray current has a great effect on corrosion, and so affects the service life and safe use of buried pipelines (Ding et al. 2010). Therefore, it is important to better understand stray current corrosion.

There are three types of stray currents: direction current (DC), alternating current (AC), and natural telluric current in the earth. Among these, the DC stray current causes most damage to a buried pipeline (Gao et al. 2010). The DC stray current mainly originates from DC electrified railways, DC electrolytic equipment grounding electrodes, and the anode bed of cathodic protection systems (Wang et al. 2010).

Numerical methods have been shown to be powerful tools to analyze corrosion problems in the last two decades. Numerical methods used for corrosion studies include the finite difference method (FDM), the finite element method (FEM) (Xu and Cheng 2013), and the boundary element method (BEM) (Metwally et al. 2007; Boumaiza and Aour 2014; Bordón et al. 2014). The BEM was applied to model cathodic protection systems in the early 1980s (Wrobel and Miltiadou 2004; DeGiorgi and Wimmer 2005; Lacerda et al. 2007; Abootalebi et al. 2010; Lan et al. 2012; Liu et al. 2013). Compared with FDM and FEM, BEM requires the meshing of the boundary only. As a result, BEM needs fewer equations resulting in a smaller matrix size than FEM and can solve both finite and infinite domain problems (Jia et al. 2004; Parvanova et al. 2014). Last but not least, the BEM has been specifically developed to calculate the DC stray current induced in pipeline networks by electric railways and can model the soil and the entire traction system consisting of rails, traction stations, overhead wires, and trains (Bortels et al. 2007; Poljak et al. 2010).

In this paper, the BEM is carried out to determine the effect of DC interference corrosion on neighboring pipelines (crossing or parallel with the cathodic protection pipeline). It focuses on the DC current produced by the auxiliary anode of the impressed current cathodic protection system.

2 Mathematical model

2.1 Governing equation

Some simplifications and assumptions are made here: the solution around the pipeline is uniform and electroneutral, and there is no concentration gradient in the solution.

Based on the above assumption and Ohm's law, the current density i can be expressed as follows (Metwally et al. 2008):

$$i = \sigma_e e, \tag{1}$$

where σ_e is the electrical conductivity of soil in S/m; i is the current density in mA/m^2; and e is the electric field in V/m. Then, the static form of the equation of continuity can be given as

$$\nabla i = \nabla(\sigma_e e) = 0. \tag{2}$$

Under static conditions, the electric potential is defined by the following equivalent equation:

$$\nabla \phi = e. \tag{3}$$

Consequently, the governing equation for the electric potential is the Laplace equation (Thamita 2012):

$$\Delta \phi = 0. \tag{4}$$

2.2 Boundary conditions

Boundary conditions can be divided into the anode boundary condition, the cathode boundary condition, and the insulation boundary condition.

2.2.1 Anode and insulation boundary conditions

In this simulation, an impressed current cathodic protection system is used and it is assumed that the output current of the auxiliary anode is constant. Thus, the anode boundary condition can be described as (Lan et al. 2012)

$$I = I_0. \tag{5}$$

The insulation boundary condition can be described as

$$\frac{\partial \phi}{\partial n} = 0. \tag{6}$$

2.2.2 Cathode boundary condition

On the surface of the cathode, many complex electrochemical reactions occur. Polarization is one such consequence of these reactions. The polarization data will be used as the cathode boundary.

The polarization curve of steel was measured in the soil environment using a conventional three-electrode cell assembly. Rectangular platinum shapes were used as the counter electrode and a saturated copper sulfate electrode as reference. The working electrode was a cuboid, and its material was Q235 steel. The steel electrode was embedded in epoxy resin so that only a 1-cm^2 area of the cuboid was exposed to the soil. Before the experiment, the working electrode was polished gradually using 600–1200 grid, waterproof abrasive paper. It was then washed with distilled water, degreased with acetone, and washed with ethanol. Finally, it was dried in an unheated air stream. Electrochemical measurements were carried out using electrochemical workstation Parstat 2273, which was specially designed for the study of the electrochemical corrosion behavior. It can be used to test the open circuit potential, electrochemical impedance spectroscopy (EIS), Tafel polarization curve, cyclic voltammograms, etc. In this paper, the Tafel curve was tested, and the polarization measurement involved a scan starting from −400 to −1200 mV at a scan rate of 0.3 mV/s.

The polarization curve is used as the cathode boundary condition, but it is a nonlinear curve, so we have to use polarization data in a piecewise linear interpolation approach (Abootalebi et al. 2010; Liang et al. 2011; Liu et al. 2013; Li et al. 2013). The polarization curve is shown in Fig. 1, and we present it as a piecewise linear curve.

2.3 Boundary element method (BEM)

When the BEM is applied, only the boundaries of the domain need to be discretized. The pipeline is discretized using the pipe element method, see Fig. 2. The number of nodes and elements is reduced a lot with this method and as a result the computation is simplified.

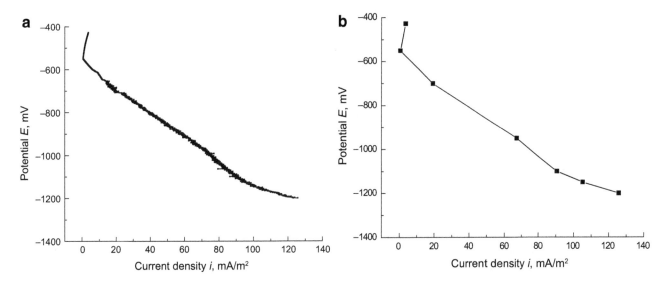

Fig. 1 The polarization curve. **a** Experimental polarization curve. **b** Piecewise linear polarization curve

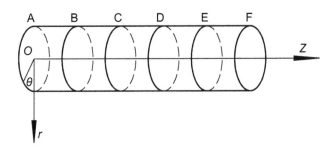

Fig. 2 Pipe surface discretization

It should be noted that certain conditions need to be met to use the pipe element method (Meng et al. 1998). (1) The geometry of the protected body needs to be suitable for cylinder unit subdivision. (2) The potential everywhere on the same cylinder unit is considered constant.

It is known that the fundamental solution of the boundary integral equation is $\frac{1}{4\pi r}$. Here "r" refers to the distance between the boundary node and the source node. Therefore, the integration will become a singularity when the boundary node coincides with the source node.

2.3.1 Computation of the nonsingular coefficient

Based on the pipe element method above together with the standard BEM formula, to make the boundaries discrete, the coefficient matrix of each element is obtained by integral transformation (Brichau and Deconinck 1994).

$$G_{i,jm}(t) = 4\frac{RL}{4\pi}|J|\phi_m(t)\frac{4K(k)}{[(R+B)^2+(Lt-Z_{rn})^2]^{1/2}}, \quad (7)$$

$$H_{i,jm}(t) = |J|\phi_m(t)$$
$$\frac{-RL}{[(R+B)^2+(Lt-Z_{rn})^2]^{1/2}}\frac{(Lt-Z_{rn})E(k)}{\pi[(R-B)^2+(Lt-Z_{rn})^2]^2}. \quad (8)$$

In Eqs. (7) and (8), $i \neq j$. $K(k)$ is the elliptic integral of the first kind, while $E(k)$ is the elliptic integral of the second kind; J and B are the results of the coordinate transformation; t is the local coordinate; and Z_{rn} is the third coordinate of the last pipe element node.

2.3.2 Computation of the singular coefficient

For the semi-infinite region (Wu 2008),

$$H_{ii} = -\sum H_{ij}, \quad i \neq j. \quad (9)$$

The analytical method to solve G_{ii} is

$$G_{ii} = \frac{L}{2\pi}\left[1 - \ln\left(\frac{L}{16r}\right)\right]. \quad (10)$$

Finally, the standard BEM formula is represented by a numerical integral, with the result

$$\{G\} \times \{Q\} = \{H\} \times \{\phi - \eta(Q)\}. \quad (11)$$

3 Experimental

To validate the mathematical model, related experiments were carried out in a controlled laboratory environment.

3.1 Experimental design

An experimental box was made of wood, 8000 mm × 6000 mm × 1500 mm (length × width × height), covered with an insulating board. We also placed a PVC board under it and around it. Two steel pipes were buried inside the box: the protected pipe and the DC-interfered pipe. The parameters of the protected pipe were material Q235 steel—no coating, outside diameter 20 mm, wall thickness 3 mm, length 6000 mm, and depth 1 m. The parameters of the DC-interfered pipe were material Q235—no coating, outside diameter 20 mm, wall thickness 3 mm, length 4000 mm, depth 0.5 m. The parameters of the auxiliary anode with a treated cylindrical surface were diameter 0.03 m, length 0.1 m, depth 1 m, distance from the pipe 0.3 m, no fillers, and the output current 1 mA.

The test points were set up on the pipes as shown in Fig. 3. Three cables of potentiostat (DJS-292) were connected to the energized point of the pipe, the auxiliary anode, and the reference electrode, respectively. The reference electrode was a saturated copper–copper sulfate electrode (CSE), which was buried near the energized point.

The cathodic protection potential data were acquired with an NIUSB6210 data acquisition module using a test block power interruption method.

3.2 Comparison between simulation and experimental results

Figure 4 shows the measured and simulated potentials at test points on the protected pipe. Experimental results for the DC-interfered pipe are shown in Fig. 5.

Figures 4 and 5 show that the experimental data are in relatively good agreement with simulation results. The biggest difference is less than 15 mV which is less than 2 %. So the mathematical model of the DC stray current interference is fairly accurate. BEM modeling proves to be very effective for simulating DC stray current interference on pipelines.

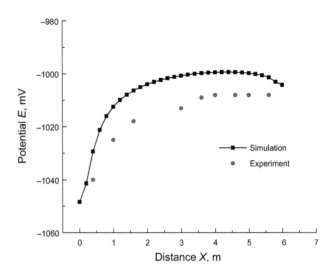

Fig. 4 Potentials on the protected pipe

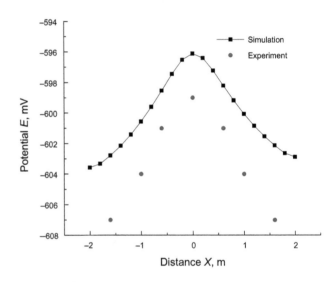

Fig. 5 Potentials on the DC-interfered pipe

4 Simulation with BEASY

In this paper, the boundary element numerical simulation software BEASY was used. The simulation model was performed in accordance with Fig. 6. For BEM simulations, the structures under study have been located inside a big cubic box. According to BEASY recommendations (BEASY 2005), the simulation box has to be 20 times the size of the model. The box surrounding the model does not need so many elements unless the model is placed very close to it. As long as the model is very far from the box walls, the current should not reach the box. Therefore, there is no need to put so many elements; 4 elements/face of the box (including the ground) would be fine (Wu et al. 2011).

The model includes the following: an impressed current cathodic protection system (ICCP) with a protected

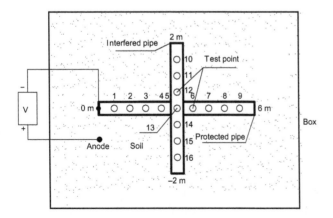

Fig. 3 Schematic diagram of the experimental setup

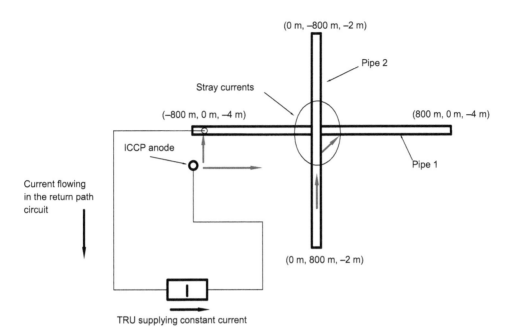

Fig. 6 Schematic of the DC interference model. The anode ground bed produces the DC stray current. Pipeline 2 is rotated anticlockwise to study four different crossing angles (30°, 45°, 60°, and 90°)

pipeline (pipeline 1). Pipeline 2 is located near pipeline 1 and has no applied cathodic protection. Therefore, part of the cathodic protection current that the auxiliary anode carries to pipeline 1 will flow into pipeline 2 as stray current and affects the corrosion of pipeline 2—see Fig. 6.

The origin is chosen as the intersection between the two pipelines, and the parameters in the model are set as follows: Pipeline 1: Two endpoint coordinates are (−800 m, 0 m, −4 m) and (800 m, 0 m, −4 m), diameter is 0.762 m, and the material is Q235; Pipeline 2: Two endpoints coordinate are (0 m, −800 m, −2 m) and (0 m, 800 m, −2 m), diameter is 0.4064 m, and the material is Q235. The coating on both pipelines is assumed to have 5 % damage. The auxiliary anode which is vertically buried has the following parameters: two endpoint coordinates (−800 m, −100 m, −1 m) and (−800 m, −100 m, −6 m), diameter 0.1 m, and constant current 2400 mA. Soil conductivity in the area of the buried pipelines is 0.005 S/m. It should be noted that all simulated potential data below are with respect to the saturated copper sulfate reference electrode.

5 Results and discussion

5.1 Effect of pipeline crossing

Considering the situation of two pipelines intersecting at an angle of 90°, the setting of each parameter is the same as in Sect. 4. Both potential distribution and current density distribution of pipeline 2 are obtained by simulation, and the results are shown in Fig. 7.

As shown in Fig. 7a, the change in the potential distribution of pipeline 2 is very large where the two pipelines intersect. The corrosion potential near the intersection is higher than the self-corrosion potential, and the potential of the intersection is then most positive. The potential at each end of the pipeline is lower than the pipeline self-corrosion potential. Considering the stray corrosion current density in Fig. 7b, in this section, the current density is positive which means that the current flows out of the pipeline. Therefore, that section of the pipeline becomes an anode, and its potential is higher than the self-corrosion potential which makes corrosion more severe. The section that shows a negative current density has current flowing into the pipeline—in other words, it becomes a cathode. Its potential is lower than the self-corrosion potential. It receives some cathodic protection which reduces corrosion.

5.2 Effect of pipeline crossing on the cathodic protection potential distribution

The following assumptions are made: Pipeline 2 also receives applied impressed current cathodic protection. The coordinates of the auxiliary anode of pipeline 2 are (100 m, −800 m, −1 m) and (100 m, −800 m, −6 m). The current is also 2400 mA, and the rest of the parameters remain unchanged. The distribution of the cathodic protection potential and current density of two pipelines are shown in Fig. 8.

According to Fig. 8, a large change of the cathodic protection potentials occurs near the intersection between the two pipelines. There is a clear potential increase with the maximum at the intersection. The current density,

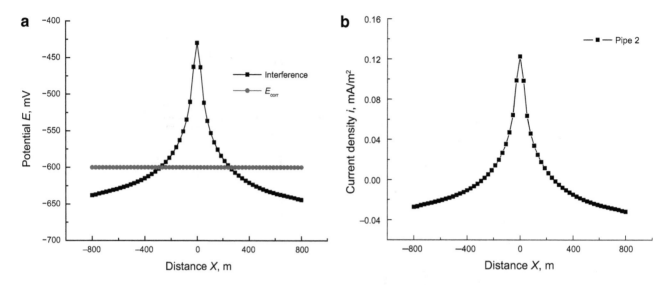

Fig. 7 Calculated effect of pipeline crossing. **a** Potential distribution. **b** Stray current density distribution

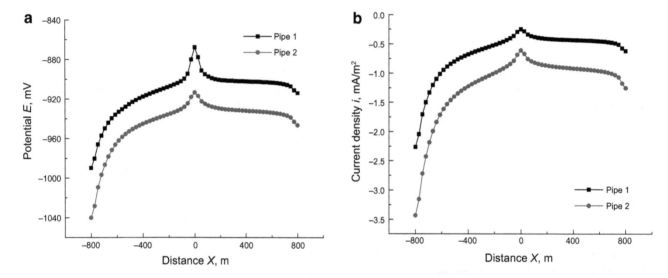

Fig. 8 Calculated effect of pipeline crossing on the cathodic protection potential (**a**) and calculated current density distribution (**b**)

however, is low near the intersection and almost zero at the intersection itself. Therefore, the degree of pipeline cathodic protection is weakened around the intersection, possibly causing insufficient protection.

5.3 Calculated effect of changing the crossing angle

We simulated both corrosion potential and current density distribution of pipeline 2 for four different crossing angles between the two pipelines: 30°, 45°, 60°, and 90°. The different angles are obtained by rotating pipeline 2 anticlockwise, while all other parameters remain unchanged. The results are shown in Fig. 9.

Figure 9 reveals that the potential near the intersection becomes increasingly negative on increasing the crossing angle from 30° to 90°. However, the change of potential

within the first 100 m of the intersection is only about 30 mV. On the other hand, the calculated potential change a few hundred meters away from the intersection is relatively large. This is because the relative position between the pipeline and the auxiliary anode has changed more significantly while the location of both anode and intersection remains unchanged. This is very important when comparing the potentials of intersections.

5.4 Calculated effect of changing the vertical crossing distance

We simulate the corrosion potential and current density distribution of pipeline 2 buried at depths 0.5, 1, 2, and 3 m—see Fig. 10. Other parameters remain unchanged.

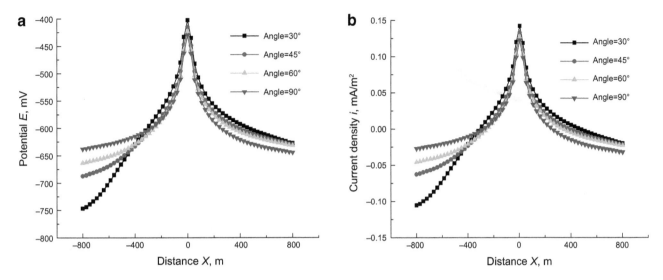

Fig. 9 Calculated effect of different crossing angles. **a** Potential distribution. **b** Stray current distribution

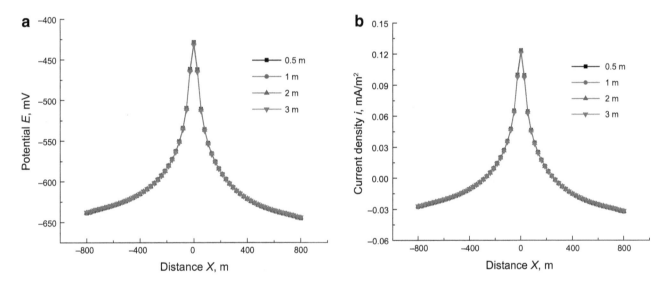

Fig. 10 Calculated effect of changing the vertical crossing distance. **a** Potential distribution. **b** Stray current distribution

Our calculated results indicate that both the pipeline corrosion potential and the current density remain practically unchanged for the different vertical crossing distances in the range investigated.

5.5 Calculated effect of the horizontal distance between parallel pipelines

The distance between parallel pipelines can vary a lot in practice. We simulate both the potential distribution and the current density distribution of pipeline 2 for distances of 0, 50, 100, 150, and 200 m with all other parameters unchanged. The simulated results are shown in Fig. 11.

Our results indicate that the potential distribution of pipeline 2 becomes more negative as the horizontal distance between two pipelines increases. Also the

potential distribution of the entire pipeline becomes more evenly distributed. From the distribution of the stray current density, we know that the current density will decrease with increasing distance and that the current density of most types of intersections is about zero within 200 m. Here the DC interference from stray current turns out to be very small. As a result, DC interference in pipeline 2 decreases with increasing horizontal distance.

5.6 Calculated effect of anode output current

The anode output current has a strong influence on the magnitude of the stray current. Here we calculate both the potential distribution and the current density distribution of pipeline 2 for three anode output currents: 1500, 2400, and

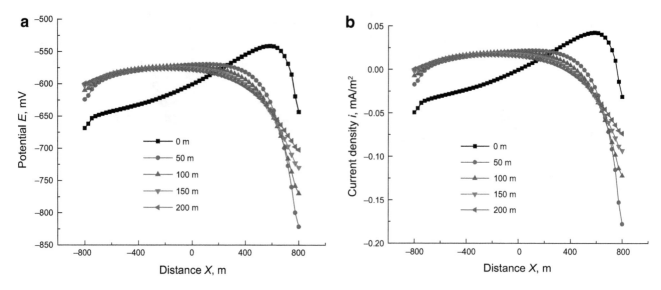

Fig. 11 Calculated effect of changing the horizontal distance between two parallel pipes. **a** Potential distribution. **b** Stray current distribution

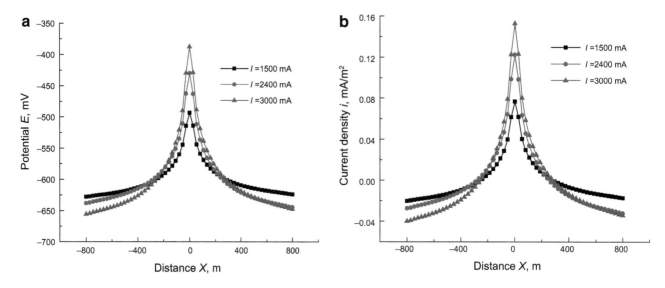

Fig. 12 Calculated effect of the anode output current. **a** Potential distribution. **b** Stray current density distribution

3000 mA. All other parameters remain unchanged. The simulation results are shown in Fig. 12.

Both the corrosion potential distribution and the current density distribution around pipeline 2 change significantly with an increase in the anode output current. When the anode output current increases, a large current flows into the lower end (-800 m) of pipeline 2 (consistent with Fig. 12b) and the potential decreases gradually. Because a large current flows into the lower end of pipeline 2, a large current will flow out of pipeline 2 from the intersection. The potential will increase and become greater than the self-potential, and the DC interference becomes more severe. Interestingly despite the increase in the anode

output current, both the potential distribution and the current density distribution for the upper end (800 m) of pipeline 2 remain unchanged.

5.7 Calculated effect of anode depth

We would expect that the anode depth can strongly affect the DC interference. We simulated the potential distribution and the current density distribution for pipeline 2 for varying anode depths (3.5, 23.5, and 53.5 m). All other parameters remain unchanged. The results are shown in Fig. 13.

To our surprise, the results indicate that both the pipeline potential distribution and the current density remain

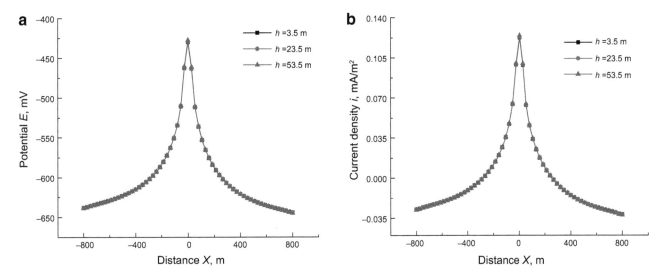

Fig. 13 Calculated effect of changing anode depths. **a** Potential distribution. **b** Stray current density distribution

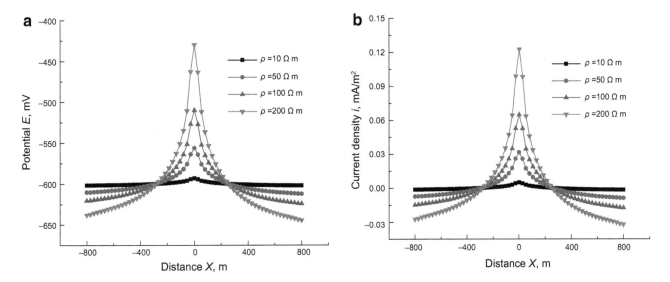

Fig. 14 Calculated effect of soil resistivity. **a** Potential distribution. **b** Stray current density distribution

practically unchanged when the anode depth changes, in the range investigated. A possible explanation may be that the distance between the anode and pipeline 2 (400 m) is much larger than the considered anode depths. Therefore, at least in our modeled scenario, the anode depth shows no significant impact on the DC interference of pipeline 2.

5.8 Calculated effect of soil resistivity

Soil resistivity has a strong influence on the magnitude of the stray current. We simulate both the corrosion potential and the current density distribution for the pipeline with soil resistivities of 10, 50, 100, and 200 Ω m. All other parameters remained unchanged. The simulation results are shown in Fig. 14.

Both the potential distribution and the current density distribution of pipeline 2 become very uneven if the soil resistivity is increased. When the soil resistivity is very small, the stray current density of pipeline 2 is near zero and the corrosion potential is approximately the self-corrosion potential. In other words, pipeline 2 is hardly affected. When the soil resistivity is very large, the current density near the pipeline intersection however becomes positive and the corrosion potential is significantly more positive than the self-corrosion potential.

As can be seen in Fig. 14, there are two common intersections between all curves. This means there are two possibilities: The current density between two intersections is positive (current flows out and the associated corrosion of the pipeline is more severe) or the current density outside two intersections is negative (current flows in and the pipeline has some cathodic protection).

As a consequence, the change of soil resistivity does not result in the change of the location and length of affected pipe sections. It only affects the degree of corrosion.

6 Conclusions

The following conclusions can be drawn based on our experiments and simulation:

(1) The mathematical model of the DC stray current interference in pipelines investigated in this paper is accurate, and BEM is confirmed to simulate DC stray current interference on pipelines very well.

(2) The potential distribution of the pipeline changes a lot when the pipelines cross. The potential at the intersection is about 200 mV higher than the self-corrosion potential which renders the corrosion of the pipeline in the intersection quite severe. The potentials at the two ends are about 50 mV lower than the self-corrosion potential. This part of the pipeline appears to receive some cathodic protection.

(3) The variation of the pipeline crossing angle, vertical crossing distance, and anode depth has little impact on the potential and the current density distribution of the DC-interfered pipeline.

(4) Upon increasing the horizontal distance between parallel pipelines, the corrosion potential of the affected section becomes negative, the potential distribution becomes more uniform, and the degree of the DC interference decreases.

(5) Upon increasing either the anode output current or the soil resistivity, the corrosion potential of the DC-interfered pipeline becomes very uneven. The corrosion potential of the pipeline near the intersection has a large positive offset, and pipeline corrosion becomes worse. However, the corrosion potential and the current density distribution at the far end (800 m) of pipeline 2 remain unchanged after increasing the anode output current. Change in resistivity just changes the degree of corrosion of the affected section but does not change the location and length of the affected pipe sections.

References

Abootalebi O, Kermanpur A, Shishesaz MR, et al. Optimizing the electrode position in sacrificial anode cathodic protection systems using boundary element method. Corros Sci. 2010;52(3):678–87. doi:10.1016/j.corsci.2009.10.025.

BEASY software user guide, computational mechanics BEASY Version 10. Southampton, UK, 2005. Available: http://www.beasy.com.

Bertolini L, Carsana M, Pedeferri P. Corrosion behavior of steel in concrete in the presence of stray current. Corros Sci. 2007;49(3):1056–68. doi:10.1016/j.corsci.2006.05.048.

Bordón JDR, Aznárez JJ, Maeso O. A 2D BEM–FEM approach for time harmonic fluid–structure interaction analysis of thin elastic bodies. Eng Anal Bound Elem. 2014;43(6):19–29. doi:10.1016/j.enganabound.2014.03.004.

Bortels L, Dorochenko A, Bossche BV, et al. Three-dimensional boundary element method and finite element method simulations applied to stray current interference problems. A unique coupling mechanism that takes the best of both methods. Corrosion. 2007;63(6):561–76. doi:10.5006/1.3278407.

Boumaiza D, Aour B. On the efficiency of the iterative coupling FEM–BEM for solving the elasto-plastic problems. Eng Struct. 2014;72(8):12–25. doi:10.1016/j.engstruct.2014.03.036.

Brichau F, Deconinck J. A numerical model for cathodic protection of buried pipes. Corrosion. 1994;50(1):39–49. doi:10.5006/1.3293492.

Brichau F, Deconinck J, Driesens T. Modeling of underground cathodic protection stray currents. Corrosion. 1996;52(6):480–8. doi:10.5006/1.3292137.

DeGiorgi VG, Wimmer SA. Geometric details and modeling accuracy requirements for shipboard impressed current cathodic protection system modeling. Eng Anal Boundary Elem. 2005;29(1):15–28. doi:10.1016/j.enganabound.2004.09.006.

Ding HT, Li LJ, Jiang MH, et al. Lab research of DC & AC stray current for the failure effect of the gas pipelines buried underground. Guizhou Chem Ind. 2010;35(5):9–12 (in Chinese).

Gao B, Shen LS, Meng XQ, et al. DC stray current corrosion and protection of oil gas pipelines. Pipeline Technol Equip. 2010;4:42–3 (in Chinese).

Guo YB, Liu C, Wang DG, Liu SH. Effects of alternating current interference on corrosion of X60 pipeline steel. Pet Sci. 2015;12(2):316–24. doi:10.1007/s12182-015-0022-0.

Jia JX, Song G, Atrens A, et al. Evaluation of the BEASY program using linear and piecewise linear approaches for the boundary conditions. Mater Corros. 2004;55(11):845–52. doi:10.1002/maco.200403795.

Lacerda LAD, Silva JMD, Jose L. Dual boundary element formulation for half-space cathodic protection analysis. Eng Anal Bound Elem. 2007;31(6):559–67. doi:10.1016/j.enganabound.2006.10.007.

Lan ZG, Wang XT, Hou BR, et al. Simulation of sacrificial anode protection for steel platform using boundary element method. Eng Anal Boundary Elem. 2012;36(5):903–6. doi:10.1016/j.enganabound.2011.07.018.

Li YT, Wu MT, Zheng F, et al. Stray current monitoring for pipelines and an application example in the Chuanxi gas field. Total Corros Control. 2010;24(1):25–8 (in Chinese).

Li ZL, Cui G, Shang XB, et al. Defining cathodic protection potential distribution of long distance pipelines with numerical simulation. Corros Prot. 2013;34(6):468–71 (in Chinese).

Liang CH, Yuan CJ, Huang NB. Defining the voltage distribution of a pipeline in frozen earth under cathodic protection with a boundary element method. J Dalian Marit Univ. 2011;37(4):109–16 (in Chinese).

Liu C, Shankar A, Orazem ME, et al. Numerical simulations for cathodic protection of pipelines. Undergr Pipeline Corros. 2014;63:85–126.

Liu GC, Sun W, Wang L, et al. Modeling cathodic shielding of sacrificial anode cathodic protection systems in sea water. Mater Corros. 2013;63(6):472–7. doi:10.1002/maco.201206726.

Meng XJ, Wu ZY, Liang XW, et al. Improvement of algorithms for a regional cathodic protection model. J Chin Soc Corros Prot. 1998;9:221–6 (in Chinese).

Metwally IA, Al-Mandhari HM, Gastli A, et al. Factors affecting cathodic-protection interference. Eng Anal Bound Elem. 2007;31(6):485–93. doi:10.1016/j.enganabound.2006.11.003.

Metwally IA, Al-Mandhari HM, Gastli A, et al. Stray currents of ESP well casings. Eng Anal Bound Elem. 2008;32(1):32–40. doi:10.1016/j.enganabound.2007.06.003.

Parvanova SL, Dineva PS, Manolis GD, et al. Dynamic response of a solid with multiple inclusions under anti-plane strain conditions by the BEM. Comput Struct. 2014;139(7):65–83. doi:10.1016/j.compstruc.2014.04.002.

Poljak D, Sesnic S, Goic R. Analytical versus boundary element modelling of horizontal ground electrode. Eng Anal Bound Elem. 2010;34(4):307–14. doi:10.1016/j.enganabound.2009.10.008.

Thamita SK. Modeling and simulation of galvanic corrosion pit as a moving boundary problem. Comput Mater Sci. 2012;65(12):269–75. doi:10.1016/j.commatsci.2012.07.029.

Wang Y, Yan YG, Dong CF, et al. Effect of stray current on corrosion of Q235, 16Mn and X70 steels with damaged coating. Corros Sci Prot Technol. 2010;22(2):117–9 (in Chinese).

Wrobel LC, Miltiadou P. Genetic algorithms for inverse cathodic protection problems. Eng Anal Bound Elem. 2004;28(3):267–77. doi:10.1016/S0955-7997(03)00057-2.

Wu HT. The application of boundary element method in heat transfer. Beijing: National Defence Industry Press; 2008. p. 24–7 (in Chinese).

Wu JH, Xing SH, Liang CH, et al. The influence of electrode position and output current on the corrosion related electro-magnetic field of ship. Adv Eng Softw. 2011;42(10):902–9. doi:10.1016/j.advengsoft.2011.06.007.

Xu LY, Cheng YF. Development of a finite element model for simulation and prediction of mechanoelectrochemical effect of pipeline corrosion. Corros Sci. 2013;73(8):150–60. doi:10.1016/j.corsci.2013.04.004.

9

Improvement of the prediction performance of a soft sensor model based on support vector regression for production of ultra-low sulfur diesel

Saeid Shokri · Mohammad Taghi Sadeghi ·
Mahdi Ahmadi Marvast · Shankar Narasimhan

Abstract A novel data-driven, soft sensor based on support vector regression (SVR) integrated with a data compression technique was developed to predict the product quality for the hydrodesulfurization (HDS) process. A wide range of experimental data was taken from a HDS setup to train and test the SVR model. Hyper-parameter tuning is one of the main challenges to improve predictive accuracy of the SVR model. Therefore, a hybrid approach using a combination of genetic algorithm (GA) and sequential quadratic programming (SQP) methods (GA–SQP) was developed. Performance of different optimization algorithms including GA–SQP, GA, pattern search (PS), and grid search (GS) indicated that the best average absolute relative error (AARE), squared correlation coefficient (R^2), and computation time (CT) (AARE = 0.0745, R^2 = 0.997 and CT = 56 s) was accomplished by the hybrid algorithm. Moreover, to reduce the CT and improve the accuracy of the SVR model, the vector quantization (VQ) technique was used. The results also showed that the VQ technique can decrease the training time and improve prediction performance of the SVR model. The proposed method can provide a robust, soft sensor in a wide range of sulfur contents with good accuracy.

S. Shokri · M. T. Sadeghi (✉)
Department of Chemical Engineering, Iran University of Science and Technology (IUST), Tehran, Iran
e-mail: sadeghi@iust.ac.ir

M. A. Marvast
Process and Equipment Technology Development Division, Research Institute of Petroleum Industry (RIPI), Tehran, Iran

S. Narasimhan
Department of Chemical Engineering, IIT Madras, Chennai, India

Edited by Xiu-Qin Zhu

Keywords Soft sensor · Support vector regression · Hybrid optimization method · Vector quantization · Petroleum refinery · Hydrodesulfurization process · Gas oil

List of symbols

w	Weight vector
b	Bias term
AARE	Average absolute relative error
R^2	Squared correlation coefficient
RBF	Gaussian radial basis kernel function
N	Sample size
g ($1/2\sigma^2$)	Hyper-parameter
C	Regularization parameter (hyper-parameter)
$\hat{X}$	Code vector
F_i	Validation data
T_i	Training data
k	Subsets (folds)
Exp_i	Actual values
Pre_i	Predicted values
$S_{\exp}$	Experimental value of sulfur content
S_{pre}	Predicted value of sulfur content
Q_{H2}	Hydrogen flow rate
Q_{gasoil}	Gas–oil flow rate

Greek symbol

σ	Width of kernel of radial basis function
ε	Precision parameter (hyper-parameter)
ξ_i, ξ_i^*	Slack variables
$\Phi(x)$	High-dimensional feature space

1 Introduction

Sulfur compounds are one of the most important impurities in crude oil and various petroleum fractions. Reduction of sulfur content of end product to the new lower limits is one of the recent challenges in the petroleum refineries. Online determination of sulfur concentration in the end product is difficult or impossible due to the limitations in process technology and measurement techniques. This index as the key indicator of process performance is normally determined by offline sample analysis in laboratories or online hardware analyzers that are mostly expensive with high maintenance costs.

Soft sensors can be a supplement to hardware process analyzers as their measurements may often be unavailable due to instrument failure, maintenance calibration necessity, insufficient accuracy, and long dead time (Kartik and Narasimhan 2011). Moreover, soft sensors can be applied for product quality estimation for industrial processes as an alternative to laboratory testing (Bolf et al. 2010). The core of a soft sensor is the construction of a soft sensing model (Yan et al. 2004). Different classes of soft sensors are (Kadlec 2009): (1) Model-driven or white-box model, (2) Data-driven or black-box model, and (3) Hybrid model or gray-box model. The Model-driven or first principle models obtained from the fundamental process knowledge require a lot of expert process knowledge, effort, and time to develop. Data-driven models are based on the data taken from the processing plants, and thus describe the real process conditions (Kadlec et al. 2009, 2011). These data-driven models can be developed more quickly with less expense. The hybrid model is a combination of both methods.

Artificial neural networks (ANNs) have been widely used as a useful tool for nonlinear soft sensing models. However, they give no guarantees of high convergence speed or of avoiding local minima, while there are no general methods to choose the number of hidden units in the networks. Moreover, they need a large number of controlling parameters, have difficulty in obtaining stable solutions with the danger of overfitting, and thus lack generalization capability (Liu et al. 2010).

In recent years, the support vector machine (SVM) technique, based on machine learning formalism and developed by Vapnik (1995), has been gaining popularity over ANN due to its many attractive features and promising empirical performance (Pan et al. 2010).

King et al. (2000) have compared SVM with ANN and concluded that SVM can provide more reliable and better performance under the same training conditions. Li and Yuan (2006) have applied SVMs to the prediction of key state variables in bioprocesses and indicated that SVM is better than ANN.

SVM can be used for classification, regression and other tasks. Applying the SVM to solve regression problems is called the support vector regression (SVR) method (Basak et al. 2007). The SVR tries to find an optimal hyper-plane as a decision function in high-dimensional space. SVR is different from conventional regression techniques, since it uses structural risk minimization (SRM), instead of empirical risk minimization (ERM) induction principles (Boser et al. 1992; Cristianini and Taylor 2000).

Hyper-parameter tuning is one of the main challenges in improving the predictive accuracy of an SVR model. Moreover, the generalization capability of SVR is highly dependent upon its learning parameters. The grid search method (GSM) is the most common method to determine appropriate values of hyper-parameters. Most researchers have followed a standard procedure using the GSM (Lu et al. 2009). This method's the computation time (CT) is too high, and it is unable to converge to the global optimum, and it is dependent on the parameters of boundary selection (Min and Lee 2005). There have been some research and development efforts into tuning of SVR hyper-parameters. Duan et al. (2003) have found a reasonably good hyper-parameter set for SVM using the Xi-Alpha bound. Some researchers have developed heuristic algorithms for the parameter optimization of SVR. Wu et al. (2009) have developed a kernel parameter-optimization technique using a hybrid model of GA and SVR. Huang (2012) has employed the hybrid GA–SVR methodology to solve an important stock selection for an investment problem. Chen and Wang (2007) have optimized the SVR parameters using metaheuristic algorithms.

Therefore, potentials of the hybrid strategies for optimization of these parameters need to be further investigated. This study proposes a novel hybrid metaheuristic approach for the SVR models to increase their performance both in accuracy and CT by hybridization of GA and SQP.

Moreover, despite having large datasets in process industries, the important issue is the need for high-speed and extensive memory capacities to process the data. The data compression phenomena provided by the VQ technique can be employed to overcome the problem (Somasundaram and Vimala 2010). Using VQ, the training time for choosing optimal parameters is greatly reduced. The most impactful gain here is the robustness of such systems.

The objectives of the present study are (1) Designing a robust and reliable data-driven soft sensor using an SVR model for prediction of sulfur content of treated gasoil. (2) Applying the VQ technique for data compression in the SVR model. This technique can simplify and compress the training set and speed up the computing time and also simultaneously improve the accuracy of the SVR model. (3) Optimizing the hyper-parameter of SVR model. An integrated hybrid GA–SQP algorithm was employed for optimizing the SVR hyper parameters using a fivefold cross-validation technique. To validate the prediction accuracy of the proposed hybrid model, the prediction performance of the proposed hybrid model was compared to those of GS–SVR, PS–SVR and GA–SVR.

2 Methodology

2.1 Support vector regression (SVR)

The basic concept of SVR is to map nonlinearly the original data x into a higher-dimensional feature space and solve a linear regression problem in this feature space (Gunn 1998). A number of loss functions such as the Laplacian, Huber's, Gaussian, and ε-insensitive can be used in the SVR formulation. Among these, the robust ε-insensitive loss function (L_ε) is more common (Vapnik et al. 1996; Si et al. 2009):

$$L_\varepsilon(f(x)-y) = \begin{cases} |f(x)-y| - \varepsilon & \text{for} \quad |f(x)-y| \geq \varepsilon \\ 0 & \text{otherwise} \end{cases} \quad (1)$$

where ε is a precision parameter representing the radius of the tube located around the regression function, $f(x)$ (see Fig. 1). The goal of using the ε-insensitive loss function is to find a function that can fit current training data with a deviation less than or equal to ε. The optimization problem can be reformulated as

$$\text{Min} \ \frac{1}{2}\|w\|^2 + C\sum_{i=1}^{l}\left(\xi_i^- + \xi_i^+\right) \quad (2)$$

subject to the following constraints:

$$y = \begin{cases} y_i - (\langle w, x_i \rangle + b) \leq \varepsilon + \xi_i \\ (\langle w, x_i \rangle + b) - y_i \leq \varepsilon + \xi_i^* \\ \xi_i, \xi_i^* \geq 0 \end{cases}$$

The positive slack variables ξ_i and ξ_i^* represent the distance from actual values and the corresponding boundary values of the ε-tube, respectively. The constant $C > 0$ is a parameter determining the trade-off between the empirical risk and the model flatness.

The basic idea in SVR is to map the dataset x_i into a high-dimensional feature space via nonlinear mapping. Kernel functions perform nonlinear mapping between the input space and a feature space. Different kernel trick functions were used (Table 1) (Yeh et al. 2011).

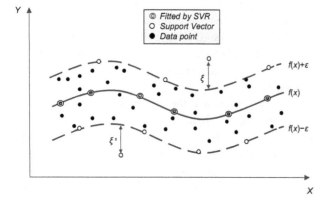

Fig. 1 A schematic diagram of SVR using an ε-sensitive loss function

Table 1 Different kernel functions

No.	Kernel type	Equation
1	Linear kernel	$K(x, y) = x_i^T \times y_i$
2	Polynomial kernel	$K(x, y) = (x_i^T y_i + t)^d$
3	Sigmoid kernel	$K(x, y) = \tanh(x_i^T \times y_i)$
4	Gaussian (radial basis function, RBF) kernel	$K(x, y) = \exp\left(-\frac{\|x_i - y_j\|^2}{2\sigma^2}\right)$

2.2 Vector quantization (VQ)

VQ is a data compression method based on the principle of block coding. VQ is applied to reduce a large dataset replacing examples by prototypes. Using VQ, the training time for choosing optimal parameters is greatly reduced. The most impactful gains here are the robustness of such systems.

The prediction speed is very important in soft sensor design. Therefore, in order to speed up the training time and reliability prediction of SVR model, the VQ technique is applied for data compression. The main goal of this method is to simplify the training set and increase the prediction accuracy. In the VQ technique, the data are quantized in the form of contiguous blocks called vectors rather than individual samples. VQ maps a K-dimensional vector x in the vector space R^k to another K-dimensional vector y that belongs to a finite set C (code book) of output vectors (code words).

In this method, K-dimensional input vectors are derived from input data $\{X\} = \{x_i : i = 1, 2, \cdots, N\}$. Data vectors are quantized into a finite set of code words $\{Y\} = \{y_j : j = 1, 2, \cdots, K\}$. Each vector y_j is called a code vector or a code word, and the set of all the code words is called a code book where the overall distortion of the system should be minimized. The purpose of the generated code book is to provide a set of vectors which generate minimal distortion between the original vector and the quantized vector.

The generation of the code book is the most important process that determines the performance of VQ. The aim of code book generation is to find code vectors (code book) for a given set of training vectors by minimizing the average pairwise distance between the training vectors and their corresponding code words (Horng 2012).

Each vector is compared with a collection of representative code vectors, $\hat{X}_i$ ($i = 1, 2, \cdots, N_c$), taken from a previously generated code book. The best-matching code vector is chosen using a minimum distortion rule (Gersho and Gray 1992). To minimize the distortion, the following formula is used to determine the distance between two code words:

$$d(X, \hat{X}) = \frac{1}{N} \sum_{i=1}^{n} (x_i - \hat{x}_i)^2 \qquad (3)$$

where $d(X, \hat{X})$ denotes the distortion incurred in replacing the original vector X with the code vector $\hat{X}$.

Therefore, VQ comprises three stages: (1) Code book generation, (2) Vector encoding, and (3) Vector decoding. It works by encoding values from a multidimensional vector space into a finite set of values from a discrete subspace of lower dimension.

2.3 K-fold cross-validation (CV)

The quality of the soft sensor model identified from data can be assessed using CV. In k-fold cross-validation, the original sample is randomly partitioned into k subsets (folds) of approximately equal size. Of the k subsets, a single subset is retained as the validation data for testing the model, and the remaining k-1 subsets are used as training data (An et al. 2007). Therefore, the training dataset X is randomly divided into k mutually exclusive folds of approximately equal size parts Z_i ($i = 1, 2, \cdots, k$). By training the model k times and leaving out one of the k subsets each time, k pairs are obtained as follows:

$$
\begin{aligned}
F_1 &= Z_1, \ T_1 = Z_2 \cup Z_3 \cup \cdots \cup Z_K \\
F_2 &= Z_2, \ T_1 = Z_1 \cup Z_3 \cup \cdots \cup Z_K \\
&\vdots \\
F_k &= Z_k, \ T_k = Z_1 \cup Z_2 \cup \cdots \cup Z_{k-1}
\end{aligned}
\qquad (4)
$$

where F_i represents the validation dataset, and T_i represents the training dataset. The k results from the folds can then be averaged to produce a single estimation. The advantage of this method over repeated random sub sampling is that all the observations are used for both training and validation, while each observation is used for validation exactly once. As k increases, the percentage of training samples increases, and a more robust estimator can be obtained; however, the validation sets become smaller.

3 Experimental set-up

As one of the vital catalytic units in oil refineries, the HDS process is very effective in sulfur removal from petroleum fractions where the molecules containing sulfur lose their sulfur atoms via hydrogenation reactions (Zahedi et al. 2011). HDS of gas–oil fractions is commonly accomplished in a trickle-bed reactor where there are three phases, namely gas (hydrogen), liquid (gasoil), and solid (catalyst particles) (Froment 2004; Korsten and Hoffmann 1996).

A pilot plant facility for HDS processing of petroleum streams has been set up at the Research Institute of Petroleum Industry of Iran (RIPI). A schematic diagram of the experimental set-up used in this work is shown in Fig. 2. The major parameters of the set-up are shown in Table 2. Gas–oil containing 7,200 ppm (by weight) of sulfur is fed into the reactor. Feedstock selected for HDS set-up is gasoil with the characteristics listed in Table 3.

Gasoil is first pumped into the unit, preheated and mixed with hydrogen. The mixture is then passed through the trickle-bed reactor. Output of the reactor is directed to the condenser in which the treated gas–oil and H_2S are separated. Co-Mo HDS catalyst on alumina support (DC-130) procured from CRITERION Company is used in the experiments.

The content of sulfur in the product depends on (1) reactor temperature, (2) reactor pressure, and (3) H_2/Oil ratio. Therefore, in order to train and test the SVR model, a set of experiments were carried out using the setup. The inlet temperature varied from 320 to 370 °C, while the reactor pressure changed from 50 to 70 bars and H_2/oil ratio from 85 to 170 Nm^3/m^3. Values of the parameters are shown in Table 4.

Only one factor was allowed to change in every test evaluating the parameters. Over 300 experiments were performed in the laboratory to find the values. Minimum and maximum contents in the products were 10 and 4,900 ppm wt, respectively. A single model capable of predicting the product sulfur concentration over the wide range is sought. The samples are collected based on 4 h of operation under nearly steady-state conditions. A time interval of 2 h between every experiment was required to reach the next steady-state condition. Treated gasoil sulfur content is collected from each experiment as the output values.

4 Development of model

The input and output variables of the SVR model were selected as shown in Table 5. In order to consider the effect of reactor (catalyst) size, the reactor outlet temperature was selected as one of the input variables of the SVR model. In this way, the trained model could be applied to industrial scale reactors independent of the (catalyst) size. A five-dimensional input vector $X = [x_1, x_2, \cdots, x_5]^T$ and the corresponding row of Y matrix denoting the one-dimensional desired (target) output vector $Y = [y_1]^T$ were employed in training the SVR model.

The SVR model is developed using the LIBSVM package (Chang and Lin 2001). Implementation of the model was carried out using MATLAB 7.10 simulation software. The experimental results were obtained using a personal computer equipped with Intel (R) Core (TM) 2 CPU (3.0 GHz) and 3.25 GB of RAM.

To build an SVR model efficiently, the SVR parameters must be specified carefully. These parameters include (1)

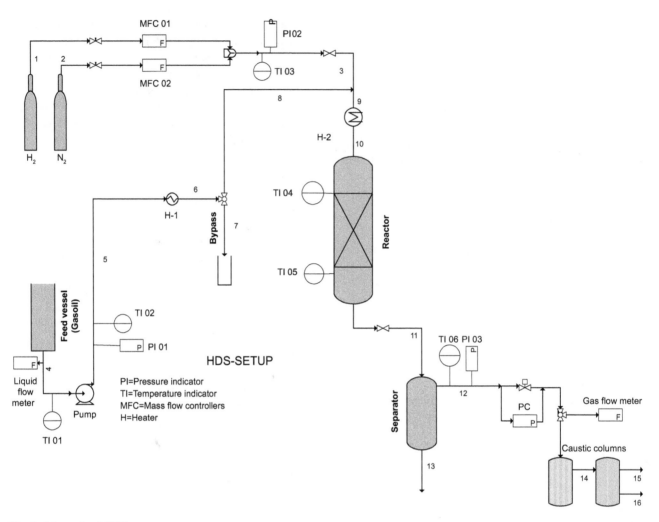

Fig. 2 Schematic of HDS set-up

Table 2 Setup specification

Reactor	
Reactor diameter, m	0.0127
Reactor length, m	0.63
Catalyst bed length, m	0.11
Catalyst	
Chemical composition, wt% dry basis	
Cobalt	3.4
Molybdenum	13.6
Physical properties	
Surface area, m²/g	235
Pore volume, cc/g (H₂O)	0.53
Flat plate crush strength, N/cm (lb/mm)	200
Attrition index	99
Compacted bulk density, g/c	0.72

Table 3 Characteristics of selected gasoil

	Temperature °C	Fraction vol %
Distillation curve	171.4	0
	204.4	5
	216.2	10
	240.8	20
	263.7	30
	281.0	40
	295.3	50
	308.7	60
	322.8	70
	338.5	80
	357.5	90
	370.6	95
	371.7	100
Specific gravity, g/cm³		0.865
Total sulfur in feed, ppm		7,200

Kernel function, (2) bandwidth of the kernel function (σ^2), (3) regularization parameter C, and (4) the tube size of ε-insensitive loss function (ε). Furthermore, in order to

Table 4 The parameter levels

Parameter	Level 1	Level 2	Level 3	Level 4	Level 5
Inlet temperature, °C	320	337	353	370	
Reactor pressure, bar	50	60	70		
H_2/Oil ratio, nm^3/m^3	85	100	120	140	170
Liquid flow rate, cc/min	0.20	0.23	0.26	0.29	0.32

Table 5 Input and output parameters for SVR model

Input variables (X)	Output variable (Y)
Hydrogen flow rate (Q_{H2})	Product sulfur content
Gasoil flow rate (Q_{gasoil})	
Reactor pressure	
Inlet temperature of reactor	
Outlet temperature of reactor	

simplify the training set and to reduce training time, the VQ technique was applied. In this study, different algorithms including GS, GA, PS, and GA–SQP were applied for optimizing the SVR hyper-parameters. The C and $g\left(\frac{1}{2\sigma^2}\right)$ hyper-parameters were selected as the optimization parameters. Figure 3 represents the structure of the proposed method and details of the parameter-optimization procedure. About 70 experiments were selected randomly for testing data, and the 230 data were used as training data. According to Fig. 3, the main steps of model development were as follows:

Step 1: Data compression: Extracting a collection of raw data and generate training and testing sets and reduce CT of the SVR model by applying the VQ technique. Thus, the SVR model was trained with low-dimensional dense datasets, which can lead to speeding up of computation with a reasonable accuracy.

Step 2: Selecting the SVM (the ε-SVR model was used); applying cross-validation technique (the fivefold cross-validation technique was used); and selecting the type of core kernel.

Step 3: Hyper-parameter optimization: Optimizing model parameters (C and $g\left(\frac{1}{2\sigma^2}\right)$) using GSM, GA, PS, and GA–SQP algorithms;

Step 4: Validating the model and predicting the sulfur content.

Hyper-parameter optimization is one of the vital challenges in SVR models. In addition to the commonly used GS, other techniques were also employed in SVR (or SVM) to correct appropriate values of hyper-parameters. Huang and Wang (2006) presented a GA-based feature selection and parameters' optimization for SVM. Also, Momma and Bennett (2002) developed a fully automated pattern search (PS) methodology for model selection of SVR.

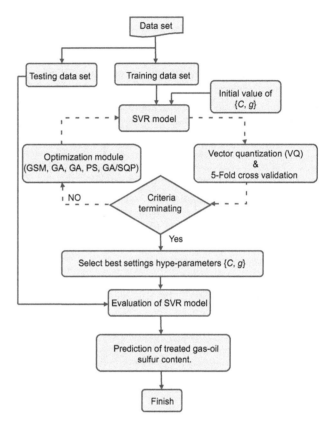

Fig. 3 The procedure of parameter tuning in SVR

4.1 Parameter tuning of SVR with GSM

The GSM is the most common method used to determine the appropriate values of hyper-parameters. This method suffers from the main drawbacks of being very time consuming, lacks guarantee of convergence to a global optimal solution, and involves dependency on the parameters' boundary selection.

In this study, two typical ranges were selected for hyper-parameters' boundary of GSM. First, $\log \frac{C}{2}$ and $\log \frac{g}{2}$ varied between $[-3\ 3]$ and $[-5\ 4]$, respectively. Then, $\log \frac{C}{2}$ and $\log \frac{g}{2}$ varied between $[-2\ 2]$ and $[-3\ 2]$, respectively. Since ε has little effect on ARRE, it was assumed to be 0.01. Typical results by this method are shown in the Table 7.

Selecting a wide range for this method can increase the accuracy but the CT would become very long. Since the accuracy of the SVR model depends on a proper setting of

SVR hyper-parameters, some optimization algorithms have been developed.

4.2 Optimizing the SVR parameters based on GA

The concept of GA was developed by Holland (1975). GA is a heuristic search method that mimics the process of natural evolution. Furthermore, it is a stochastic search technique that can be used in finding the global optimum solution in a complex multidimensional search space. It can search large and complicated spaces using ideas from natural genetics and the evolutionary principle (Goldberg 1989). In this work, the procedure for hyper-parameter optimization with GA method is summarized in the following steps:

(1) *Start* Initialize the parameters for GA and choose a randomly generated population, population size, the number of subpopulations and individuals per subpopulation, the type of kernel function, and the range of the SVR parameters. The SVR hyper-parameters $\{C, g, \text{ and } \varepsilon\}$ are directly coded to generate the chromosome randomly.

(2) *Calculating the fitness* The fitness function is defined as the AARE cross-validation on the training dataset as follows:

$$\text{Min}(Fitness) = \text{Min}(AARE \text{ of } CV)$$
$$= \text{Min} \frac{1}{k} \sum_{i=1}^{k} \left(\frac{k}{m} \sum_{i=1}^{m/k} \left| \frac{(Exp_i - Pre_i)}{Exp_i} \right| \right)$$
(5)

where Exp_i and Pre_i are the actual values and the predicted values, respectively. In this research, a fivefold cross-validation method was being used ($k = 5$). m denotes the total number of training sets ($m = 230$).

(3) *Creating the offspring by genetic operators* To select the subpopulation individuals for the mating pool. The integration of discrete recombination and line recombination is applied to randomly paired chromosomes, which determines whether a chromosome should be mutated in the next generation.

(4) *Elitist strategy* Elitist reinsertion is used to prevent losing good information and is a recommended method.

(5) *Migration* The migration model is used to divide the population into multiple subpopulations.

(6) *Check the termination condition* If the executed generation number equals the special generation number, the algorithm ends; otherwise, it goes back to step 2. The GA creates generations by selecting and reproducing parents until termination criteria are met.

4.3 Parameters tuning of SVR based on the PS algorithm

The PS method is a class of direct search methods to solve nonlinear optimization problems. The PS algorithm can calculate the function values of a pattern and tries to find the minimum value. For the hyper-parameter optimization with the PS algorithm, the procedures are summarized in the following steps:

(1) Parameters setting, set iteration $i = 0$
(2) Set iteration $i = i + 1$
(3) Model training: Hyper-parameter optimization, five-fold CV
(4) Fitness definition and evaluation
(5) Termination: The evolutionary process proceeds until a stopping criterion is met (maximum iterations predefined or the error accuracy of the fitness function). Otherwise, we go back to step (2).

4.4 Parameter tuning of SVR based on GA–SQP hybrid algorithm

This method relies on both local search and global search techniques. The SQP method is a deterministic method, while the GA is a stochastic method. The SQP method is one of the most effective gradient-based algorithms for constrained nonlinear optimization problems. The method is sensitive to initial point selection. It can guarantee local optima as it follows a gradient search direction from the starting point toward the optimum point. GA is efficient for global optimization by finding the most promising regions of search space. Hybridization of GA and SQP can complement the qualities of GA by focusing on accuracy and solution time. The GA is first applied to produce the proper estimation point for SQP. In other words, GA and SQP were used in series.

The algorithm starts with the GA, since the SQP is sensitive to the initial point. Therefore, GA is the main optimizer, and the SQP is used to fine tune for improvement of the every solution of the GA. GA has shown to be efficient on global optimization by finding the most promising regions of search space; however, it suffers from excessive solution time and low accuracy. On the other hand, the SQP can complement the qualities of GA by focusing on accuracy and solution time. GA can be applied first in order to refine the initial point, and then the SQP will be able to reach the solution fast. In other words, the calculation continues with the GA for a specific number of generations or a user-specified number for stall generation during which the approximate solution becomes closer to the real solution. The algorithm then shifts to the SQP which is a faster method. Details of the procedure are illustrated in the flowchart shown in Fig. 4.

Fig. 4 Flow diagram of the combined GA–SQP and SVR for parameter optimization

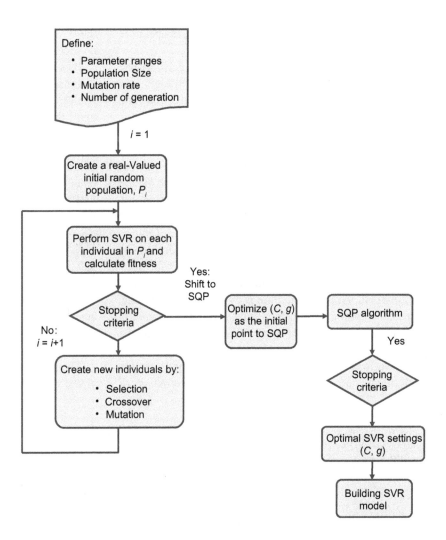

According to Fig. 4, the procedure of hyper-parameter optimization with GA–SQP method is summarized as follows:

(1)   *Start* To define the parameters for GA and choose a randomly generated population, population size, the number of subpopulations and individuals per sub-population, the mutation rate, type of kernel function, and the range of the SVR parameters.

(2)   *Calculating the fitness* The fitness function is defined as the AARE cross-validation on the training dataset as per Eq. 5.

(3)   *Creating the offspring by genetic operators* The GA uses selection, crossover, and mutation operators to generate the offspring of the existing population. Offspring replaces the old population and forms a new population in the next generation. The evolutionary process proceeds until a stopping criterion is satisfied.

(4)   *Shift to the SQP* The GA creates generations by selecting and reproducing parents until a stopping

criterion is met. One of the stopping criteria is a specified maximum number of generations. Another stopping strategy involves population convergence criteria. After satisfying a stopping criterion, the algorithm shifts to the SQP method. The search will continue until a stopping criterion is satisfied.

5 Results and discussion

In this research, over 300 experiments were conducted on a pilot scale hydro-desulfurization set up. Gas–oil flow rate, H_2 flow rate, reactor pressure, and inlet temperature were chosen as the different operating parameters in the experiments. The gas–oil sulfur contents used in experiments varied from 10 to 4,900 ppm. Besides the mentioned parameters, the reactor outlet temperature was also selected as an input parameter of the SVR model. This facilitates the application of the developed SVR model to simulate the behavior of industrial reactors. Note that when catalyst deactivation occurs during the time, the outlet temperature would change for the same input conditions.

One of the important factors in forecasting performance of SVR is the kernel function. In this work, different kernels namely linear kernel, polynomial kernel, sigmoid, and radial basis function (RBF) kernel were used, and the effects of these kernel functions on SVR model based on GS-optimization method are summarized in Table 6. The results show that SVR model with Gaussian (RBF) kernel provides a lower AARE. Furthermore, in order to obtain better accuracy of SVR model and data compression, the VQ technique was employed. The impacts of VQ on CT and prediction accuracy of SVR model are shown in Table 6. The VQ technique can reduce the CT and simultaneously improve the accuracy of the SVR model.

The most important factor influencing the efficiency and robustness of the SVR algorithm is hyper-parameter tuning. Hence, the optimization method is the most critical factor to determine the convergence speed of the SVR model and the ability to search for the global optimal solution.

The effects of different optimization methods on the SVR model are shown in Table 7. The performance of

these methods was evaluated by the statistical criteria (AARE and R^2).

It is seen that the GS results depend completely on the boundary value of C and g. PS gives a better result of AARE and R^2 than GA; however, the GA–SQP algorithm gives the best AARE, R^2, and CT.

From the results, it can be concluded that the performances of the PS, GA, and GA–SQP integrated with SVR are relatively superior to GSM integrated with SVR. On the other hand, integrating these methods (PS, GA, and GA–SQP) with SVR presented attractive advantages compared with GSM and SVR, as follows:

(1) Optimization of the SVR parameters without drawbacks of GSM.

(2) Reduction of computational time.

Some of the results from the hybrid GA–SQP algorithm are shown in Table 8. As seen in this table, integration of GA–SQP with SVR model has good accuracy for prediction of sulfur content of the treated gas–oil in a wide range.

The parity plots for different optimization algorithms integrated with SVR model are shown in Fig. 5. It shows that the

Table 6 The impact of kernel function and VQ on prediction by SVR model

	Kernel type	AARE (with VQ)	AARE (without VQ)	CT (with VQ)	CT (without VQ)
1	Linear	1.295	4.274	228	327
2	Polynomial	0.227	2.894	264	368
3	Sigmoid	0.246	2.928	243	339
4	Gaussian	0.083	0.546	312	595

Table 7 Optimal SVR hyper-parameters obtained by different algorithms ($\varepsilon = 0.1$)

No.	Method	Boundary	C	g	Type of kernel	AARE (training)	AARE (test)	R^2 (training)	R^2 (test)	CT(S)
1	GS–SVR	C:2^[−2 2], g:2^[−3 2]	4.0	0.125	RBF	0.0978	0.1063	0.985	0.983	291
2	GS–SVR	C:2^[−3 3], g:2^[−5 4]	8.0	0.125	RBF	0.0885	0.0893	0.988	0.986	323
3	PS–SVR	C:[0.01 1e4], g:[0.01 1e4]	52.0	0.100	RBF	0.0723	0.0828	0.996	0.995	107
4	GA–SVR	C:[0.01 1e4], g:[0.01 1e4]	48.9	0.030	RBF	0.0734	0.0844	0.995	0.994	112
5	GA–SQP–SVR	C:[0.01 1e4], g:[0.01 1e4]	50.5	0.099	RBF	0.0652	0.0745	0.998	0.997	56

Table 8 Typical input and output data for the SVR testing with GA–SQP method

Test no.	Q_{gasoil}, kg/s	Q_{H2}, m³/h	P, kPa	T_{in}, °C	T_{out}, °C	S_{exp}, ppm	S_{pre}, ppm
1	2.85E−06	3.54E−05	5,000	320	322.27	3,660	3,652
2	2.85E−06	3.54E 05	5,000	337	339.24	1,598	1,586
3	2.85E−06	3.54E−05	5,000	353	355.76	313	273
4	2.85E−06	6.90E−05	5,000	370	373.28	15	20
5	3.71E−06	6.90E−05	5,000	320	320.66	4,333	4,354
6	3.71E−06	6.90E−05	5,000	337	338.47	2,346	2,354
7	3.71E−06	6.90E−05	5,000	353	355.99	675	670

Table 8 continued

Test no.	Q_{gasoil}, kg/s	Q_{H2}, m³/h	P, kPa	T_{in}, °C	T_{out}, °C	S_{exp}, ppm	S_{pre}, ppm
8	4.56E−06	6.90E−05	5,000	337	338.92	2,950	2,921
9	2.85E−06	6.90E−05	5,000	320	321.27	3,644	3,634
10	2.85E−06	6.90E−05	5,000	353	356.44	320	297
11	2.85E−06	3.54E−05	7,000	320	321.18	3,737	3,705
12	2.85E−06	3.54E−05	7,000	337	338.59	1,733	1,819
13	2.85E−06	3.54E−05	7,000	353	356.43	384	369
14	2.85E−06	3.54E−05	7,000	370	373.56	18	22
15	4.56E−06	6.90E−05	7,000	320	320.60	4,885	4,859
16	4.56E−06	6.90E−05	7,000	337	338.44	3,097	3,054
17	3.71E−06	1.13E−04	7,000	320	320.60	4,377	4,343
18	3.71E−06	1.13E−04	7,000	337	338.87	2,477	2,429
19	3.71E−06	1.13E−04	7,000	353	356.33	805	7,83
20	2.85E−06	1.13E−04	7,000	337	339.33	1,731	1,691

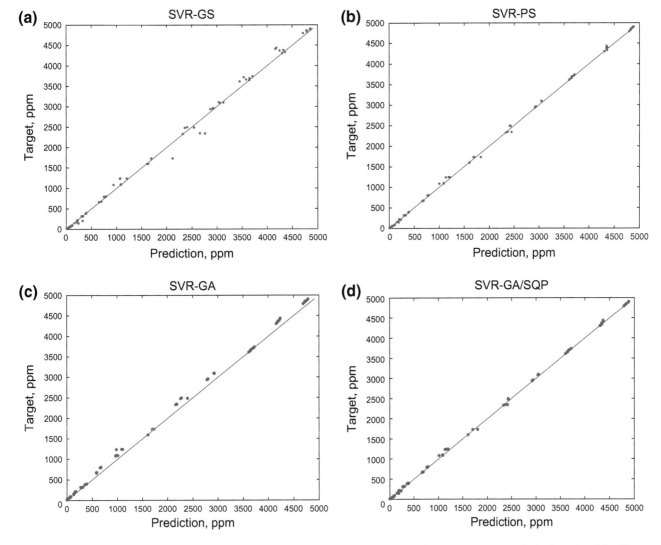

Fig. 5 The parity plot for different algorithms. **a** Hyper-parameter optimization using GS. **b** Hyper-parameter optimization using PS. **c** Hyper-parameter optimization using GA. **d** Hyper-parameter optimization using GA–SQP

SVR model is a robust and reliable model to predict the treated gas–oil sulfur content, no matter what algorithm is being selected for hyper-parameter optimization. Consequently, the model can be applied with good confidence to predict sulfur content in the industrial plants with any characteristics.

6 Conclusion

The aim of this study was to improve the prediction performance and the CT of a data-driven, soft sensor used in the production of ultra-low sulfur diesel. A novel, soft sensor model integrating VQ technique with SVR model was proposed. Selection of optimal parameters of the model is a vital challenge directly affecting prediction accuracy. An integrated GA and SQP (GA–SQP) optimization procedure that is a relatively a fast alternative to the time-consuming GS approach was employed.

The other important factor in the predictive performance of SVR model is kernel function. Four different kernels, namely, linear, polynomial, sigmoid, and Gaussian kernels were evaluated. Results show that the SVR model with a Gaussian (RBF) kernel gives a lower AARE. The model was validated against a wide range of experimental data taken from the gas–oil HDS set-up. The results revealed that the proposed VQ–SVR model coupled with hybrid GA–SQP optimization algorithm is superior to other methods and gives the best prediction for the sulfur content with the highest accuracy (AARE = 0.0745, $R^2 = 0.997$) and the lowest computation time (CT = 56 s).

The proposed approach can pave the way for design of reliable data-driven soft sensors in petroleum industries.

References

An S, Liu W, Venkatesh S. Fast cross-validation algorithms for least squares support vector machine and kernel ridge regression. Pattern Recogn. 2007;40(8):2154–62.

Basak D, Pal S, Patranabis DC. Support vector regression. Neural Inf Process. 2007;11(10):203–25.

Boser B, Guyon I, Vapnik V. A training algorithm for optimal margin classifiers. In: Proceedings of 5th Annual Workshop on Computer Learning Theory, Pittsburgh: ACM. 1992;144–152.

Bolf N, Galinec G, Baksa GT. Development of soft sensor for diesel fuel quality estimation. Chem Eng Technol. 2010;33(3):405–13.

Chang CC, Lin CJ. LIBSVM: a library for support vector machines; software. http://www.csie.ntu.edu.tw/cjlin/libsvm. 2001.

Chen K, Wang C. Support vector regression with genetic algorithms in forecasting tourism demand. Tour Manag. 2007;28(1):215–26.

Cristianini N, Taylor JS. An introduction to support vector machines. London: Cambridge University Press; 2000.

Duan K, Keerthi S, Poo A. Evaluation of simple performance measures for tuning SVM hyperparameters. Neuro Comput. 2003;51:41–59.

Froment GF. Modeling in the development of hydrotreatment processes. Catal Today. 2004;98(1–2):43–54.

Gunn SR. Support vector machines for classification and regression, ISIS Technical Report. Southampton: University of Southampton; 1998.

Gersho A, Gray RM. Vector quantization and signal compression. Dordrecht: Kluwer Academic Publishers; 1992.

Goldberg DE. Genetic algorithms in search, optimization and machine learning. Boston: Addison-Wesley Longman Publishing Co, Inc.; 1989.

Holland JH. Adaptation in Natural and Artificial Systems. Michigan: University of Michigan Press; 1975.

Horng MH. Vector quantization using the Firefly algorithm for image compression. Expert Syst Appl. 2012;39(1):1078–91.

Huang CL, Wang CJ. A GA-based feature selection and parameters optimization for support vector machines. Expert Syst Appl. 2006;31(2):231–40.

Huang CF. A hybrid stock selection model using genetic algorithms and support vector regression. Appl Soft Comput. 2012;12(2):807–18.

Kartik CKN, Narasimhan S. A theoretically rigorous approach to soft sensor development using principle components analysis. In: 21st European Symposium on Aided Process Engineering-ESCAPE2. 2011;793–798.

Kadlec P. On robust and adaptive soft sensors. PhD Thesis, Bournemouth: Bournemouth University; 2009.

Kadlec P, Gabrys B, Strandt S. Data-driven soft sensors in the process industry. Comput Chem Eng. 2009;33(4):795–814.

Kadlec P, Grbic R, Gabrys B. Review of adaptation mechanisms for data-driven soft sensors. Comput Chem Eng. 2011;35(1):1–24.

Korsten H, Hoffmann U. Three-phase reactor model for hydrotreating in pilot trickle-bed reactors. AIChE J. 1996;42(5):1350–60.

King SL, Bennett KP, List S. Modeling noncatastrophic individual tree mortality using logistic regression, neural networks, and support vector methods. Comput Electron Agric. 2000;27(1–3):401–6.

Liu G, Zhou D, Xu H, et al. Model optimization of SVM for a fermentation soft sensor. Expert Syst Appl. 2010;37(4):2708–13.

Li Y, Yuan J. Prediction of key state variables using support vector machines in bioprocesses. Chem Eng Technol. 2006;29(3):313–9.

Lu CJ, Lee TS, Chiu CC. Financial time series forecasting using independent component analysis and support vector regression. Decis Support Syst. 2009;47(2):115–25.

Min JH, Lee YC. Bankruptcy prediction using support vector machine with optimal choice of kernel function parameters. Expert Syst Appl. 2005;28(4):603–14.

Momma M, Bennett KP. A pattern search method for model selection of support vector regression. In: Second SIAM International Conference on Data Mining. 2002.

Pan M, Zeng D, Xu G. Temperature prediction of hydrogen producing reactor using SVM regression with PSO. J Comput. 2010;5(3):388–93.

Si F, Romero CE, Ya Z, et al. Inferential sensor for on-line monitoring of ammonium bisulfate formation temperature in coal-fired power plants. Fuel Process Technol. 2009;90(1):56–66.

Somasundaram K, Vimala S. Fast encoding algorithm for vector quantization. Int J Eng Sci Technol. 2010;2(9):4876–9.

Vapnik V, Golowich S, Smola A. Support vector method for function approximation, regression estimation and signal processing. In: Mozer M, Jordan M, Petsche T, editors. Advances in neural information processing systems. 9th ed. Cambridge: MIT Press; 1996. p. 281–7.

Vapnik VN. The nature of statistical learning theory. New York: Springer; 1995.

Wu CH, Tzeng GH, Lin RH. A novel hybrid genetic algorithm for kernel function and parameter optimization in support vector regression. Expert Syst Appl. 2009;36(3):4725–35.

Yan W, Shao H, Wang X. Soft sensing modeling based on support vector machine and bayesian model selection. Comput Chem Eng. 2004;28(8):1489–98.

Yeh CY, Huang CW, Lee SJ. A multiple-kernel support vector regression approach for stock market price forecasting. Expert Syst Appl. 2011;38:2177–86.

Zahedi S, Shokri S, Baloochi B, et al. Analysis of sulfur removal in gasoil hydrodesulfurization process by application of response surface methodology. Korean J Chem Eng. 2011;28(1):93–8.

10

From molecular dynamics to lattice Boltzmann: a new approach for pore-scale modeling of multi-phase flow

Xuan Liu[1,2] · Yong-Feng Zhu[3] · Bin Gong[1] · Jia-Peng Yu[1] · Shi-Ti Cui[3]

Abstract Most current lattice Boltzmann (LBM) models suffer from the deficiency that their parameters have to be obtained by fitting experimental results. In this paper, we propose a new method that integrates the molecular dynamics (MD) simulation and LBM to avoid such defect. The basic idea is to first construct a molecular model based on the actual components of the rock–fluid system, then to compute the interaction force between the rock and the fluid of different densities through the MD simulation. This calculated rock–fluid interaction force, combined with the fluid–fluid force determined from the equation of state, is then used in LBM modeling. Without parameter fitting, this study presents a new systematic approach for pore-scale modeling of multi-phase flow. We have validated this approach by simulating a two-phase separation process and gas–liquid–solid three-phase contact angle. Based on an actual X-ray CT image of a reservoir core, we applied our workflow to calculate the absolute permeability of the core, vapor–liquid H_2O relative permeability, and capillary pressure curves.

Keywords Molecular dynamics · Lattice Boltzmann · Multi-phase flow · Core simulation

✉ Bin Gong
 gongbin@pku.edu.cn

[1] College of Engineering, Peking University, Beijing 100871, China

[2] Sinopec Petroleum Exploration and Production Research Institute, Beijing 100083, China

[3] Petrochina Tarim Oilfield Exploration and Production Research Institute, Korla 841000, Xinjiang, China

Edited by Yan-Hua Sun

1 Introduction

Multi-phase flow in porous media is a common process in production of oil, natural gas, and geothermal fluids from natural reservoirs and in environmental applications such as waste disposal, groundwater contamination monitoring, and geological sequestration of greenhouse gases. Conventional computational fluid dynamics (CFD) methods are not adequate for simulating these problems as they have difficulty in dealing with multi-component multi-phase flow systems, especially phase transitions. In addition, the complex pore structure of rocks is a big challenge for conventional grid generation and computational efficiency (Guo and Zheng 2009).

In hydrology and the petroleum industry, it is common practice to model multi-phase flow using Darcy's law and relative permeability theory. The relative permeability curve, usually obtained from laboratory experiments, is the key to calculate flow rates of different phases. Although measuring approaches have been widely used and results have been largely accepted for many years in most cases, laboratory experiments are usually expensive, not robust especially for low and ultra-low permeability core measurements, can damage the cores, and cannot always be repeated for different fluids or under different flow scenarios.

It is desirable to obtain the core properties through numerical modeling based on actual pore structure characterizations. Recently, the lattice Boltzmann (LB) method, which is based on a molecular velocity distribution function, has been proposed as a feasible tool for simulation of multi-component multi-phase flow in porous media (Huang et al. 2009, 2011; Huang and Lu 2009). In 1991, Chen et al. proposed the first immiscible LB model that uses red and blue-colored particles to represent two types of fluids

(Chen et al. 1991). The phase separation is produced by the repulsive interaction based on the color gradient. In 1993, Shan and Chen proposed to impose nonlocal interactions between fluid particles at neighboring lattice sites by adding an additional force term to the velocity field (Shan and Chen 1993, 1994; Shan and Doolen 1995). The potentials of the interaction control the form of the equation of state (EOS) of the fluid, and phase separation occurs naturally once the interaction potentials are properly chosen. In 1995, Swift et al. (1995, 1996) proposed a free-energy model, in which the description of non-equilibrium dynamics, such as the Cahn–Hilliard approach, is incorporated into the LB model using the concept of the free-energy function. However, the free-energy model does not satisfy Galilean invariance, and the temperature dependence of the surface tension is incorrect (Nourgaliev et al. 2003). In 2003, Zhang and Chen proposed a new model, in which the body force term was directly incorporated in the evolution equation (Zhang and Chen 2003). Compared with the Shan and Chen (SC) model, the Zhang and Chen (ZC) model avoids negative values of effective mass. However, simulation results from the Zhang and Chen model show that the spurious current gets worse and the temperature range that this model can deal with is much smaller than the SC model (Zeng et al. 2009). In 2004, by introducing the explicit finite difference (EFD) method to calculate the volume force, Kupershtokh developed a single-component Lattice Boltzmann model (LBM) (Kupershtokh and Medvedev 2006; Kupershtokh et al. 2009; Kupershtokh 2010). Compared to previous models, this model has a significant improvement in parameter ranges of temperature and density ratio. Our work in this paper is partially based on this model.

In conventional LB models, the force between fluid and rock is supposed to be proportional to the fluid density. This assumption lacks theoretical support and cannot describe the true physical phenomena under certain circumstances. As an improvement, we propose to simulate the force between the fluid component and rock for different fluid density using the molecular dynamics (MD) method.

Molecular dynamics simulation is an effective method for investigating microscopic interactions and detailed governing forces that dominates the flow. Among the MD studies, various issues in multi-phase processes were paid close attention. Ten Wolde and Frenkel studied the homogeneous nucleation of liquid phase from vapor (Ten Wolde and Frenkel 1998). Wang et al. studied thermodynamic properties in coexistent liquid–vapor systems with liquid–vapor interfaces (Wang et al. 2001). A sharp peak and a small valley at the thin region outside the liquid–vapor interface were found to be evidence of a non-equilibrium state at the interface. In our work, we established a similar system to simulate forces between the fluid and solid components for different fluid densities.

Previous work in combining LB and MD methods together can be classified into two types: one is conducted by Succi, Horbach, and Sbragalia (Chibbaro et al. 2008; Horbach and Succi 2006; Sbragaglia et al. 2006; Succi et al. 2007). They applied MD and LB methods for the same problem and then compared the results. The second type is conducted by Duenweg, Ahlrichs, Horbach, and Succi (Ahlrichs and Dünweg 1998, 1999; Fyta et al. 2006). They applied the two methods to the motion simulation of polymer, DNA, or other macromolecules in water. The coupling of the MD calculation for the macromolecule part and the LB modeling for the solvent is achieved via a friction ansatz, in which they assumed the force exerted by the fluid on one monomer was proportional to the difference between the monomer velocity and the fluid velocity at the monomer's position.

To be exact, the first type summarized above is not the coupling of LB and MD. The second approach is considered as multi-scale coupling of LB and MD, where MD is used for the focus part such as the polymer or the fluid–gas interface, while LB is used for other parts of the system, such as the solvent or the fluid flow. LB and MD simulations are conducted at the same time step, and the variables are exchanged between these two simulation domains under certain boundary constraints. Such a synchronous calculation method is extremely time-consuming in porous media flow simulation because of the large amount of calculation of MD simulation on both gas–liquid and rock–fluid interfaces.

Our proposed method integrates, rather than couples simultaneously, the LB and MD models efficiently. In this approach, the interaction forces between rock and fluid of different density are firstly calculated by MD simulation. Combined with the fluid–fluid force determined from the EOS, the two types of interaction forces are then accurately described for LBM modeling. We validated our integrated model by simulating a two-phase separation process and gas–liquid–solid three-phase contact angle. The success of MD–LBM results in agreement with published EOS solution, and experimental results demonstrated a breakthrough in pore-scale, multi-phase flow modeling. Based on an actual X-ray CT image of a reservoir core, we applied our workflow to calculate the absolute permeability of the core, the vapor–liquid H_2O relative permeability, and capillary pressure curves.

2 Methodology

2.1 The lattice Boltzmann model

The Boltzmann equation describes the evolution with regard to a space–velocity distribution function from motions of microscopic fluid particles (Atkins et al. 2006).

$$\frac{\partial f}{\partial t} + \boldsymbol{\xi} \cdot \nabla_x f + \boldsymbol{a} \cdot \nabla_{\xi} f = \Omega(f), \tag{1}$$

where t is time, vector $\boldsymbol{x}$ is location, vector $\boldsymbol{\xi}$ is the fluid molecular velocity at time t and location $\boldsymbol{x}$, and f is the velocity distribution function of the fluid molecules, which is equivalent to the density of the fluid molecules whose velocity is $\boldsymbol{\xi}$ at time t on location $\boldsymbol{x}$, $\boldsymbol{a}$ is the acceleration of the fluid molecules, and $\Omega(f)$ is the velocity distribution function change caused by collision between fluid molecules.

It is not realistic to solve the integral–differential Boltzmann equation directly. An alternative approach is to solve the discrete form of the Boltzmann equation. The most widely used approach is the LBM. The key idea of LBM is both the location and velocity of the particles which are discretely characterized (Fig. 1). A typical LB equation can be written as (Shan and Chen 1993; Qian et al. 1992)

$$f_i(\boldsymbol{x} + \boldsymbol{e}_i \delta t, t + \delta t) - f_i(\boldsymbol{x}, t)$$
$$= \frac{1}{\tau}(f_i^{\text{eq}}(\boldsymbol{x}, t) - f_i(\boldsymbol{x}, t)) + \psi_i(\boldsymbol{x}, t), \tag{2}$$

where $\boldsymbol{x}$ denotes the position vector, $\boldsymbol{e}_i$ ($i = 0, 1,\ldots, q - 1$) is the particle velocity vector to the neighbor sites, q is the number of neighbors, which depends on the lattice geometry, f_i is the particle velocity distribution function along the ith direction, f_i^{eq} is the corresponding local equilibrium distribution function satisfying the Maxwell distribution, τ is the collision relaxation time, and ψ_i is the change in the distribution function due to the body force. The LB

equation implies two kinds of particle operations: streaming and collision. The term on the left side of Eq. (2) describes particles moving from the local site $\boldsymbol{x}$ to one of the neighbor sites $x + e_i \delta t$ within each time step. The first term on the right side of Eq. (2) describes the collisions contributing to loss or gain of the particles with a velocity of e_i. After collision, the velocity distribution will relax to an equilibrium distribution, f_i^{eq}.

The fluid density ρ and its velocity $\boldsymbol{u}$ at one node are calculated in Eqs. (3) and (4) (Qian et al. 1992). The relationship between τ and fluid kinematic viscosity v can be described as $v = (2\tau - 1)\delta x^2 / (6\delta t)$, where δx is the lattice constant and δt is the lattice time (Qian et al. 1992; Swift et al. 1996).

$$\rho = \sum_{i=0}^{q-1} f_i \tag{3}$$

$$\rho \boldsymbol{u} = \sum_{i=0}^{q-1} f_i \boldsymbol{e}_i \tag{4}$$

We use the equilibrium distribution function f_i^{eq} in the standard form (Yuan and Schaefer 2006)

$$f_i^{\text{eq}}(\boldsymbol{x},\ t) = f_i^{\text{eq}}(\rho,\ \boldsymbol{u})$$
$$= \rho \omega_i \left(1 + \frac{\boldsymbol{e}_i \cdot \boldsymbol{u}}{\theta} + \frac{(\boldsymbol{e}_i \cdot \boldsymbol{u})^2}{2\theta^2} - \frac{\boldsymbol{u}^2}{2\theta}\right), \tag{5}$$

where ρ and $\boldsymbol{u}$ are for $\rho(\boldsymbol{x}, t)$ and $\boldsymbol{u}(\boldsymbol{x}, t)$, respectively, corresponding to local fluid density and macro velocity. With this equilibrium distribution function, the "kinetic temperature" in standard LB models, such as "D2Q9" and "D3Q19," is equal to $\theta = \delta x^2 / (3\delta t^2)$ (Yuan and Schaefer 2006).

In this work, we use the D2Q9 model for the 2D simulations. The weighting factor and discrete velocity for this model are given below:

$$\boldsymbol{e}_i = \begin{cases} (0,\ 0), & i = 0 \\ (\pm 1,\ 0)\ (0,\ \pm 1), & i = 1,\ 2,\ 3,\ 4 \\ (\pm 1,\ \pm 1), & i = 5,\ 6,\ 7,\ 8 \end{cases}$$

$$\omega_i = \begin{cases} 4/9,\ i = 0 \\ 1/9,\ i = 1,\ 2,\ 3,\ 4 \\ 1/36,\ i = 5,\ 6,\ 7,\ 8 \end{cases}$$

For the implementation of the body force term, ψ_i, it is appropriate to use the exact difference method (Kupershtokh and Medvedev 2006; Kupershtokh 2010)

$$\psi_i(\vec{x},\ t) = f_i^{\text{eq}}(\rho,\ \vec{u} + \Delta \vec{u}) - f_i^{\text{eq}}(\rho,\ \vec{u}), \tag{6}$$

where the change of the velocity is determined by the force $\boldsymbol{F}$ acting on the node

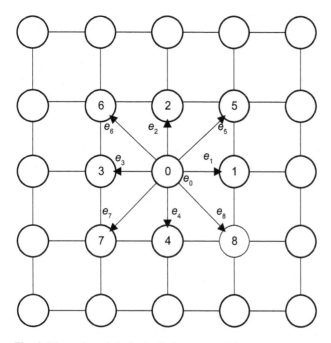

Fig. 1 Illustration of the lattice Boltzmann model

$$\Delta \boldsymbol{u} = \boldsymbol{F}\Delta t/\rho \qquad (7)$$

$\boldsymbol{F}$ is identified in three categories: attractive (or repulsive) force between local fluid and the fluid of their neighbor lattices $\boldsymbol{F}_{\mathrm{ff}}$, attractive force between the local lattice fluid and the boundary wall $\boldsymbol{F}_{\mathrm{sf}}$, and the macro body force, e.g., the gravity, $\boldsymbol{F}_{\mathrm{g}}$.

With the body force $\boldsymbol{F}$, the real fluid velocity $\boldsymbol{v}$ should be evaluated at half the time step (Kupershtokh 2010)

$$\rho \boldsymbol{v} = \sum_{i=0}^{q-1} f_i \boldsymbol{e}_i + \frac{1}{2}\boldsymbol{F}\Delta t \qquad (8)$$

Thus, the most important part for LB modeling is the determination of the interaction forces between neighboring fluids $\boldsymbol{F}_{\mathrm{ff}}$ and the fluid and rock wall $\boldsymbol{F}_{\mathrm{sf}}$.

2.2 Application of the equation of state to obtain fluid–fluid interaction forces $\boldsymbol{F}_{\mathrm{ff}}$

It is suggested that the fluid–fluid interacting force can be obtained by solving the state equations (Shan and Chen 1993, 1994; Kupershtokh and Medvedev 2009; Yuan and Schaefer 2006)

$$\boldsymbol{F}_{\mathrm{ff}}(\boldsymbol{x},\ t) = -\nabla U(\boldsymbol{x},\ t), \qquad (9)$$

where $U(\boldsymbol{x},\ t)$ is a function of state that can be expressed as (Kupershtokh 2010)

$$U(\boldsymbol{x},\ t) = p(\rho_{\mathrm{m}}(\boldsymbol{x},\ t),\ T(\boldsymbol{x},\ t)) - \rho_{\mathrm{m}}(\boldsymbol{x},\ t)\theta, \qquad (10)$$

where ρ_{m} is the specific density of the fluid, which can be expressed as $\rho = \rho_{\mathrm{m}}M$ where M is the standard molar quality of fluid component. T is temperature. $p(\rho_{\mathrm{m}},\ T)$ is a certain state equation. Replacing U with the effective mass density φ, we get

$$\varphi^2(\boldsymbol{x},\ t) = |U(\boldsymbol{x},\ t)| \qquad (11)$$

Then the interacting force between fluids can be written as

$$\boldsymbol{F}_{\mathrm{ff}}(\boldsymbol{x},\ t) = 2\varphi(\boldsymbol{x},\ t)\nabla\varphi(\boldsymbol{x},\ t) \qquad (12)$$

In our work, the Peng–Robinson (P–R) state equation is used in the expression

$$p = \frac{\rho_{\mathrm{m}}RT}{1 - b\rho_{\mathrm{m}}} - \frac{a\alpha(T)\rho_{\mathrm{m}}^2}{1 + 2b\rho_{\mathrm{m}} - (b\rho_{\mathrm{m}})^2} \qquad (13)$$

where $a = 0.458R^2T_{\mathrm{c}}^2/p_{\mathrm{c}}$, $b = 0.0778RT_{\mathrm{c}}/p_{\mathrm{c}}$, $\alpha(T) = [1 + f(\omega)(1-(T/T_{\mathrm{c}})^{0.5})]^2$ where T_{c} and p_{c} are the critical temperature and pressure of fluid components, ω is the acentric factor which depends on the fluid composition itself.

According to the correspondence principle, the molar density ρ_{m}, viscosity μ, etc. are the same when the two fluids have the same values of T/T_{c} and p/p_{c}. Following this principle, the lattice fluid and the real fluid should satisfy the following equations for the P–R equation to be used in the LB model (Yuan and Schaefer 2006)

$$\frac{T^{\mathrm{l}}}{T_{\mathrm{c}}^{\mathrm{l}}} = \frac{T^{\mathrm{r}}}{T_{\mathrm{c}}^{\mathrm{r}}} \quad \frac{p^{\mathrm{l}}}{p_{\mathrm{c}}^{\mathrm{l}}} = \frac{p^{\mathrm{r}}}{p_{\mathrm{c}}^{\mathrm{r}}} \quad \frac{\rho_{\mathrm{m}}^{\mathrm{l}}}{\rho_{\mathrm{m,c}}^{\mathrm{l}}} = \frac{\rho_{\mathrm{m}}^{\mathrm{r}}}{\rho_{\mathrm{m,c}}^{\mathrm{r}}}, \qquad (14)$$

where the superscript l and r stand for the lattice and real systems, respectively, and c means the critical state.

2.3 Application of the molecular dynamics simulation to obtain rock–fluid interaction forces $\boldsymbol{F}_{\mathrm{sf}}$

It is clear that the force acting between fluid components and the rock boundary wall has a significant effect on flow modeling with LBM. In fact, this force decides the capillary force and relative permeability curve for multi-phase systems.

In previous studies (Huang et al. 2009, 2011; Huang and Lu 2009; Martys and Chen 1996; Hatiboglu and Babadagli 2007, 2008), the force between the fluid and boundary wall is simplified as

$$\boldsymbol{F}_{\mathrm{sf}}(\boldsymbol{x},\ t) = -\varphi(\rho(\boldsymbol{x},\ t))G_{\mathrm{sf}}\sum_{i=0}^{q}\omega_i\delta(\boldsymbol{x}+\boldsymbol{e}_i)\boldsymbol{e}_i, \qquad (15)$$

where ρ is the fluid component density, boolean $\delta = 0,\ 1$ to indicate if it is the fluid or solid lattice, respectively. G_{sf} is the force strength factor between fluid and solid, and it is usually determined by fitting the macroscopic fluid–solid contact angle.

There are two disadvantages of this method. One is that G_{sf} cannot always be obtained because of the lack of contact angle data for some common fluids such as methane, nitrogen, carbon dioxide, and polymers for instance. Another disadvantage is that the assumption that $\boldsymbol{F}_{\mathrm{sf}}$ is directly proportional to the fluid effective mass density yields significant error under actual conditions. For example, this assumption cannot represent the adsorption force for coal gas and shale gas to the formation rock.

To overcome these limitations, we propose to obtain the force between fluid components and the rock through molecular dynamic simulation.

There are a variety of minerals in the reservoir rock. But it is mainly composed of minerals including quartz (SiO_2), calcite ($CaCO_3$), dolomite ($CaMg(CO_3)_2$), feldspar ($KAlSi_3O_8$, $NaAlSi_3O_8$, $CaAl_2Si_2O_8$ mixture), and clay ($Al_3Si_2O_5(OH)_4$) with a mixing ratio. In our preliminary investigation, we chose monocrystalline silicon to represent the rock solid, although elemental silicon does not occur in reservoir rocks. It is however convenient for model validation against theoretical and experimental results. In our on-going research, we now use the more realistic assumption that the rock grains are composed of SiO_2 and other components.

The SPC/E model (Jorgensen et al. 1983) was used to model water molecules. This model specifies a 3-site rigid water molecule with charges and Lennard-Jones parameters assigned to each of the 3 atoms. The bond length of O–H is 1.0 Angstrom, and the bond angle of H–O–H is 109.47°.

The well-known LJ potential and the standard Coulombic interaction potential were applied to calculate the intermolecular forces between all the molecules. The parameters are listed in Table 1. The cross parameters of LJ potential were obtained by Lorentz Berthelot combining rules. As originally proposed, this model was run with a 9 Angstrom cut-off for both LJ and Coulombic terms.

LAMMPS code was used to construct the simulation system.

As shown in Fig. 2, the simulation domain was a box with periodic boundary conditions applied in the x and

Table 1 Parameters for LJ potential and Coulombic potential

Parameters	σ, Angstroms	ε, kcal/mole	q, e
Silicon	3.826	0.4030	0
Oxygen	3.166	0.1553	−0.8476
Hydrogen	0.0	0.0	0.4238

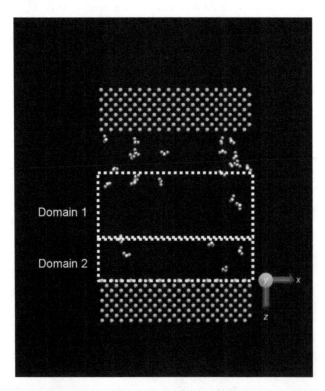

Fig. 2 The molecular dynamics model of the Si–H$_2$O system. $L_x = L_y = 43.45$ Angstrom, $L_z = 60.14$ Angstrom. *Green* atom is for Si, *pink* for O, and *white* for H. The density of water phase is determined by H$_2$O molecules in domain 1. Force between water and the boundary is determined by H$_2$O molecules in domain 2

y directions. The length along these two directions is both 43.45 Angstrom, and the length along the z direction is 60.14 Angstroms. The solid walls were represented by layers of DM (face-centered cubic) silicon atoms (1152 × 2). The speed and the force applied on the Si atoms were both set as zero to make them represent boundary walls and to keep a constant volume of the system.

For different cases, at the beginning of simulation, water molecules of different numbers were sandwiched by the two solid walls.

Firstly, using the constant NVT (certain number, volume, and temperature–canonical ensemble) time integration via the Nose/Hoover method, the whole system was set at a uniform temperature of 373.15 K (110 °C).

Then an annealing schedule of 340,000 steps was applied to allow the system to reach equilibrium (there were four cycles altogether in the annealing schedule. In one single cycle, the temperature is raised from 373.15 to 423.15 K in 20,000 steps, to 473.15 K in 20,000 steps, and then reduced to 423.15 K in 20,000 steps, to 373.15 K in 20,000 steps). After that, the equilibrium state of the system was reached at 373.15 K (110 °C). Then the NVT ensemble was changed into NVE (certain number, volume, and energy– micro-canonical ensemble) and run for 20,000 steps. In these two processes, the equivalent length scale for the simulations was 0.5 feets to ensure energy conservation.

With the equilibrium system, the force between Si and H$_2$O can be calculated. As Fig. 2 shows, since the distance between domain 1 and the solid wall is larger than the cut-off radius (9 Angstroms), the water molecules in domain 1 can be considered as free molecules. So the mass densities of free water in different cases were calculated based on the molecules in this domain. For the water molecules in domain 2, the force F applied by the silicon solid wall was directly calculated according to the potential field in the simulation. Then we can get the stress $\sigma = F/A$, where A is the area of the wall.

2.4 Integrated workflow from MD to LBM simulation

For the first step, we build a MD model as presented in Sect. 2.3 to calculate the rock–fluid interaction forces $\boldsymbol{F}_{sf}$ between any solid component of the wall and the fluid of different density. Input parameters for this step include molecular species of the fluid and boundary solid and the potential parameters. The result of this step is the relationship between the solid–fluid interface and the fluid density.

Then we calculate $\varphi(\rho)$ of different fluid density ρ from P–R EOS calculations as presented in Sect. 2.2. The fluid–fluid interaction force $\boldsymbol{F}_{sf}$ is then determined from Eq. (12).

Input parameters for this step include fluid density, critical pressure and temperature, and acentric factor of the fluid component.

With an actual core X-ray scan image for lattice grid construction and the interaction forces being determined from MD and EOS calculation, we use the LB method to simulate the fluid flow in porous media and to obtain the absolute permeability, relative permeability, and capillary pressure curves.

The integrated workflow is illustrated in Fig. 3.

3 Results

3.1 LBM simulation on liquid–vapor phase transition process of water

To validate the EOS model, the single-component fluid phase transition process was simulated using the model described in Sect. 2.2.

Water is the fluid we choose in our simulation. Its critical temperature is 373.99 °C, critical pressure is 22.06 MPa, critical density is 322.0 kg/m^3, and its acentric factor is 0.344.

A 2D 200 × 200 square lattice was used in this simulation, and the periodic boundary conditions were applied on the four boundaries. The average mass density of the computational domain was set to be 500 kg/m^3. At the beginning of simulation, the mass density was homogenously initialized with a small (0.1 %) random perturbation.

Figure 4 shows the mass density contour in the computation domain at different time steps ($t = 250, 500, 1000$ and 2000) when the temperature is 110 °C. The density in the pink area is high (793 kg/m^3), which indicates it is saturated with water liquid. The density in the green area is

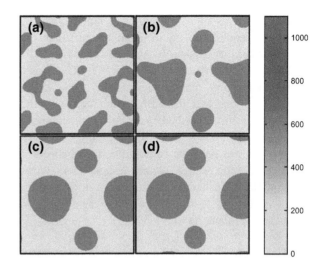

Fig. 4 Mass density (kg/m^3) distribution of water vapor and liquid at different time steps t. **a** $t = 250$. **b** $t = 500$. **c** $t = 1000$. **d** $t = 2000$

low (about 0.76 kg/m^3), which indicates it is saturated with water vapor. It is clearly shown in this figure that the small liquid droplets aggregate to form bigger ones as time increases. Eventually, all these small droplets coalesce to form bigger droplets and the rest of the space of the computational domain is occupied by vapor only. All these simulation results for the single-component phase transition process are consistent with those reported by Zeng et al. (2009) and Qin (2006).

We then repeated the simulation at different temperatures for model validation. Figure 5 shows the density

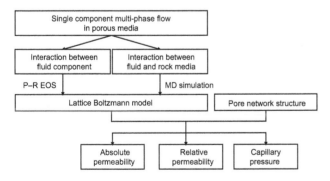

Fig. 3 Schematic illustrations of the workflow. First, the interaction forces between fluids are determined from P–R EOS calculation, and the rock–fluid interaction force is determined from MD simulation. Then we run the lattice Boltzmann simulation based on the grid constructed from the core X-ray scan image to obtain the absolute permeability, relative permeability, and capillary pressure curves

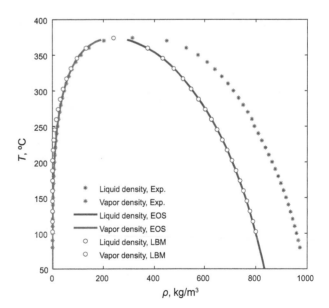

Fig. 5 Saturated water (in *blue*) and vapor (in *red*) densities versus temperature. "o" is the result of LB simulation. The *solid line* is the solution of the P–R equation using the Maxwell equal-area construction. *asterisk* Denotes the experimental data from Handbook of Chemical Engineering (Liu et al. 2001)

distribution curves of both vapor and liquid when they reach phase equilibrium at different temperatures. This figure indicates that the density curve calculated with LBM is nearly identical to that obtained from P–R EOS using the Maxwell equal-area construction (Yuan and Schaefer 2006). It suggests that the LBM, combined with EOS, provides an accurate method to model single-component gas–fluid two-phase flow.

It is noted that there are some differences between the theoretical solution using EOS and the experimental results for liquid density calculations, and this is originated from the inadequacy of P–R EOS and is in agreement with (Atkins and William 2006).

3.2 MD simulation in determination of force between fluid components and the boundary wall

In the MD simulation, 42 sets of systems with different H_2O densities were generated. We select 4 cases for illustration in Fig. 6.

As shown in Fig. 5, at the simulation temperature (110 °C), the force calculation is only meaningful when the

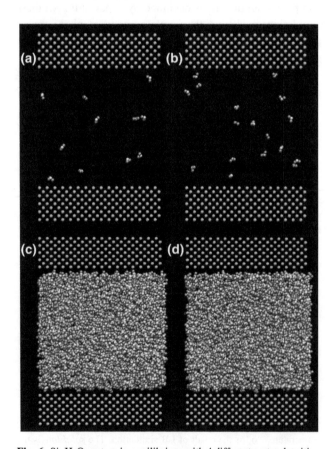

Fig. 6 Si–H_2O system in equilibrium with 4 different water densities. In vapor phase: **a** $\rho = 0.43$ kg/m^3 and **b** $\rho = 0.66$ kg/m^3. In liquid phase: **c** $\rho = 842$ kg/m^3 and **d** $\rho = 911$ kg/m^3

vapor density is smaller than 0.787 kg/m^3 or the liquid density is greater than 793 kg/m^3. The relationship between density and stress is plotted in Fig. 7. It clearly shows that the stress varies with the fluid density nonlinearly, thus demonstrates the assumption that "the force between fluid and rock is supposed to be proportional to the fluid density" used by former studies (Martys and Chen 1996; Hatiboglu and Babadagli 2007, 2008) is not appropriate.

3.3 Calculation of monocrystalline silicon–water contact angle

Contact angle reflects the interaction forces at the phase interfaces and the wettability. Thus, it is a significant parameter to describe the interaction between oil, gas, water, and rock in petroleum reservoirs (Wolf et al. 2009).

We simulated the contact angle of the monocrystalline Si–water liquid–water vapor system by applying our integrated MD–LBM approach and then compared the value with literature for validation.

Figure 8 demonstrates the equilibrium state of the system after LBM simulation. In the simulation, the computational domain is 100×50, where the upper and lower boundaries are solid walls, and the east and west boundary is periodic. For the solid node, before the streaming step, a bounce–back algorithm was implemented to mimic the non-slip wall boundary condition.

In Fig. 8, the gray grid on the top and bottom is for Si, the red is for water liquid, and the blue is for water vapor. The interaction force between water fluids was calculated by EOS, and the interaction force between water and Si was obtained by MD. This figure shows that the macro contact angel is approximate 101°, which is consistent with the result given by (Williams and Goodman 1974) in which it states that the contact angle is near 90°. Therefore, the method of MD used in fluid–solid interaction force calculation incorporated into LBM is reasonable in porous media flow simulation.

3.4 Rock permeability determination based on an actual X-ray CT image of a reservoir core

Similar to the methodology in previous literature (Guo and Zheng 2009; Huang et al. 2009, 2011; Huang and Lu 2009; Hatiboglu and Babadagli 2007, 2008; Jorgensen et al. 1983), we calculated the relative permeability curves by simulating a two-phase flow in the porous media. Figure 9 is a post-processed 2D X-ray image of a reservoir core in Tarim Basin conglomerate with a length $L = 2.7$ mm and width $A = 1.35$ mm. The domain is gridded into $540 \times 270 = 145,800$ cells with each one being a 5×5 μm square. In this image, the part in black

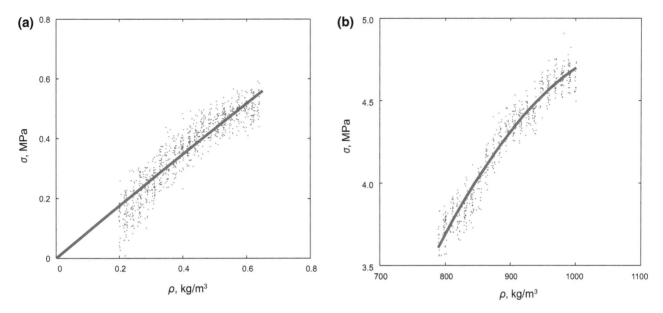

Fig. 7 The stress–density relationship. **a** in the vapor region, linear fitted by the least squares method and **b** in the liquid region, quadratic fitted by the least squares method

represents pores and white for the rock grains. Porosity ϕ = 43.5 %.

We applied a single-vapor phase flow of water to calculate the absolute permeability of the core represented by

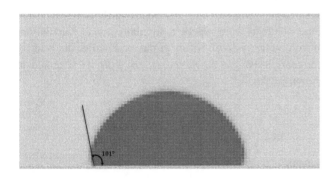

Fig. 8 Equilibrium state simulated from LBM for a water liquid drop in contact with the Si surface. The *gray* grid on the top and bottom is for Si, the *red* for water liquid, and the *blue* for water vapor

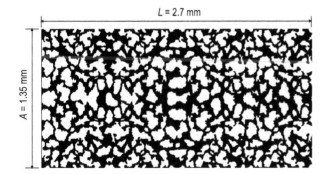

Fig. 9 A 2D CT image of a reservoir rock in Tarim Basin conglomerate

the image in Fig. 9. The system is at temperature $T = 110$ °C. At the initial state, each cell was saturated with H_2O steam with a density of 0.754 kg/m^3. In order to use the MD simulation results on interaction forces between H_2O and Si consistently, we assumed that the rock grain is made of Si (in our on-going research, we use the more realistic assumption that the rock grains are composed of quartz (SiO_2) and other minerals).

As in Fig. 10, we applied a virtual body force g and periodic boundary conditions in calculating permeability based on LBM solution of the water flow problem. The body force is equivalent of adding a pressure drop $\Delta p = \bar{\rho}gL$ ($\bar{\rho} = \sum_{i=1}^{n} \rho_i/n$ is the average density of water in the

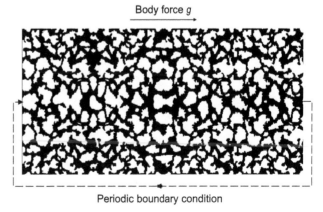

Fig. 10 The flow problem in calculating absolute permeability. Closed boundaries on the *top* and *bottom*. A virtual body force g towards right is applied to the simulation domain, and periodic boundary condition is set on *left* and *right* sides of the simulation domain

simulation domain, n is the total number of the grids filled with water) between the left and right side of the core. By applying different values of g, we can calculate flow velocity under different pressure drops.

For a given pressure drop Δp, the flow velocity of H_2O steam reaches equilibrium after some steps of simulation. The volumetric velocity of H_2O steam flow is calculated as $v_g = \phi \sum_{i=0}^{n} v_i/n$ (Cihan et al. 2009). Figure 11 depicts the calculated relationship between v_g and Δp.

The linearity between v_g and Δp shown in Fig. 11 is in agreement with Darcy's law $v = K\Delta p/(\mu L)$, where μ is the viscosity of water vapor, L is the length of the core, and K is permeability of the core.

Plug $L = 2.7$ mm and $\mu_g = 12.4 \times 10^{-6}$ Pa s into Darcy's law and use least square linear fitting of calculated points in Fig. 11, we calculate the permeability of the rock as 636 mD.

3.5 Vapor–liquid two-phase relative permeability determination based on an actual X-ray CT image of a reservoir core

Relative permeability, as a function of phase saturation, could be modeled by multi-phase flow simulation using the proposed workflow. Similar to the conditions set for absolute permeability simulation, virtual body force g and periodic boundary condition were applied to the simulation domain, and temperature was set at 110 °C.

At initial state, the core was randomly filled with vapor–liquid two-phase water at different mixing ratios to

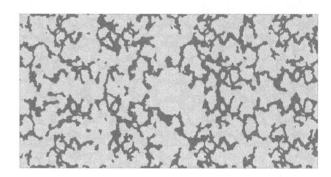

Fig. 12 Distribution of vapor (in *pink*) and liquid (in *blue*) phases of H_2O in rock pores

represent different saturations (Fig. 12). At the simulation temperature, 110 °C, saturated water vapor density is 0.787 kg/m³, and saturated water liquid density is 793 kg/m³. Then the average density of water in simulation domain, $\bar{\rho}$, was calculated for different saturations. Given that $\Delta p = \bar{\rho} g L$, the pressure drop for any values of g can be obtained.

At the simulation temperature, the viscosities of saturated water vapor and liquid are $\mu_g = 12.4 \times 10^{-6}$ Pa s and $\mu_w = 252 \times 10^{-6}$ Pa s.

We calculated the permeability of the vapor and liquid phases using Darcy's law for different phases, $K_g = v_g \mu_g L/\Delta p$, $K_w = v_w \mu_w L/\Delta p$. K_g and K_w at different saturations were obtained from repeated simulations and calculations of volumetric velocity of each phase. Finally, the relative permeability $K_{rg} = K_g/K$ and $K_{rw} = K_w/K$ is obtained as shown in Fig. 13.

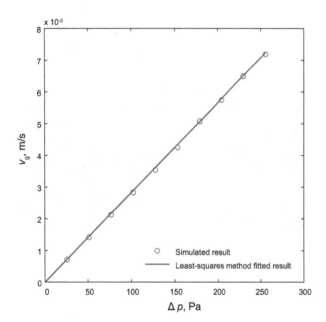

Fig. 11 Calculated water flow velocity versus pressure drop

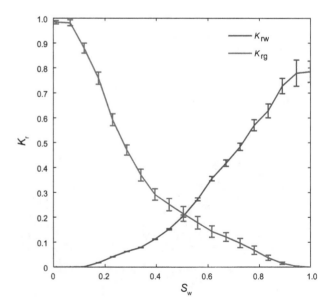

Fig. 13 Calculated vapor–liquid two-phase water relative permeability curves

Different from the simulation of absolute permeability, the multi-phase flow modeling results using a statistical method such as LBM could be unstable, thus causing the volumetric velocity obtained at different times be slightly different. This means that the relative permeability of each phase at a given saturation could be a distribution as shown in Fig. 13. The relative permeability curves were constructed by taking the average values of K_r at each saturation.

3.6 Capillary pressure curve determination based on an actual X-ray CT image of a reservoir core

We also applied our methodology to calculate the capillary pressure curve. As shown in Fig. 14, constant pressure boundary conditions (p_1 on the left and p_2 on the right) and closed boundary conditions on the top and bottom were applied to the simulation domain.

In the initial state, the domain was filled with saturated water vapor. We injected saturated water liquid from the left side under constant pressure p_1. Since $p_1 > p_2$, water vapor will flow out of the domain from the right side. By applying different pressure drops $\Delta p = p_1 - p_2$, the injected liquid flowed into the pores to yield different phase saturations. At an equilibrium state as shown in Fig. 15, the capillary pressure is equal to the pressure drop, i.e., $p_c(S_g) = \Delta p$. The calculated capillary pressure curve is shown in Fig. 16.

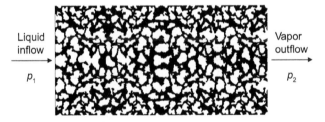

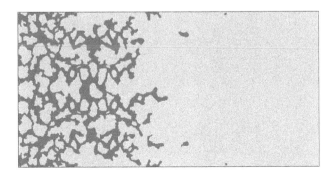

Fig. 14 The flow problem in calculating capillary pressure curve. A constant pressure boundary is set on both sides of the simulation domain

Fig. 15 The viscous–capillary equilibrium of steam flooding

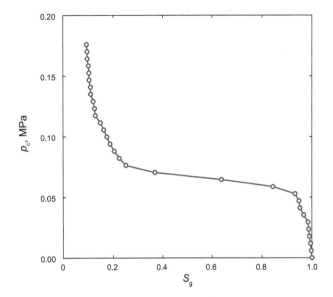

Fig. 16 Calculated vapor–liquid H$_2$O capillary pressure curve

4 Conclusions

In this work, we proposed a new systematic workflow to integrate MD simulation with the LB method to model multi-phase flow in porous media. As an improvement, this new approach avoids parameter fitting or incorrectly assuming a linear relationship between the rock–fluid interaction force and fluid density. We have validated this approach by simulating a two-phase separation process and a gas–liquid–solid three-phase contact angle. The success of MD–LBM results in agreement with published EOS solution, and experimental results demonstrated a breakthrough in pore-scale, multi-phase flow modeling. Based on an actual X-ray CT image of a reservoir core, we applied our workflow to calculate absolute permeability of the core, vapor–liquid water relative permeability, and capillary pressure curves. With the application of this workflow to a more realistic model considering actual reservoir rock and fluid parameters, the ultimate goal is to develop an accurate method for prediction of permeability tensor, relative permeability, and capillary curves based on 3D CT image of the rock, actual fluid, and rock components.

References

Ahlrichs P, Dünweg B. Lattice-Boltzmann simulation of polymer-solvent systems. Int J Mod Phys C-Phys Comput. 1998;9(8): 1429–38.

Ahlrichs P, Dünweg B. Simulation of a single polymer chain in solution by combining lattice Boltzmann and molecular dynamics. J Chem Phys. 1999;111:8225.

Atkins P, William P, Paula JD. Atkins' physical chemistry. Oxford: Oxford University Press; 2006.

Chen S, Chen H, Martnez D, et al. Lattice Boltzmann model for simulation of magnetohydrodynamics. Phys Rev Lett. 1991;67(27):3776–9.

Chibbaro S, Biferale L, Diotallevi F, et al. Evidence of thin-film precursors formation in hydrokinetic and atomistic simulations of nano-channel capillary filling. Europhys Lett. 2008;84:44003.

Cihan A, Sukop MC, Tyner JS, et al. Analytical predictions and lattice Boltzmann simulations of intrinsic permeability for mass fractal porous media. Vadose Zone J. 2009;8(1):187–96.

Fyta MG, Melchionna S, Kaxiras E, et al. Multiscale coupling of molecular dynamics and hydrodynamics: application to DNA translocation through a nanopore. Multiscale Model Simul. 2006;5:1156.

Guo Z, Zheng C. Theory and applications of the lattice Boltzmann method. Beijing: Science Press; 2009 (in Chinese).

Hatiboglu CU, Babadagli T. Lattice-Boltzmann simulation of solvent diffusion into oil-saturated porous media. Phys Rev E. 2007;76(6):066309.

Hatiboglu CU, Babadagli T. Pore-scale studies of spontaneous imbibition into oil-saturated porous media. Phys Rev E. 2008;77(6):066311.

Horbach J, Succi S. Lattice Boltzmann versus molecular dynamics simulation of nanoscale hydrodynamic flows. Phys Rev Lett. 2006;96(22):224503.

Huang H, Li Z, Liu S, et al. Shan-and-Chen-type multiphase lattice Boltzmann study of viscous coupling effects for two-phase flow in porous media. Int J Numer Methods Fluids. 2009;61(3):341–54.

Huang H, Lu X. Relative permeabilities and coupling effects in steady-state gas–liquid flow in porous media: a lattice Boltzmann study. Phys Fluids. 2009;21(9):092104.

Huang H, Wang L, Lu X. Evaluation of three lattice Boltzmann models for multiphase flows in porous media. Comput Math Appl. 2011;61(12):3606–17.

Jorgensen W, Chandrasekhar J, Madura J, et al. Comparison of simple potential functions for simulating liquid water. J Chem Phys. 1983;79(2):926–35.

Kupershtokh AL, Medvedev DA. Lattice Boltzmann equation method in electrohydrodynamic problems. J Electrost. 2006;64:581–5.

Kupershtokh AL, Medvedev DA, Karpov DI. On equations of state in a lattice Boltzmann method. Comput Math Appl. 2009;58(5):965–74.

Kupershtokh AL. Criterion of numerical instability of liquid state in LBE simulations. Comput Math Appl. 2010;59(7):2236–45.

Liu G, Ma L, Liu J, et al. Handbook of chemical engineering (inorganic volume). Beijing: Chemical Industry Press; 2001 (in Chinese).

Martys NS, Chen H. Simulation of multicomponent fluids in complex three-dimensional geometries by the lattice Boltzmann method. Phys Rev E. 1996;53(1):743.

Nourgaliev RR, Dinh TN, Theofanous TG, et al. The lattice Boltzmann equation method: theoretical interpretation, numerics and implications. Int J Multiph Flow. 2003;29(1):117–69.

Qian Y, Humires D, Lallemand P. Lattice BGK models for Navier–Stokes equation. Europhys Lett. 1992;17:479–84.

Qin R. Mesoscopic interparticle potentials in the lattice Boltzmann equation for multiphase fluids. Phys Rev E. 2006;73(6):066703.

Sbragaglia M, Benzi R, Biferale L, et al. Surface roughness-hydrophobicity coupling in microchannel and nanochannel flows. Phys Rev Lett. 2006;97(20):204503.

Shan X, Chen H. Lattice Boltzmann model for simulating flows with multiple phases and components. Phys Rev E. 1993;47(3):1815–9.

Shan X, Chen H. Simulation of nonideal gases and liquid-gas phase transitions by the lattice Boltzmann equation. Phys Rev E. 1994;49(4):2941–8.

Shan X, Doolen G. Multicomponent lattice–Boltzmann model with interparticle interaction. J Stat Phys. 1995;81(1/2):379–93.

Succi S, Benzi R, Biferale L, et al. Lattice kinetic theory as a form of supra-molecular dynamics for computational microfluidics. Tech Sci. 2007;55(2):151–8.

Swift MR, Osborn WR, Yeomans JM. Lattice Boltzmann simulation of nonideal fluids. Phys Rev Lett. 1995;75(5):830–3.

Swift MR, Orlandini E, Osborn WR, et al. Lattice Boltzmann simulations of liquid–gas and binary fluid systems. Phys Rev E. 1996;54(5):5041–52.

Ten Wolde PR, Frenkel D. Computer simulation study of gas–liquid nucleation in a Lennard–Jones system. J Chem Phys. 1998;109:9901.

Wang ZJ, Chen M, Guo ZY, et al. Molecular dynamics study on the liquid–vapor interfacial profiles. Fluid Phase Equilib. 2001;183–184:321–9.

Williams R, Goodman AM. Wetting of thin layers of SiO_2 by water. Appl Phys Lett. 1974;25:531.

Wolf FG, Santos LOE, Philippi PC. Modeling and simulation of the fluid–solid interaction in wetting. J Stat Mech. 2009;2009(06):P06008.

Yuan P, Schaefer L. Equations of state in a lattice Boltzmann model. Phys Fluids. 2006;18:042101.

Zeng J, Li L, Liao Q, et al. Simulation of phase transition process using lattice Boltzmann method. Chin Sci Bull. 2009;54(24):4596–603.

Zhang R, Chen H. Lattice Boltzmann method for simulations of liquid–vapor thermal flows. Phys Rev E. 2003;67(6):066711.

11

Analysis of spudcan–footprint interaction in a single soil with nonlinear FEM

Dong-Feng Mao · Ming-Hui Zhang ·
Yang Yu · Meng-Lan Duan · Jun Zhao

Abstract The footprints that remain on the seabed after offshore jack-up platforms completed operations and moved out provide a significant risk for any future jack-up installation at that site. Detrimental horizontal and/or rotational loads will be induced on the base cone of the jack-up platform leg (spudcan) in the preloading process where only vertical loads are normally expected. However, there are no specific guidelines on design of spudcan re-installation very close to or partially overlapping existing footprints. This paper presents a rational design approach for assessing spudcan–footprint interaction and the failure process of foundation in a single layer based on nonlinear finite element method. The relationship between the distance between the spudcan and the footprint and the horizontal sliding force has been obtained. Comparisons of simulation and experimental results show that the model in this paper can deal well with the combined problems of sliding friction contact, fluid–solid coupling, and convergence difficulty. The analytical results may be useful to jack-up installation workovers close to existing footprints.

Keywords Jack-up · Existing footprint · Spudcan–footprint interaction · Numerical simulation · Nonlinearity

1 Introduction

With an increase in frequency of operations, the situation that installation of jack-up platforms on sites which contains old footprints is becoming more common and inevitable. According to van den Berg's statistics (Van den Berg et al. 2004), within Shell EP Europe alone roughly 1,200 footprint points had been registered in geotechnical and footprint datasets. In addition, there are approximately 80 new single footprint points added to the existing datasets every year. Thus, it can be seen that footprints are not rare and they pose a serious and growing threat to operational safety of jack-up drilling platforms. Figure 1 shows when a leg is close to an existing footprint, the non-uniform bearing load caused by the footprint will make the spudcan slide into the footprint in the jacking process, which was proven by Gaudin et al. (2007), Leung et al. (2007). The sliding trend is affected by the leg stiffness, connection between leg and hull, and in-place condition of other two legs, and the size of the trend is measured by the horizontal sliding force and overturning moment (McClelland et al. 1982; Hossain and Randolph 2007; Bouwmeester et al. 2009). If a slide occurs, the legs will incline in different directions, so that the legs may become stuck in the platform and this would mean the platform cannot be raised. The potential risk of slipping is a serious threat to the operational safety of platforms.

Re-installing a spudcan very close to or partially overlapping existing footprints is generally not recommended in the guidelines (SNAME OC-7 panel. 2007; Hossain and Randolph 2008). In a situation where this is inevitable, the guidelines recommend the use of an identical jack-up (same footing geometries and leg spacing) and locating it in exactly the same position as the previous unit, where possible. However, it is unlikely that two jack-up units have an identical design because the structures of most units are often custom-made and the deployments of units are subject to availability. It is evident that existing guidelines are not adequate for rig operators to install jack-up units in close proximity to existing footprints safely.

D.-F. Mao (✉) · M.-H. Zhang · Y. Yu · M.-L. Duan · J. Zhao
College of Mechanical and Transportation Engineering, China University of Petroleum, Beijing 102249, China
e-mail: maodf@cup.edu.cn

Edited by Yan-Hua Sun

Fig. 1 Schematic diagram of existing footprint problems

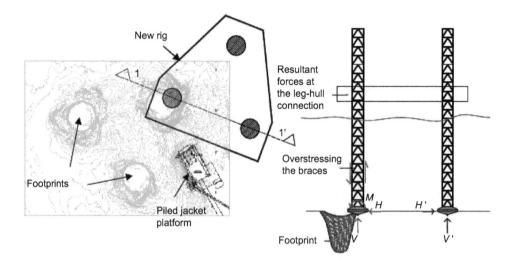

Footprint issues involve soil elastoplasticity, material and geometric nonlinearities, fluid–solid coupling, friction contact during spudcan preloading, and difficult convergence of numerical solutions (Hanna and Meyerhof 1980; Kellezi and Stromann 2003; DeJong et al. 2004; Deng and Kong 2005; Leung et al. 2008). Previous research mainly focuses on the spudcan–footprint interaction through the centrifuge model test. Murff et al. (1991), Hossain et al. (2005), Cassidy et al. (2004, 2009), Teh et al. (2010), Gan (2009), Gan et al. (2012), Kong et al. (2010, 2013), Xie et al. (2012) conducted a series of drum centrifuge model tests to investigate spudcan–footprint interaction and the effect of leg stiffness on spudcan–footprint interaction. With the centrifuge model tests, Stewart and his coworkers (Stewart 2005; Stewart and Finnie 2001) studied the effect of bending rigidity of legs on spudcan–footprint interaction and the influence of the distance between the spudcan and the footprint on sliding. Dean and Serra (2004) discussed the effect of equivalent stiffness of legs on spudcan–footprint interaction. Teh et al. (2006) reported a set of test results investigating the effects of sloping seabed (30° inclined to the horizontal) and footprint on loads developed in jack-up legs. They found that the effect of the footprint is much greater than that of the seabed slope. This indicates that the footprint problem is more serious than a sloping seabed. Other researchers have tried to investigate the footprint problem with numerical simulation (Zhang et al. 2011, 2014). Jardine et al. (2002) simplified a three-dimensional model to a plane strain one to deal with footprint issues. The current understanding of this topic is still insufficient, and only a small number of studies of the footprint problem are available in the public domain. Although it is a great challenge to obtain a converged numerical solution, a good numerical model and solution is very important because it is able to achieve more accurate estimation of carrying capacity of spudcans and better

explanations for tests. This paper takes various factors including failure process of foundation, nonlinearity, sliding friction contact, and fluid–solid coupling into account. It discusses the finite element model of spudcan–footprint interaction in spudcan re-installation near an existing footprint as well as handling relative parameters. With the model of the spudcan–footprint interaction, the changes of horizontal sliding force on the spudcan at different offset distances between the spudcan and the footprint were analyzed with ABAQUS software. The finite element model was validated by comparing the simulation result with experimental results.

2 Analytical methods and computing model

During jacking, the deformation of the surrounding soil is very large, which results in changes in pore pressure and then a reduction in the effective strength of the soil. To analyze spudcan–footprint interaction, the coupling of stress/fluid flow in soil should be considered. Undrained total stress analysis is used in the computing model, i.e., the total stress is the sum of effective stress and hydrostatic pressure. Thus, the equilibrium equation in the vertical direction is as follows (Houlsby and Martin 2003):

$$\frac{d\bar{\sigma}_z}{dz} = \begin{cases} \rho g - \gamma_w \left(S_r(1 - n^0) - \frac{dS_r}{dz}(z_w^0 - z) \right); & z \leq z_w^0, \\ \rho g; & z_w^0 \leq z \leq z^0 \end{cases}$$

(1)

where $\bar{\sigma}_z$ is the vertical stress, Pa; ρ is the soil dry density, kg/m³; γ_w is the water gravity density, N/m³; S_r is the soil saturation, %; z_w^0 is the free water surface elevation, m; z^0 is the elevation of interface between dry soil and partially saturated soil, m; and n^0 is porosity, %; when $z \leq z_w^0$ in completely saturated, $S_r = 1$, and when $z_w^0 \leq z \leq z^0$, in partially saturated, $S_r < 1$.

The advantage of ABAQUS in soil engineering is that it provides not only various elastic/plastic constitutive models for soil but also coupled analysis of stress/fluid flow in soil. In numerical computation, the finite element mesh is fixed on the soil skeleton, and fluid may flow through the mesh and satisfy the fluid continuity equation. The Forchheimer equation (Zeng and Grigg 2006) is adopted to describe nonlinear flow in soil (porous medium). Since less relative parameters in calculation are needed, the Mohr–Coulomb constitutive model is used (Li 2004), i.e., the soil is considered as a perfect elastic–plastic material, and obeys the noncorrelation flow rule. The Mohr–Coulomb yield criterion is as follows:

$$s + \sigma_m \sin \phi - c \cos \phi = 0, \tag{2}$$

where $s = (\sigma_1 - \sigma_3)/2$ is half of the difference of maximum and minimum principal stresses, kPa; $\sigma_m = (\sigma_1 + \sigma_3)/2$ is the average value of maximum and minimum principal stresses, kPa; c is cohesion, kPa; and ϕ is the internal friction angle, °. Except for over-consolidated soil, clay always shows little dilatancy, and thus the dilatancy angle $\phi = 0$. Assume that the deformation modulus is approximately proportional to the undrained shear strength, then $E = 500s_u$ (s_u is the undrained shear strength, kPa).

A vertical plane containing the line connecting the spudcan and the footprint center is chosen and a finite element model is established, as shown in Fig. 2. The diameter and depth of the footprint are D and d, respectively. In order to reduce the boundary effect on accuracy of the numerical simulation, the width and depth of the surrounding soil are taken as $15D$ and $7d$, respectively. The offset distance between the spudcan and the footprint center is denoted as S. The 8-node plane strain and pore pressure element, CPE8PR, is used to simulate the soil element to avoid self-locking phenomena and to increase the computational accuracy in numerical simulation. The active–passive surface contact algorithm is used to deal with the contact interaction and relative displacement

between the spudcan and the surrounding soil, and the spudcan surface is taken as the active surface and the soil surface as the passive surface (Zhuang et al. 2005). The principle for choosing an active or passive surface is that the mesh of the passive surface should be finer, and if both mesh densities are similar to each other, the surface of the softer material should be passive. The tangential contact obeys the Coulomb friction law, and the normal contact follows the hard touching mode, i.e., penetration is not allowed between the spudcan element and the soil element, but they are allowed to separate (Zhuang et al. 2005). In order to obtain the correct horizontal sliding force–displacement curve, the displacement control method is used to load. A simplified spudcan, with its side friction ignored because of its relative smaller area, is adopted to reduce the difficulty of convergence in calculation. The friction coefficients for undrained clay and drained granular soil are 0.2–0.3 and $\tan \delta$, respectively, where δ is the friction angle between the spudcan and the soil. It must be pointed out that whether setting a reasonable degree of spudcan–soil contact will lead to the calculation converging or not.

Since the ultimate bearing capacity would be underestimated if the initial geo-stress equilibrium were not considered in numerical simulation, this paper deals with the initial geo-stress equilibrium first and imports a stress file with an 'initial conditions' method. This is instead of the 'Geostatic' way, a commonly used geo-stress equilibrium analysis method in general simulation involving in soil that is difficult to deal with for such a complex problem as spudcan–soil interaction with an existing footprint. In addition, because of serious soil deformation under a large spudcan penetration depth, in order to avoid huge warping and ensure accuracy of calculation, ALE self-adaptive meshes are employed.

3 Spudcan–footprint interaction in clay

3.1 Failure process of clay foundations

Let $S = 0.75D$ ($D = 6$ m, $d = 6$ m). The mechanical characteristics of uniform soil such as clay are shown in Table 1.

The gradual failure process of clay foundation occurs in three stages: elastic balance, plastic expansion, and complete plastic damage (Fig. 3). Figure 3a shows that plastic damage first appears at the bottom edge of the footprint

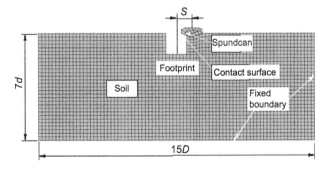

Fig. 2 Schematic diagram of the finite element model

Table 1 Material parameters of single-layer foundation

Effective density ρ, kg/m^3	Cohesion C, kPa	Internal friction angle ϕ, °
860	20	0

close to the spudcan. Figure 3b shows the expansion of the soil foundation plastic zone from the bottom edge of the footprint toward the farther edge of the spudcan with load increasing. Figure 3c indicates that when the complete plastic damage of clay foundation appears, the plastic zones have expanded to form a continuous sliding surface.

3.2 Clay foundation yield at different S

Changing only S while keeping other parameters constant, the situations of clay foundation yield at different S are

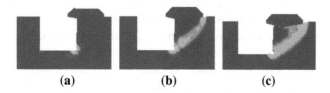

(a)   **(b)**   **(c)**

Fig. 3 Plastic zone of clay foundation in loading (*part around the footprint*)

shown in Fig. 4. This indicates that the plastic zone becomes larger with an increase in S and the failure pattern of soil around the spudcan gradually changes from asymmetric to symmetric.

3.3 Soil movement patterns at different S

When the spudcan arrives at the designed depth, the soil displacement vectors under different S are shown in Fig. 5, from which we see that there is an obvious uplift trend at the bottom of the footprint and the soil near the footprint clearly migrates toward the footprint. The bulge on the farther side surface of the clay foundation changes little with an increase in S. However, the apophysis on the footprint bottom increases significantly and the soil movement patterns on the closer side to the spudcan and below the spudcan change greatly. When S is small, part of the soil below the spudcan moves to the footprint, while another part migrates downward with the spudcan. With the S increasing, the soil under the spudcan bottom

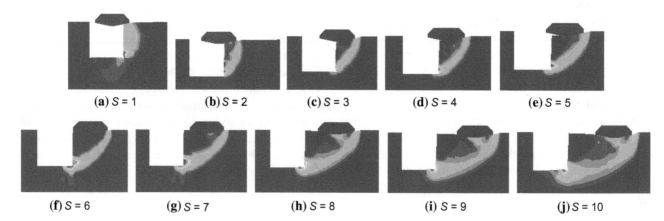

(a) $S = 1$ **(b)** $S = 2$ **(c)** $S = 3$ **(d)** $S = 4$ **(e)** $S = 5$

(f) $S = 6$ **(g)** $S = 7$ **(h)** $S = 8$ **(i)** $S = 9$ **(j)** $S = 10$

Fig. 4 The complete plastic damage zone at different S (*part around the footprint*)

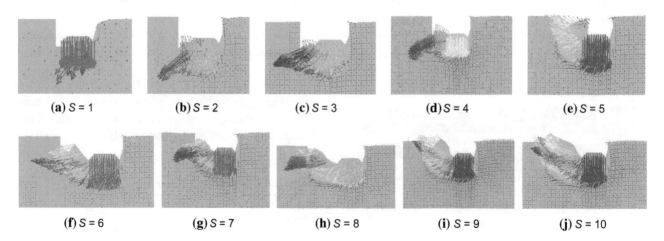

(a) $S = 1$ **(b)** $S = 2$ **(c)** $S = 3$ **(d)** $S = 4$ **(e)** $S = 5$

(f) $S = 6$ **(g)** $S = 7$ **(h)** $S = 8$ **(i)** $S = 9$ **(j)** $S = 10$

Fig. 5 The displacement vector of clay at different S (*part around the spudcan*)

basically migrates downward, while most of the soil on the closer side of the footprint moves into the footprint and only a little moves downward with the spudcan edge. This may provide a coping idea for jack-up re-installation close to footprint (which will be discussed in a separate paper).

3.4 Influence of S on horizontal slip force

The relation between the horizontal slipping force on the spudcan and the spudcan vertical displacement, i.e., depth at different S is displayed in Fig. 6. This shows that at any S, with the depth increasing, the horizontal force on the spudcan increases initially then decreases after it reaches a peak value. The peak values at different S appear at a depth from 2.5 to 4.5 m, and the maximum peak horizontal force is about 0.7 MN when $S = 4$ m. This indicates that the most potentially dangerous situation is when the spudcan partially overlaps the existing footprint. In order to investigate the overall relationship between the peak horizontal force on the spudcan and S, the peak horizontal forces are sorted at different S in dimensionless form (Table 2).

For the problem with a 'footprint,' the horizontal slip force on the spudcan varies with soil strength, footprint dimension, diameter of the spudcan, and the offset distance between the spudcan and the footprint center. Taking these factors into consideration, the expression of the peak horizontal force on the spudcan in dimensionless form can be summarized as

$$H_{max} = f' \left(\frac{D_s}{D_f}, \frac{S}{D_f}, \frac{d}{D_f} \right) \times s_u D_s^2, \qquad (3)$$

where H_{max} is the peak horizontal force on the spudcan, MN; S_u is the soil undrained shear strength; D_f is the diameter of the footprint, m; D_s is the diameter of the

Table 2 Peak horizontal forces at different 'S'

S, m	S/D	Peak horizontal force, MN
1	0.166	0.264
2	0.333	0.497
3	0.498	0.593
4	0.664	0.698
5	0.834	0.653
6	1.000	0.561
7	1.166	0.504
8	1.333	0.443
9	1.498	0.383
10	1.664	0.326

spudcan in future operations, m; S is the distance between the spudcan and the footprint center, m; and d is the depth of the footprint, m.

In this paper, only the influence of the offset distance on the peak horizontal slip force on the spudcan is considered, as given in Table 2. The horizontal force on the spudcan will be zero when $S = 0$ as the spudcan is located exactly in the footprint. Using Matlab to fit the numerical simulation results, the peak horizontal force on the spudcan is obtained as follows:

$$H_{max} = 4.1248 \cdot \left(\frac{S}{D_f} \right)^{1.3439} \times \exp\left(-1.9555 \frac{S}{D_f} \right), \qquad (4)$$

The fitting curve of Eq. (4) and the numerical simulation results are shown in Fig. 7. This demonstrates that the curvature tolerance of Eq. (4) is very small and it could reliably represent the relationship between the peak horizontal sliding force on the spudcan and the offset distance S. The peak horizontal force reaches a maximum value

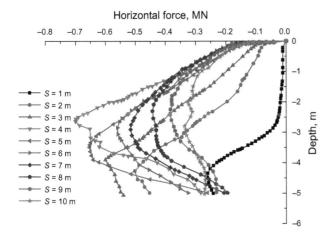

Fig. 6 The horizontal force–depth diagram at different S

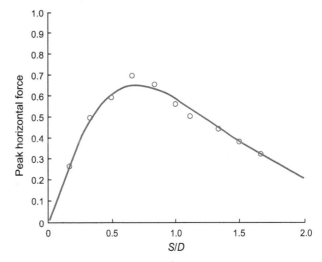

Fig. 7 The fitted curve between the peak horizontal force and S

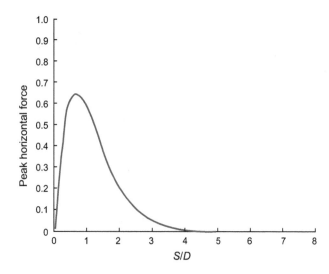

Fig. 8 The whole relation between the peak horizontal force and S

when $S/D = 0.6$. The horizontal force increases quickly before it reaches the maximum value and then gradually decreases. The rate of decrease is far less than the rate of increase. In order to observe the successive change of the peak horizontal force, the horizontal force is calculated at larger 'S according to Eq. (4), and the whole relation between the peak horizontal sliding force and the offset distance is given in Fig. 8. When $S/D \geq 5$, the peak

horizontal force becomes almost zero, which means in this case that the influence of the existing footprint could be ignored.

4 Verification of numerical simulation results

Based on the University of Western Australia centrifuge model test (Table 3; Gan 2009), we built 2-dimensional and 3-dimensional simulation models (Fig. 9) to conduct finite element simulation. Results at different S (0.25D, 0.50D, 0.75D, 1.0D) are shown in Figs. 10 and 11. Comparisons of results from the 2-dimensional or 3-dimensional simulation models and from the experiments indicate that the simulation results are in good agreement with experimental results, and the results from the 3-dimensional model are a little closer to the test results than those from the 2-dimensional model. However, with the 3-dimensional model, not only the computing time needed is much longer, but also the calculation is much more difficult to converge. Using the 2-dimensional model built in this paper would significantly reduce the necessary computing time, and the simulation results are in good agreement with experimental results, which shows that the 2-dimensional model built in this paper is feasible and reliable.

Table 3 List of major experimental parameters (after Gan 2009)

Test No.	Spudcan diameter			Initial penetration					Re-penetration		Remarks
	Initial penetration D_f, m	Re-penetration D_s, m	Size ratio D_f/D_s	Soil strength profile			Preload pressure q_0, kPa	Penetration depth d_0, m	Radial distance R_d, m	R_d/D_f	
				s_{um}, kPa	k, kPa/m	kD_f/s_{um}					
OA1	6	6	1	25	5	1.20	460	5.84	0.0	0.00	Tests done
OA2	6	6	1	28	5	1.07	460	5.61	1.5	0.25	in NUS
OA3	6	6	1	28	5	1.07	460	5.30	3.0	0.50	
OA4	6	6	1	28	5	1.07	460	5.19	4.5	0.75	
OA5	6	6	1	28	5	1.07	460	5.19	6.0	1.00	
OA6	6	6	1	30	5	1.00	460	4.70	9.0	1.50	

Test No.	Size ratio D_f/D_s	R_d/D_f	Depth ratio d_s/D_f	Re-penetration							
				Maximum horizontal load, H_{max}				Maximum moment, M_{max}			
				d/D_s	H_{max}, MN	θ, degree	$H/s_u D_s^2$	d/D_s	M_{max}, MN	e/D_s	$M/s_u D_s^3$
OA1	1	0.00	0.97	1.02	0.11	0.54	0.06	0.98	0.31	0.005	0.03
OA2	1	0.25	0.94	0.75	0.41	2.76	0.20	0.78	1.81	0.033	0.14
OA3	1	0.50	0.88	0.84	0.49	2.32	0.23	0.44	1.91	0.047	0.15
OA4	1	0.75	0.87	0.52	0.72	4.29	0.34	0.10	2.29	0.109	0.18
OA5	1	1.00	0.86	0.78	0.63	2.69	0.30	0.27	2.13	0.047	0.17
OA6	1	1.50	0.78	0.88	0.30	1.15	0.14	0.44	0.45	0.007	0.03

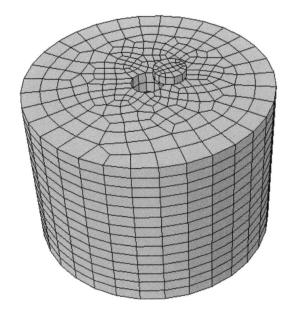

Fig. 9 3-dimensional finite element model

5 Conclusions

1. In the initial loading stage, plastic damage first appears at the bottom edge of the footprint close to the spudcan. Then the plastic zone expands with increasing load and finally it forms a continuous sliding surface.

2. With an increase in the distance between the spudcan and the footprint, the soil failure pattern gradually changes from asymmetric to symmetric.

3. The soil migration patterns on the closer side of the footprint and below the spudcan change greatly at different offset distances. With the distance increasing, the soil on the spudcan bottom basically migrates downward, while most of the soil on the closer side of the footprint moves into the footprint, and only a little moves downward with the spudcan edge. This means "stomping" (repeated raising and lowering of the jackup leg) may be a successful solution for the jack-up installation close to a footprint.

4. The peak horizontal sliding forces on spudcan at different offset distances modeled with Matlab to fit the numerical simulation results and the possible dangerous ranges during re-installation have been obtained. The peak horizontal force reaches its maximum value when $S/D = 0.6$. When $S/D \geq 5$, the horizontal sliding force becomes almost zero, which means in this case that the influence of the footprint could be ignored.

5. The numerical simulation results show good agreement with experimental results, indicating clearly that the finite element model built in this paper can be used to solve the problems of spudcan–footprint interaction

with sliding friction contact, fluid–solid coupling, nonlinear elastic–plastic deformation, and convergence problems.

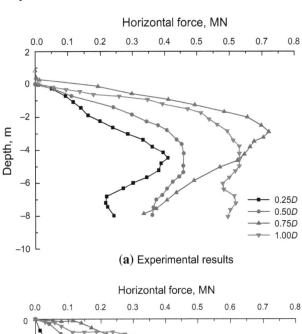

(a) Experimental results

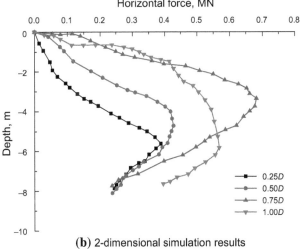

(b) 2-dimensional simulation results

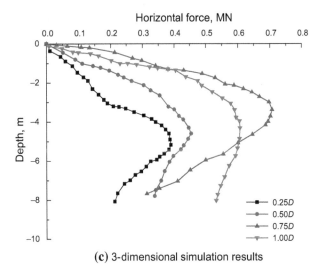

(c) 3-dimensional simulation results

Fig. 10 Simulation and experimental results

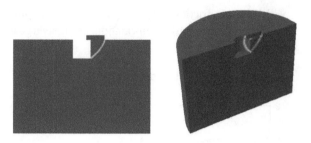

(a) The plastic zone in the loading of clay foundation

(b) The displacement vector of clay foundation

Fig. 11 2-dimensional and 3-dimensional numerical simulation results

Acknowledgments This work is financially supported by the National Natural Science Foundation of China (Grant No. 51379214) and the National Science and Technology Major Project (Grant No. 2011ZX05027-005-001).

References

Bouwmeester D, Peuchen J, Van der Wal T, et al. Prediction of breakout forces for deepwater seafloor objects. In: Offshore technology conference, OTC-19925-MS. Houston; 4–7 May 2009.

Cassidy MJ, Byrne BW, Randolph MF. A comparison of the combined load behaviour of spudcan and caisson foundations on soft normally consolidated clay. Géotechnique. 2004;54(2):91–106.

Cassidy MJ, Quah CK, Foo KS. Experimental investigation of the reinstallation of spudcan footings close to existing footprints. J Geotech Geoenviron Eng (ASCE). 2009;135(4):474–86.

Dean ER, Serra H. Concepts for mitigation of spudcan-footprint interaction in normally consolidated clay. In: Proceedings of the 14th international offshore and polar engineering conference. (ISOPE ISOPE-I-04-248), Toulon; 23–28 May 2004.

DeJong JT, Yafrate NJ, DeGroot DJ, Jakubowski J. Evaluation of the undrained shear strength profile in soft layered clay using full-flow probes. In: Proceedings of the 2nd international conference on site characterisation. Porto; 2004. pp. 679–86.

Deng CJ, Kong WX. Analysis of ultimate bearing capacity of foundations by elastoplastic FEM through step loading. Rock Soil Mech. 2005;26(3):500–4 (in Chinese).

Gan CT. Centrifuge model study on spudcan-footprint interaction doctoral dissertation. Perth: The University of Western Australia; 2009.

Gan CT, Leung CF, Cassidy MJ, Gaudin C, Chow YK. Effect of time on spudcan-footprint interaction in clay. Géotechnique. 2012;62(5):402–13.

Gaudin C, Cassidy MJ, Donovan T. Spudcan reinstallation near existing footprints. In: Proceedings of the 6th international offshore site investigation and geotechnics conference: confronting new challenges and sharing knowledge. London: Society of Underwater Technology; 2007. pp. 285–92.

Hanna AM, Meyerhof G. Design chart for ultimate bearing capacity of foundation on sand overlying soft clay. Can Geotech J. 1980;17(2):300–3.

Hossain MS, Hu Y, Randolph MF, White DJ. Limiting cavity depth for spudcan foundations penetrating clay. J Geotech. 2005;55(9):679–90.

Hossain MS, Randolph MF. Investigating potential for punch-through for spud foundations on layered clays. In: Proceedings of the 17th ISOPE. Lisbon; 1–6 July 2007. pp. 1510–17.

Hossain MS, Randolph MF. Overview of spudcan performance on clays: current research and SNAME. In: Proceedings of the 2nd jack-up Asia conference & exhibition. Singapore; 2008.

Houlsby GT, Martin CM. Undrained bearing capacity factors for conical footings on clay. Géotechnique. 2003;53(5):513–20.

Jardine RJ, Kovacevic N, Hoyle MJR, Sidhu HK, Letty A. Assessing the effects on jack-up structures of eccentric installation over infilled craters. In: Society for underwater technology (SUT), proceedings of international conference. Offshore site investigation and geotechnics—diversity and sustainability, London; 26–28 Nov 2002. pp. 307–24.

Kellezi L, Stromann H. FEM analysis of jack-up spud penetration for multi-layered critical soil conditions. Lyngby: GEO-Danish Geotechnical Institute; 2003. pp. 411–20.

Kong VW, Cassidy MJ and Gaudin C. (2010) Jack-up reinstallation near a footprint cavity. In: Proceedings of the 7th international conference on physical modelling in geotechnics, (ICPMG). Zurich; 28 Jun–1 Jul 2010. pp. 1033–38.

Kong VW, Cassidy MJ, Gan CT. Experimental study of the effect of geometry on the reinstallation of a jack-up next to a footprint. Can Geotech J. 2013;50(5):557–73.

Leung CF, Gan CT, Chow YK. Shear strength changes within jack-up spudcan footprint. In: Proceedings of the 17th international offshore and polar engineering conference, ISOPE-I-07-485. Lisbon; 1–6 Jul 2007.

Leung CF, Xie Y, Chow YK. (2008) Use of PIV to investigate spudcan-pile interaction. In: Proceedings of the international offshore and polar engineering conference. Vancouver; 2008. p. 721.

Li G. Advanced Soil Mech. Beijing: Tsinghua University Press Ltd; 2004.

McClelland B, Young A, Remmes BD. Avoiding jack-up rig foundation failure. Geotech Eng. 1982;13(2):151–88.

Murff JD, Himilton JM, Dean ETR, James RG, Kusakabe O, Schofield AN. Centrifuge testing of foundation behaviour using full jack-up rig models. In: Proceedings of 32nd offshore technology conference, OTC 6516. Houston; 1991.

SNAME recommended practice for site specific assessment of mobile jack-up units. Technical secretary to SNAME OC-7 panel. 2007.

Stewart DP, Finnie IMS. Spudcan-footprint interaction during jack-up workovers. In: Proceedings of the 11st of international offshore and polar engineering conference, (ISOPE-I-01-011). Stavanger; 17–22 Jun 2001.

Stewart DP. Influence of jack-up operation adjacent to a piled structure. Frontiers in offshore geotechnics: ISFOG. Perth; 2005. pp. 516–23.

Teh KL, Byrne BW, Houlsby GT. Effects of seabed irregularities on loads developed in legs of jack-up units. In: Jack up Asia conference and exhibition, Singapore; 2006.

Teh KL, Leung CF, Chow YK, Chow MJ. Centrifuge model study of spudcan penetration in sand overlying clay. Géotechnique. 2010;60(11):825–42.

Van den Berg B, Hulshof B, Tijssen T, van Oosterom P. Harmonisation of distributed geographic datasets: a model driven approach for geotechnical & footprint data. Delft: Delft University of Technology; 2004.

Xie Y, Leung CF, Chow YK. Centrifuge model study of spudcan-pile interaction. Géotechnique. 2012;62(9):799–810.

Zeng Z, Grigg R. A criterion for non-darcy flow in porous media. Transp Porous Media. 2006;63(1):57–69.

Zhang Y, Bienen B, Cassidy MJ, Gourvenec S. The undrained bearing capacity of a spudcan foundation under combined loading in soft clay. Mar Struct. 2011;24(4):459–77.

Zhang Y, Wang D, Cassidy MJ, Bienen B. Effect of installation on the bearing capacity of a spudcan under combined loading in soft clay. J Geotech Geoenviron Eng, ASCE. 2014. doi:10.1061 GT. 1943-5606.0001126.

Zhuang Z, Zhang F, Cen S. ABAQUS nonlinear finite element analysis and samples. Beijing: Science Press; 2005 (in Chinese).

12

Seismic fluid identification using a nonlinear elastic impedance inversion method based on a fast Markov chain Monte Carlo method

Guang-Zhi Zhang[1] · Xin-Peng Pan[1] · Zhen-Zhen Li[1] · Chang-Lu Sun[1] · Xing-Yao Yin[1]

Abstract Elastic impedance inversion with high efficiency and high stability has become one of the main directions of seismic pre-stack inversion. The nonlinear elastic impedance inversion method based on a fast Markov chain Monte Carlo (MCMC) method is proposed in this paper, combining conventional MCMC method based on global optimization with a preconditioned conjugate gradient (PCG) algorithm based on local optimization, so this method does not depend strongly on the initial model. It converges to the global optimum quickly and efficiently on the condition that efficiency and stability of inversion are both taken into consideration at the same time. The test data verify the feasibility and robustness of the method, and based on this method, we extract the effective pore-fluid bulk modulus, which is applied to reservoir fluid identification and detection, and consequently, a better result has been achieved.

Keywords Elastic impedance · Nonlinear inversion · Fast Markov chain Monte Carlo method · Preconditioned conjugate gradient algorithm · Effective pore-fluid bulk modulus

1 Introduction

Compared to amplitude versus offset (AVO) inversion based on common mid-point (CMP) gathers, elastic impedance inversion based on partial angle-stack gathers has the advantages of high computational efficiency, high stability, high noise immunity, and low dependence on the quality of seismic data. This has been widely used in reservoir fluid identification and detection, and has become one of the main directions of pre-stack inversion (Downton 2005; Yin et al. 2014).

Connolly (1999) proposed the concept of elastic impedance on the basis of acoustic impedance for the first time. Cambois (2000) considered that the high noise immunity of elastic impedance could avoid "leakage" between the various AVO attributes generated by noise. This is more advantageous in the extraction of pre-stack parameters. Whitcombe (2002) first applied a new normalized form of elastic impedance to improve the stability of parameter extraction. Additionally, Martins (2006) and Cui et al. (2010) introduced P- and P-SV wave elastic impedance in weakly anisotropic media, respectively. In recent years, fluid indicators estimated from seismic data play important roles in reservoir characterization and prospect identification, so many methods, such as elastic impedance inversion, have been introduced to extract a variety of fluid factors directly to avoid the cumulative error generated by indirect combination of parameters in the process of reservoir fluid identification (Ma 2003; Peng et al. 2008; Zong et al. 2011, 2012; Yin et al. 2013b; Zhang et al. 2013; Chen et al. 2014a, b; Li et al. 2014). Goodway et al. (1997) proposed $\lambda\rho$ and $\mu\rho$ as fluid indicators. Russell et al. (2011) used f as a fluid indicator based on the poroelastic theory. However, the sensitivity of these fluid indicators is dependent on the mixed effect of pore fluid and rock matrix. To improve the sensitivity of reservoir fluid identification, we use the effective pore-fluid bulk modulus as the fluid indicator, which is related only to pore fluid and may diminish the rock-matrix effect (Han and Batzle 2004; Yin and Zhang 2014).

✉ Xin-Peng Pan
  panxinpeng1990@gmail.com

[1] School of Geosciences, China University of Petroleum (East China), Qingdao 266580, Shandong, China

Edited by Jie Hao

The present elastic impedance inversion techniques are mostly based on linear or quasi-linear inversion methods, not only losing the precision of inversion in the process of linearization but also strongly relying on the accuracy of initial model, while convergence to the global optimum is difficult in the usage of these techniques (Su et al. 2014). However, the nature of most inverse problems is nonlinear and multi-extremum, so in terms of the complex pre-stack reservoir elastic parameter inversion, the development of elastic impedance with high efficiency and high stability based on the nonlinear inversion method has become more significant. We propose a nonlinear elastic impedance inversion method based on a fast Markov chain Monte Carlo (MCMC) method, and validate the feasibility and robustness of the method by testing the noise immunity of well data. Meanwhile, on the basis of two-phase medium theory for elastic impedance equation, we extract the effective pore-fluid bulk modulus from seismic data to apply to reservoir fluid identification and detection (Russell et al. 2003; Yin et al. 2013a).

2 Fast MCMC method

Hastings (1970) proposed an extended form of the Metropolis algorithm (Metropolis et al. 1953), the Metropolis–Hastings algorithm, laying the foundation for the development of the MCMC method. The MCMC method was first applied to fully nonlinear inverse problems by Malinverno (2002). Zhang et al. (2011a, b) studied post-stack and pre-stack seismic inversion methods based on the MCMC method. In recent years, the MCMC method has been applied to sample the posterior distribution of reservoir parameters for identifying reservoir lithology and fluid on the basis of a Bayesian framework (Bachrach et al. 2009; Grana and Rossa 2010; Rimstad and Omre 2010; Ulvmoen and Omre 2010; Ulvmoen et al. 2010; Rimstad et al. 2012). However, the conventional MCMC method has the defects of low computational efficiency and a low convergence rate toward the global optimum for multi-extremum inverse problems, so we propose a faster MCMC method than the conventional MCMC method in this paper.

2.1 Metropolis–Hastings algorithm

MCMC can sample posterior probability distribution converging to inverted parameters in Bayesian inference,

and then make some statistical analysis of random samples to obtain useful properties of parameter posterior distribution.

The construction methods of the transition kernel play important roles in the MCMC method, including the Metropolis–Hastings (M–H) algorithm and Gibbs algorithm. We choose the M–H algorithm, and after a sufficiently long iteration, stable Markov chains form, which can be used in making the statistical analysis for random samples of parameter posterior distribution, and which satisfy the detailed balance condition.

2.2 Principle of fast MCMC method

Conventional MCMC can be applied efficiently to nonlinear and single-extremum inverse problems, but it is difficult to converge to global optimums of multi-extremum inverse problems. Thus, we propose a fast MCMC method, which integrates an efficient optimization algorithm into the MCMC method to improve the convergence rate to the globally optimal solution, and greatly raises the computational efficiency of the MCMC method.

The PCG algorithm is considered an efficient and stable optimization algorithm (Stefano et al. 2013), so we combine the MCMC method based on the global optimal M–H algorithm with the PCG algorithm equipped with local search ability. Overall, the central idea of the fast MCMC method proposed in this paper is the integration of the more efficient PCG algorithm in the process of random search in whole space based on the global optimal MCMC method. The method can search the solution of inverted parameters for fast objective function convergence, which saves the huge computational cost of large numbers of iterations compared to the conventional MCMC method. We call the more efficient algorithm the "fast MCMC method", and its specific process is shown in Fig. 1.

To test the feasibility of the fast MCMC method, we designed a simple bimodal probability density function (PDF) $p(x) = 0.3 \times e^{\alpha x^2} + 0.7 \times e^{\alpha(x-10)^2}$, and the estimated parameter is α, whose true value is -0.2. Figure 2 shows two inverted results of 250 iterations from the conventional MCMC method (blue solid line) and the fast MCMC method (red solid line), respectively. From Fig. 2, we find that the fast MCMC method converges to a global optimal solution quickly, but conventional MCMC method has not reached convergence condition after 250

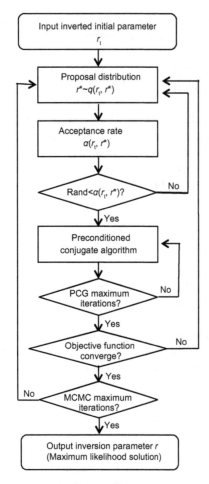

Fig. 1 Process of the fast MCMC method

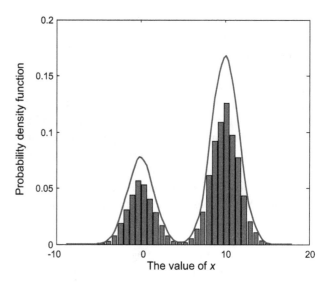

Fig. 3 Statistical characteristics of bimodal PDF sampled by the fast MCMC method

The maximum likelihood solution of unknown parameter α is regarded as the estimated value, and then, we sample the bimodal PDF $p(x)$. As shown in Fig. 3, the statistical characteristics of random samples sampled by the fast MCMC method are quite consistent with the characteristics of bimodal PDF, further verifying the feasibility and reliability of the fast MCMC method and laying the foundation for the fast algorithm of the nonlinear elastic impedance inversion method.

3 Nonlinear elastic impedance inversion method based on the fast MCMC method

The elastic inversion method used mostly now is the constrained sparse spike linear inversion method developed in the 1980s, so we propose a nonlinear elastic impedance inversion method based on the fast MCMC method to improve the accuracy of elastic impedance inversion and the reliability of reservoir prediction and fluid identification.

3.1 Elastic impedance equation based on two-phase medium theory

To improve the reliability of reservoir pore-fluid discrimination, we choose the normalized elastic impedance equation on the basis of two-phase medium theory, which highlights the effective pore-fluid bulk modulus K_f (Yin et al. 2013a; Yin and Zhang 2014):

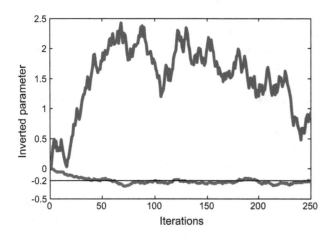

Fig. 2 Comparison of the fast MCMC result (*red*) and the conventional MCMC result (*blue*) within 250 iterations

iterations, verifying that the fast MCMC method is more efficient than conventional MCMC method to converge to a global optimum.

$$EI(\theta) = EI_0 \left(\frac{K_f}{K_{f0}}\right)^{a(\theta)} \left(\frac{f_m}{f_{m0}}\right)^{b(\theta)} \left(\frac{\rho}{\rho_0}\right)^{c(\theta)} \left(\frac{\phi}{\phi_0}\right)^{d(\theta)}, \quad (1)$$

where

$$a(\theta) = \left(1 - \frac{\gamma_{dry}^2}{\gamma_{sat}^2}\right)\frac{\sec^2\theta}{2} \quad (2a)$$

$$b(\theta) = \frac{\gamma_{dry}^2}{2\gamma_{sat}^2}\sec^2\theta - \frac{4}{\gamma_{sat}^2}\sin^2\theta \quad (2b)$$

$$c(\theta) = 1 - \frac{\sec^2\theta}{2} \quad (2c)$$

$$d(\theta) = \frac{\sec^2\theta}{2} - \frac{\gamma_{dry}^2}{\gamma_{sat}^2}\sec^2\theta + \frac{4}{\gamma_{sat}^2}\sin^2\theta, \quad (2d)$$

and

$$EI_0 = K_f^{-1/2}f_m\phi^{-3/2}\rho^{1/2}\gamma_{sat}\left(\gamma_{sat}^2 - \gamma_{dry}^2\right)^{1/2}, \quad (2e)$$

where K_f is the effective pore-fluid bulk modulus term, $f_m = \phi\mu$ is the dry rock matrix term, ρ is density, and ϕ is porosity value; K_{f0}, f_{m0}, ρ_0, and ϕ_0 are the average effective pore-fluid bulk modulus, dry rock matrix term, density, and porosity value of well data; EI_0 is the normalization coefficient; θ is the average of the incident and refracted angles; γ_{dry}^2 and γ_{sat}^2 are the square of the P- to S-wave velocities of the dry rock and saturated rock, respectively.

3.2 Nonlinear elastic impedance inversion method based on the fast MCMC method

Based on Bayes' theorem, the posterior probability density of inverted parameter r is expressed as

$$p(r|d) \propto p(r) \cdot p(d|r), \quad (3)$$

where d is the seismic observation data, r is the reflection coefficient sequence, $p(r|d)$ is the posterior probability of the reflection coefficient, $p(r)$ is the prior information, and $p(d|r)$ is a likelihood function.

Assuming seismic background noise obeys an independent Gaussian distribution with a zero mean and σ_n^2 variance, the seismic observation likelihood function can be expressed as

$$p(d|I) = \frac{1}{(2\pi\sigma_n^2)^{\frac{N}{2}}}\exp\left(-\sum\frac{(d - Gr)^T(d - Gr)}{2\sigma_n^2}\right). \quad (4)$$

Meanwhile, based on the assumption that the prior information obeys the Cauchy distribution to stand out weak reflection of the underground media, the prior information is expressed as

$$p(r) = \frac{1}{(\pi\sigma_r)^M}\prod_{i=1}^{M}\left(\frac{1}{1 + r^2/\sigma_r^2}\right). \quad (5)$$

To obtain the posterior probability distribution $p(r|d)$, we apply the Metropolis–Hastings algorithm to generate stable Markov chains converging to inverted parameter r.

However, the conventional MCMC method results in huge computational cost and there is likely to be instability in the inverted results. Therefore, having taken the computational efficiency and inversion stability into account, we apply the fast MCMC method (shown as Fig. 1) to generate stable Markov chains converging to the posterior probability density distribution of inverted parameter and invert the elastic impedance using a nonlinear method based on the fast MCMC method to identify fluid.

In the Metropolis–Hastings algorithm, we choose a symmetrical distribution satisfying a symmetric random walk as the proposal distribution of the algorithm, so the proposal distribution and acceptance rate are expressed as

$$q(r_t, r*) \sim U(r_t - delta, r_t + delta) \quad (6a)$$

and

$$\alpha(r_t, r*) = \min\left\{1, \frac{\pi(r*)q(r*, r_t)}{\pi(r_t)q(r_t, r*)}\right\} = \min\left\{1, \frac{\pi(r*)}{\pi(r_t)}\right\}. \quad (6b)$$

In the PCG algorithm, on the Bayesian framework, we construct the objective function expressed as

$$J(r) = (d - Gr)^T(d - Gr) + \mu\sum_{i=1}^{M}\ln\left(1 + r^2/\sigma_r^2\right)$$
$$+ \alpha(\eta - Cr)^T(\eta - Cr), \quad (7)$$

where $\mu = 2\frac{\sigma_n^2}{\sigma_r^2}$ is the sparse constraint factor, and the larger its value is, the more sparse the reflection coefficient is; the last term in the equation is the elastic impedance constraint term, C and η are integral operator matrix and relative impedance value, respectively, and α is the elastic impedance constraint factor, and the larger the value is, the more stable and accurate the inverted results are. Optimizing the objective function, we get the ultimate equation expressed as

$$\left(G^{\mathrm{T}}G + \mu Q + \alpha C^{\mathrm{T}}C\right)r = \left(G^{\mathrm{T}}d + \alpha C^{\mathrm{T}}\eta\right), \tag{8}$$

where $Q = \mathrm{diag}\left[\cdots, \frac{1}{\left(1+r_i^2/\sigma_r^2\right)^2}, \cdots\right]$, and it adds the denominator squared term to reduce the effect of strength contrast and to highlight weak reflection of the underground media, and it can be termed the modified Cauchy constraint (Alemie and Sacchi 2011). Eventually, we apply the fast MCMC method to invert the reflection coefficient of different angles based on the convergence judgment of Eq. (7), and then, we obtain the elastic impedance of different angles by using the path integral method or the recursion method.

3.3 Direct extraction of the effective pore-fluid bulk modulus parameter

Expecting to apply elastic impedance volume of at least four angles to extract the effective pore-fluid bulk modulus directly following Eq. (1), we must turn the equation into a log-domain equation at first, and then obtain 16 fitted regression coefficients $(a(\theta_i), b(\theta_i), c(\theta_i), d(\theta_i), i = 1, 2, 3, 4)$ via the least square method or the singular value decomposition method by using the elastic impedance data and well log data from nearby wells, and finally the inverted elastic impedance data volume of four angles is put into the equation expressed as

3.4 Process of fluid identification using the nonlinear elastic impedance inversion method based on the fast MCMC method

Synthesizing the research and analysis above, we conclude that the whole process of fluid identification using the nonlinear elastic impedance inversion method based on the fast MCMC method proposed in this paper is as follows:

1. Pretreatment of pre-stack seismic data and well log data;
2. Accurate extraction of four angle wavelets from pre-stack seismic data;
3. Inversion of elastic impedance of four angles based on the fast MCMC method by using four abstracted angle-stack seismic gathers and four angles extracted from seismic data from nearby wells;
4. Extraction of effective pore-fluid bulk modulus from inverted elastic impedance of four angles on the basis of two-phase medium theory for the elastic impedance equation as the representation of reservoir pore-fluid information;
5. Application of extracted effective pore-fluid modulus to identify fluid and predict reservoir parameters.

4 Model test

To test the feasibility of fluid identification using elastic impedance based on the fast MCMC method, we carry

$$\begin{cases} \ln\dfrac{EI(t,\theta_1)}{EI_0} = a(\theta_1)\ln\dfrac{K_f(t)}{K_{f0}} + b(\theta_1)\ln\dfrac{f_m(t)}{f_{m0}} + c(\theta_1)\ln\dfrac{\rho(t)}{\rho_0} + d(\theta_1)\ln\dfrac{\phi(t)}{\phi_0} \\[2mm] \ln\dfrac{EI(t,\theta_2)}{EI_0} = a(\theta_2)\ln\dfrac{K_f(t)}{K_{f0}} + b(\theta_2)\ln\dfrac{f_m(t)}{f_{m0}} + c(\theta_2)\ln\dfrac{\rho(t)}{\rho_0} + d(\theta_2)\ln\dfrac{\phi(t)}{\phi_0} \\[2mm] \ln\dfrac{EI(t,\theta_3)}{EI_0} = a(\theta_3)\ln\dfrac{K_f(t)}{K_{f0}} + b(\theta_3)\ln\dfrac{f_m(t)}{f_{m0}} + c(\theta_3)\ln\dfrac{\rho(t)}{\rho_0} + d(\theta_3)\ln\dfrac{\phi(t)}{\phi_0} \\[2mm] \ln\dfrac{EI(t,\theta_4)}{EI_0} = a(\theta_4)\ln\dfrac{K_f(t)}{K_{f0}} + b(\theta_4)\ln\dfrac{f_m(t)}{f_{m0}} + c(\theta_4)\ln\dfrac{\rho(t)}{\rho_0} + d(\theta_4)\ln\dfrac{\phi(t)}{\phi_0} \end{cases} \tag{9}$$

So we can obtain the effective pore-fluid bulk modulus at any sampling point.

out the feasibility and noise immunity test of a well in one work area in the eastern China. The target reservoir

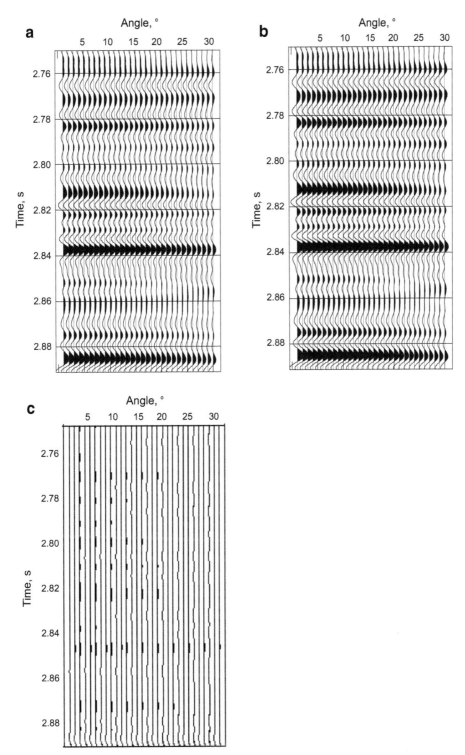

Fig. 4 Comparison of synthetic angle gathers using well log data and gathers by using inverted results in a noise-free situation. **a** synthetic angle gathers by using well data, **b** synthetic angle gathers by using inverted results in noise-free situation, **c** residual error between two synthetic angle gathers

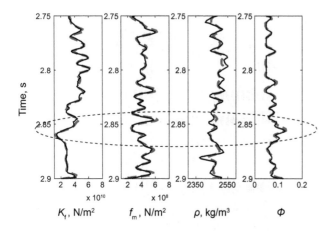

K_f, N/m² f_m, N/m² ρ, kg/m³ Φ

Fig. 5 Inverted parameter results from using the nonlinear elastic impedance inversion method based on the efficient MCMC in a noise-free situation (*black* and *red solid lines* indicate true values and inverted results, respectively; *black dashed ellipse* shows the oil-bearing reservoir from logging interpretation)

is a clastic reservoir at about 2.85 s in the seismic profile.

As shown in Figs. 4a and 6a, we apply well log data to make a synthetic seismogram in noise and noise-free situations, and then generate high-accuracy and high-resolution elastic impedance of four angles based on the fast MCMC method for the direct extraction of effective pore-fluid bulk modulus parameters, testing the feasibility and noise immunity of the method.

From Fig. 5, we find that inverted effective pore-fluid bulk modulus, dry rock matrix term, density, and porosity values in the noise-free situation are consistent with the true values, and synthetic angle gathers using these inverted results produce fewer errors compared with real angle gathers, so the inverted parameters can obviously reflect the major characteristics of the oil-bearing reservoir, agreeing with the results of oil-bearing reservoir from logging interpretation.

From Fig. 7, we find that when adding random noise in SNR = 3 to the synthetic seismograms, inverted effective pore-fluid modulus, dry rock matrix term, density, and porosity values are also consistent with the true values,

reflecting the major characteristics of the oil-bearing reservoir well. Therefore, it validates the robustness and noise immunity of the nonlinear elastic impedance inversion method based on the fast MCMC in fluid identification, and the inverted parameters are relatively accurate, which can reflect the major characteristics of the oil-bearing reservoir well and be applied to identify fluid and predict reservoirs.

5 Application of real seismic data

The real work area is selected from an exploration area in eastern China, and the target is a clastic reservoir. As shown in Fig. 8, the logging interpretation results indicate that the clastic reservoir at 2.85 s shows an oil layer with thickness up to 13 m. To verify the application effect of the method of fluid identification using the nonlinear elastic impedance inversion based on the fast MCMC method, we apply this method to real seismic data.

First of all, we invert elastic impedance based on the fast MCMC method using angle-stack seismic profiles with four different angles, and the inverted elastic impedance profiles of four angles are shown in Fig. 9, in which the well logs are elastic impedance curves of four different angles. Then, we extract the effective pore-fluid bulk modulus based on inverted elastic impedance data volume of four angles directly, and apply them to fluid identification and reservoir prediction. With the two effective pore-fluid bulk modulus logging curves in Fig. 10, the figure shows the extracted effective pore-fluid bulk modulus profile and enlarged partial profile.

From Fig. 9, we find that the inverted elastic impedance profiles based on the fast MCMC method with four different angles are highly precise, which agree with the results from the oil-bearing reservoir logging interpretation.

Figure 10 shows that the extracted effective pore-fluid bulk modulus is also consistent with the logging interpretation results, presenting low values and reflecting the characteristics of the oil-bearing reservoir, so it further

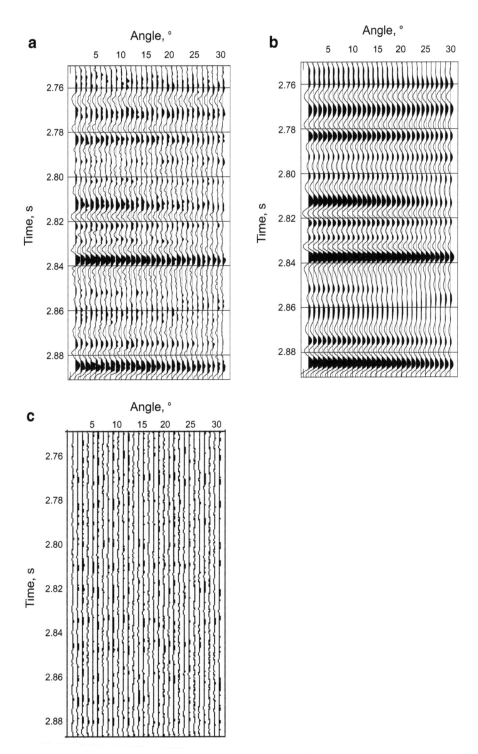

Fig. 6 Comparison of synthetic angle gathers using well data and gathers by using inverted results in a noise situation (SNR = 3). **a** synthetic angle gathers using well data, **b** synthetic angle gathers by using inverted results in the noise situation, **c** residual error between two synthetic angle gathers

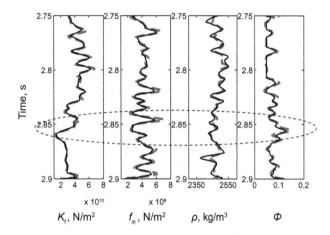

Fig. 7 Inverted parameter results from the nonlinear elastic impedance inversion method based on the efficient MCMC in the noise situation (SNR = 3) (*black* and *red solid lines* indicate true values and inverted results respectively; the *black dashed ellipse* shows the oil-bearing reservoir from the logging interpretation)

validates the reservoir fluid detection and identification from seismic data by using the nonlinear elastic impedance inversion method based on the fast MCMC method, which is promising in practical applications.

6 Conclusions

We introduce a more reliable fluid identification method using nonlinear elastic impedance inversion based on the fast MCMC in this paper, and the fast MCMC method is the key of foundation for the nonlinear elastic impedance inversion algorithm. Based on the Bayesian framework, we can identify the reservoir hydrocarbon more accurately owing to the more sensitive reservoir fluid indicator, that is the effective pore-fluid bulk modulus, and this method may improve the reliability and stability of fluid identification. Application research based on actual logging and

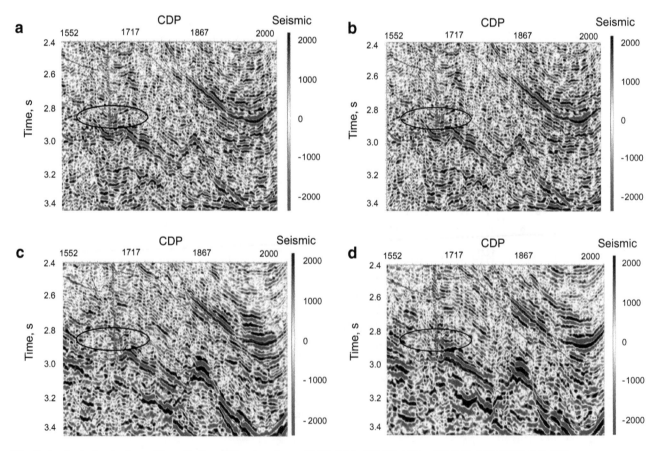

Fig. 8 Angle-stack seismic profiles with four different angles. **a** 6° (3°–9°) **b** 12° (9°–15°) **c** 18° (15°–21°) **d** 24° (21°–27°)

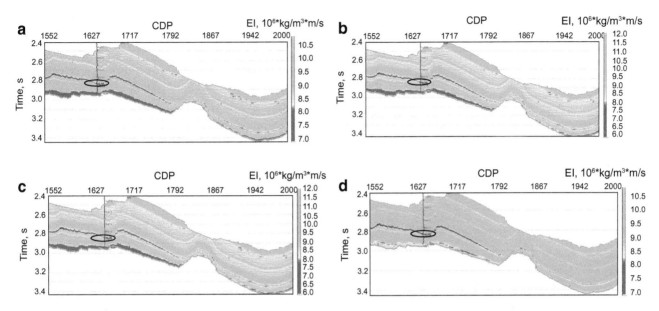

Fig. 9 Inverted elastic impedance profiles based on the fast MCMC method with four different angles. **a** 6° elastic impedance profile, **b** 12° elastic impedance profile, **c** 18° elastic impedance profile, **d** 24° elastic impedance profile

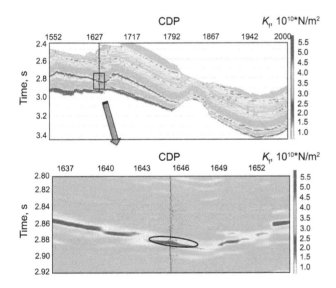

Fig. 10 Extracted effective pore-fluid bulk modulus profile and enlarged partial profile

seismic data shows that the nonlinear elastic impedance fluid identification method based on the fast MCMC is a practical method for reservoir fluid identification. However, the method has a limitation of obtaining only one inverted result of the effective pore-fluid bulk modulus, and may lack the uncertainty evaluation about reservoir fluid identification. So further in-depth research and analysis is needed, and we are intending to continue our research in this field.

Acknowledgments We would like to express our gratitude to the sponsorship of the National Basic Research Program of China (973 Program, 2013CB228604, 2014CB239201) and the National Oil and Gas Major Projects of China (2011ZX05014-001-010HZ, 2011ZX05014-001-006-XY570) for their funding of this research. We also thank the anonymous reviewers for their constructive suggestions.

References

Alemie W, Sacchi MD. High-resolution three-term AVO inversion by means of a Trivariate Cauchy probability distribution. Geophysics. 2011;76(3):R43–55.

Bachrach R, Sengupta M, Salama A, et al. Reconstruction of the layer anisotropic elastic parameter and high resolution fracture characterization from P-wave data: a case study using seismic inversion and Bayesian rock physics parameter estimation. Geophys Prospect. 2009;57(2):253–62.

Cambois G. AVO inversion and elastic impedance. 70th annual international meeting, SEG, Expanded Abstract. 2000. pp. 142–5.

Chen HZ, Yin XY, Gao CG, et al. AVAZ inversion for fluid factor based on fracture anisotropic rock physics theory. Chin J Geophys. 2014a;57(3):968–78 (**in Chinese**).

Chen HZ, Yin XY, Zhang JQ, et al. Seismic inversion for fracture rock physics parameters using azimuthally anisotropic elastic impedance. Chin J. Geophys. 2014b;57(10):3431–41 (**in Chinese**).

Connolly P. Elastic impedance. Lead Edge. 1999;18(4):438–52.

Cui J, Han LG, Liu QK, et al. P-SV wave elastic impedance and fluid identification factor in weakly anisotropic media. Appl Geophys. 2010;7(2):135–42.

Downton JE. Seismic parameter estimation from AVO inversion. Ph.D. Thesis, University of Calgary. 2005.

Goodway WN, Chen T, Downton J. Improved AVO fluid detection and lithology discrimination using Lamé petrophysical parameters; "λ", "μ", and "λ/μ fluid stack", from P and S inversions. 67th annual international meeting, SEG, Expanded Abstracts. 1997. pp. 183–6.

Grana D, Rossa ED. Probabilistic petrophysical-properties estimation integrating statistical rock physics with seismic inversion. Geophysics. 2010;75(3):O21–37.

Han DH, Batzle ML. Gassmann's equation and fluid-saturation effects on seismic velocities. Geophysics. 2004;69(2):398–405.

Hastings WK. Monte Carlo sampling methods using Markov chains and their applications. Biometrika. 1970;57(1):97–109.

Li C, Yin XY, Zhang GZ, et al. Two-term elastic impedance inversion based on the incident-angle approximation. Chin J Geophys. 2014;57(10):3442–52 (in Chinese).

Ma JF. Forward modeling and inversion method of generalized elastic impedance in seismic exploration. Chin J Geophys. 2003;46(1):118–24 (in Chinese).

Malinverno A. Parsimonious Bayesian Markov chain Monte Carlo inversion in a nonlinear geophysical problem. Geophys J Int. 2002;151(3):675–88.

Martins JL. Elastic impedance in weakly anisotropic media. Geophysics. 2006;71(3):D73–83.

Metropolis N, Rosenbluth AW, Rosenbluth MN, et al. Equations of state calculations by fast computing machines. J Chem Phys. 1953;21(6):1087–92.

Peng ZM, Li YL, Wu SH, et al. Discriminating gas and water using multi-angle extended elastic impedance inversion in carbonate reservoirs. Chin J Geophys. 2008;51(3):881–5 (in Chinese).

Rimstad K, Omre H. Impact of rock-physics depth trends and Markonv random fields on hierarchical Bayesian lithology/fluid prediction. Geophysics. 2010;75(4):93–108.

Rimstad K, Avseth P, Omre H. Hierarchical Bayesian lithology/prediction: a North Sea case study. Geophysics. 2012;77(2):B39–85.

Russell BH, Hedlin K, Hilterman FJ, et al. Fluid-property discrimination with AVO: a Biot–Gassmann perspective. Geophysics. 2003;68(1):29–39.

Russell B, Gray D, Hampson D. Linearized AVO and poroelasticity. Geophysics. 2011;76(3):C19–29.

Stefano MD, Andreasi FG, Secchi A. Towards preconditioned nonlinear conjugate gradient for generic geophysical inversions. 83th annual international meeting, SEG, Expanded Abstract. 2013. pp. 3226–30.

Su JL, Mi H, Wang YC, et al. Non-linear elastic impedance inversion method supported by vector machines. Oil Geophys Prospect. 2014;39(4):751–8 (in Chinese).

Ulvmoen M, Omre H. Improved resolution in Bayesian lithology/fluid inversion from pre-stack seismic data and well observations: part 1—Methodology. Geophysics. 2010;75(2):R21–35.

Ulvmoen M, Omre H, Buland A. Improved resolution in Bayesian lithology/fluid inversion from pre-stack seismic data and well observations: part 2—real case study. Geophysics. 2010;75(2):B73–82.

Whitcombe DN. Elastic impedance normalization. Geophysics. 2002;67(1):60–2.

Yin XY, Li C, Zhang SX. Seismic fluid discrimination based on two-phase media theory. J China Univ Petrol. 2013a;37(5):38–43 (in Chinese).

Yin XY, Zhang SX, Zhang FC, et al. Two-term elastic impedance inversion and Russell fluid factor direct estimation method for deep reservoir fluid identification. Chin J Geophys. 2013b;56(7):2378–90 (in Chinese).

Yin XY, Cao DP, Wang BL, et al. Research progress of fluid discrimination with pre-stack seismic inversion. Oil Geophys Prospect. 2014. 49(1):22–34, 46 (in Chinese).

Yin XY, Zhang SX. Bayesian inversion for effective pore-fluid bulk modulus based on fluid-matrix decoupled amplitude variation with offset approximation. Geophysics. 2014;79(5):R221–32.

Zhang GZ, Chen HZ, Wang Q, et al. Estimation of S-wave velocity and anisotropic parameter using fractured carbonate rock physics model. Chin J Geophys. 2013;56(5):1707–15 (in Chinese).

Zhang GZ, Wang DY, Yin XY, et al. Study on pre-stack seismic inversion using Markov chain Monte Carlo. Chin J Geophys. 2011a;54(11):2926–32 (in Chinese).

Zhang GZ, Wang DY, Yin XY. Seismic parameter estimated using Markov chain Monte Carlo method. Oil Geophys Prospect. 2011b;46(4):605–9 (in Chinese).

Zong ZY, Yin XY, Wu GC. Fluid identification method based on compressional and shear modulus direct inversion. Chin J Geophys. 2012;55(1):284–92 (in Chinese).

Zong ZY, Yin XY, Zhang FC. Elastic impedance Bayesian inversion for Lamé parameters extraction. Oil Geophys Prospect. 2011; 46(4):598–604, 609 (in Chinese).

13

Investment in deepwater oil and gas exploration projects: a multi-factor analysis with a real options model

Xin-Hua Qiu[1] · Zhen Wang[2] · Qing Xue[1]

Abstract Deepwater oil and gas projects embody high risks from geology and engineering aspects, which exert substantial influence on project valuation. But the uncertainties may be converted to additional value to the projects in the case of flexible management. Given the flexibility of project management, this paper extends the classical real options model to a multi-factor model which contains oil price, geology, and engineering uncertainties. It then gives an application example of the new model to evaluate deepwater oil and gas projects with a numerical analytical method. Compared with other methods and models, this multi-factor real options model contains more project information. It reflects the potential value deriving not only from oil price variation but also from geology and engineering uncertainties, which provides more accurate and reliable valuation information for decision makers.

Keywords Investment decision · Real options · Multi-factor model · Option pricing · Deepwater oil and gas

1 Introduction

Deepwater petroleum investment has attracted much attention as offshore oil and gas resources are making up a large portion of worldwide energy potentials. However,

due to the marine geographical environment, deepwater oil, and gas development projects contain higher geology and engineering risk than onshore or continental shelf projects. This situation increases the total amount of investment and the complexity of decision-making process. On one hand, the volatility of oil price causes more flexibility value for the deepwater projects which demand a longer duration for exploration. On the other hand, the technical risk of deepwater projects under development is much higher than that of onshore or continental shelf ones, and the effects of engineering and technological uncertainties on the value of deepwater projects are more significant. Under this background, the traditional theory of net present value cannot provide sufficient reliable reference for the decision making of deepwater oil and gas investment, because the value of flexibility from the uncertainties in oil price, engineering, and technology cannot be measured under the rigid assumptions. Therefore, the real options method based on uncertainty analysis is more suitable to evaluate deepwater oil and gas projects than the conventional ones.

The real options theory, originating from the financial option, regards the value from management flexibility as an option which could generate revenue. Myers (1977) analyzed the value of real options of additional investment opportunities for the first time. Since then, the flexibility value of real investment has received ongoing attention. Dixit and Pindyck (1994) summarized the research achievements of the real options theory and presented a systematic exposition of its construction and application. Firstly, they described the statistical characteristics of the uncertainty factor which influences the cash flow significantly. Secondly, they determined the functional relationship between uncertainty factors and the revenue and established the equation by non-arbitrage portfolio.

✉ Zhen Wang
wangzhen@cup.edu.cn

[1] School of Business Administration, China University of Petroleum, Beijing 102249, China

[2] Academy of Chinese Energy Strategy, China University of Petroleum, Beijing 102249, China

Edited by Xiu-Qin Zhu

Thirdly, they derive the equation for the real options model based on the assumptions and boundary conditions.

Since Brennan and Schwartz (1985) and Paddock et al. (1988) evaluated natural resources investment by adopting the real options method, research using the real options method has gradually increased. Dias (2004) presented an overview of real options models to evaluate investments in petroleum exploration and production projects. He pointed out that oil price was the only random variable in almost all the real option models and empirical studies, and other technical factors are assumed to be constant which could be obtained from engineers before the evaluation. In his review, a petroleum project was considered as a long-term investment and production process, during which the fluctuation of oil price could influence its economic value significantly. It could increase the flexibility value by adjusting production according to the oil price fluctuation. Due to the importance and financial attribute of oil price, simulation for the stochastic characteristics of oil price was a focus in real option research.

However, factors affecting the flexibility value of oil and gas projects are not limited to oil price, especially in the case of deepwater oil and gas exploration and development projects. The flexibility of geological understanding and engineering technology also has an important influence on the projects. The geological and technological information will be more accurate with increasing investment. The investors could make better decisions with the additional knowledge to realize flexibility value which may be ignored by the net present value method. In the theory of real options, the additional information and flexible management are valuable. If the flexible value of an underlying project is larger than its required investment, the project will be profitable. In order to evaluate the comprehensive flexibility value, a multi-factor real options model should be established with geological and technological factors in addition to the price factor.

Attempts to set up a multi-factor real options model have been made recently. Cortazar et al. (2001) added the information of geology and technology to the model of Brennan and Schwartz (1985) to evaluate a copper mine, but the study was not intensive. Cortazar et al. (2001) did not analyze the relationship between the information uncertainty and the flexibility value, and failed to describe the establishment or application of the model. The uncertainty factors and flexibility need further investigation. Fan and Zhu (2010) built a multi-factor real options model and applied it to an oil investment decision. However, the research did not consider the two important factors of geological and technological uncertainties. It introduced the exchange rate and resource tax rate to the model, none of which has a significant effect on the flexibility value for deepwater oil and gas projects. They proposed to adjust oil

price to exchange rate and tax rate, and then substituted the volatility of oil price in the conventional real options model for integration of the three volatilities. However, the integration is meaningless because their integrated factor has no difference from a single factor in essence. Furthermore, the tax rate does not have a stochastic characteristic. Schmit et al. (2011) built a two factors real options model to estimate the influence of U.S. ethanol policy on plant investment decisions. Similarly, this research focused on the financial aspect only, which defined the two variables as revenue and cost, ignoring the important influence from technology and engineering.

Our multi-factor real options model and its application to deepwater projects will make several contributions. In the aspect of random factors, we analyze three of the most important factors: oil price, geological information, and engineering information, based on the characteristics of deepwater oil and gas projects. We also integrate the three factors based on the stochastic process theory. In the aspect of real options model, we extend the single-factor model with geological and technical factors to better describe the flexibility value of deepwater oil and gas projects on the basis of the integration model, because the partial differential equation for three factors is too complex to be solved. In the aspect of application, we provide an example to show the practical significance of key parameters and introduce the method of parameter assignment. We also apply the real options model to value a deepwater project under a typical production-sharing contract.

This paper is organized as follows: In the second section, we will describe and integrate the variables with stochastic process theory. In Sect. 3, the multi-factor real options model will be established based on the integration model and the non-arbitrage approach. In Sect. 4, we will discuss the parameter assignment. In Sect. 5, we will apply the model to a deep water oil and gas project and analyze the optimal investment decision. Section 6 is the conclusions.

2 Three factors affecting the value of flexibility in deepwater oil and gas projects

Considering the characteristics of marine geographical environments, the flexibility value of deepwater oil and gas projects is determined not only by the volatility of oil price but also by the uncertainty of geology and engineering technology. If the exploration and development scheme is adjusted based on the additional information, the economy value of project could be increased.

The flexibility value of geology conditions implies the unremitting objectives of minimization of investment and maximization of profit since the geology information updated as the project proceeds helps make the investment budget more precise. In the oil and gas industry, the investment in exploration gives investors priority over next stage's activities. So these investors have more prominent opportunities due to their information privilege. The flexibility of technology implies potential cost savings and production increases in the process of exploration and production with the uncertainties being gradually clarified and problems solved. On the other hand, under the background of whole block development, the flexibility of technology implies the value maximization for all projects located in the same area since investors could properly design the overall development program and share the facilities among different projects in that area. Besides, the flexibility value of oil price changes could never be neglected since deep water oil and gas development projects always take many years. Investors and their management team can adjust their actual production according to the price at the time under specific technology and engineering conditions, so as to realize the best economic value of oil and gas reserves.

Geological conditions, technologies, and oil price are the main factors that affect the real options value of deepwater oil and gas projects. This work aims to study these three factors first, and build a multi-factor real options model on such basis.

There are two requirements in simulation of random factors. Firstly, the model should be concise enough for practical use and could accurately render the dynamic characteristics of random factors since the purpose of simulation is to construct financial models and realize necessary computation rather than to make predictions on future situation. Secondly, many financial models are built on the basis of Ito's lemma by now, however the random factor models are also needed to follow Brownian movements as basic variables do (Itō 2010, 2011).

2.1 Oil price simulation

There has been a lot of research conducted on simulation and prediction of oil prices because of the importance of petroleum in the world economy and international relationships. It has been shown that the fluctuation of oil price follows a random walk process with sudden increases or jumps at certain periods. Dixit and Pindyck (1994) described the oil price with several models. They pointed out that the geometric Brownian motion (GBM) should be a foundational model and the mean reverting model could describe the stochastic characteristics of oil price more accurately since oil price fluctuated around the cost of oil production which was stable. But the difficulty and the cost of oil exploitation have increased rapidly with soaring demand during the past two decades, and this change has

been reflected in the oil price fluctuation (see Fig. 1), so the mean reverting model is not as accurate to describe the characteristics of oil price.

Compared with the mean reverting model, GBM is more appropriate to embody oil price movements, as GBM conforms to both the stochastic characteristics of oil prices and the two requirements mentioned above.

Then, the simulation model of oil price will be:

$$dP_t = \mu P_t dt + \sigma_P P_t dW_t, \tag{1}$$

where P_t is the oil price at time t, W_t is a Wiener process, μ and σ_P are constants.

Assuming an initial price of P_0, use the Ito stochastic integral to solve the equation and we will get the following:

$$P_t = P_0 \exp\left(\left(\mu - \frac{\sigma_P^2}{2}\right)t + \sigma_P W_t\right). \tag{2}$$

2.2 Simulation of technological and geological factors in engineering

Technological and geological factors in engineering can substantially affect the overall economic value of deepwater oil and gas exploration projects as economic factors on the product market do, and must be included in the evaluation model.

Deepwater exploration technology is rapidly evolving. Specialists and technicians are exploring better methods to describe geology, technological conditions, and risks in the seabed. Their ideas and models can be quite different, and most of them end up in describing various geological and technological factors in the form of probabilities. Therefore, probability theory can be used to study the geological and technological uncertainties in the evaluation model for deepwater oil and gas exploration projects. However, most technological models in engineering involve various parameters, which are complicated and confidential, and cannot be directly used in economic models. Thus the model with geological and technological uncertainties must be designed with comprehensive study of technological methods in engineering to meet the requirements of an evaluation model.

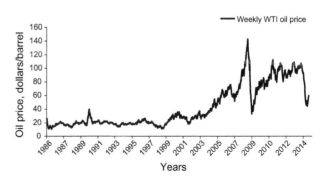

Fig. 1 The fluctuation of WTI oil price

All the random variables in a stochastic system can be described by a stochastic process. According to research findings (Clapp and Stibolt 1991), engineering and technology data follow normal distribution with certain features, and can be described by a stochastic differential equation. Engineering technology covers a wide range of factors with correlations among some of them, and it is impractical to clarify each of them in an economic model. Thus geological and technological factors in engineering will be taken as one comprehensive factor in simulation and analysis.

We hereby define a one dimension geology technology factor G, which follows Brownian movement with zero drift and constant volatility.

$$dG = G\sigma_G dW_G, \tag{3}$$

where σ_G is volatility, dW_G is the standard Wiener increment.

2.3 Variables integration

The above analysis and treatment for oil prices and geological and technological factors with a stochastic model is to investigate how to put the uncertain technological factors separately and directly into the evaluation model, so as to provide precise evaluation results for deepwater oil and gas assets. However, it is over sophisticated to put both factors P and G into a proper model together with the application of Ito's lemma. Therefore factor P and factor G will be technically integrated into one factor to build a three-factor model (Itō 2010).

Both oil price (P) and geology technology (G) impact on the value of deepwater oil and gas projects, where P is mainly affected by market dynamics, and G affected by geological conditions under the sea floor and by achievements in technological development. Therefore assume factors P and G are independent, which means:

$$dW_P dW_G = 0 \tag{4}$$

and let

$$Z \equiv F(P, G). \tag{5}$$

According to Ito's lemma, we substitute Eq. (1) and Eq. (3) into Eq. (5) and get:

$$dZ = \left(F_P P\mu + \frac{1}{2}F_{PP}P^2\sigma_P^2 + \frac{1}{2}F_{GG}G^2\sigma_G^2\right)dt + F_P P\sigma_P dW_P + F_G G\sigma_G dW_G. \tag{6}$$

In order to make Eq. (6) solvable and ensure the effectiveness of the evaluation model, we apply the principle of value additivity of stochastic process by Itō's (2010) and get:

$$Z \equiv F(P, G) = PG. \tag{7}$$

Given that factors P and G following standard Wiener process, and they both are independent and unrelated incremental variables, Eq. (7) is correct according to Ito's theory and the two stochastic processes can be superimposed to get:

$$\frac{dZ}{Z} = \mu dt + \sigma_P dW_P + \sigma_G dW_G. \tag{8}$$

Hence the new variable Z embodies the same drift rate with oil price P but larger volatility than P:

$$\sigma_Z = \sqrt{\sigma_P^2 + \sigma_G^2}. \tag{9}$$

3 Modeling

3.1 Hypothesis in modeling

Actual economic problems are far more diversified and complicated. A series of hypothesis are established in the light of the problems' particularity and research targets for simulation and computation, so as to better describe and solve problems with mathematic modeling. The fundamental hypotheses in real options modeling for deepwater oil and gas exploration projects are as follows:

(1) Oil price P follows GBM process, and its convenience yield is the function of oil price;
(2) The geological technological variables follow Brownian movement;
(3) Investment return r is known and constant;
(4) The reproduction cost of the investment portfolio is negligible;
(5) The real options value $V(Z, t)$ in the form of variable Z and time t is second order differentiable, and follows Ito's lemma;
(6) The compound option is perpetual since oil and gas exploration contracts last for many years.

3.2 Model establishment

The evaluation model is established on the basis of no-arbitrage portfolio theory. Assume $F(Z, \tau)$ is the price function of petroleum futures bought at time t with expiration time at T, where $\tau = T - t$. According to Ito's lemma, the instant yield of the futures is:

$$dF = \left(-F_\tau + \frac{1}{2}F_{ZZ}\sigma^2 Z^2\right)dt + F_Z dZ, \tag{10}$$

where F_Z and F_{ZZ} are first and second order partial derivatives.

And with Eq. (7) we have:

$$F_P = F_Z \times G \quad F_{PP} = F_{ZZ} \times G^2 \tag{11}$$

Then we generalize such an investment portfolio: An investor goes long on one unit of crude oil in the spot market and goes short on $(F_P)^{-1}$ unit of crude oil as underlying asset in the future market. Suppose no dividend is to be paid. According to Eqs. (10 and 11), the rate of return for this portfolio is:

$$\frac{dP}{P} + \frac{C(Z)dt}{P} - (PF_P)^{-1}dF$$
$$= (PF_P)^{-1}\left[F_P C(Z) - \frac{1}{2}F_{PP}\sigma^2 P^2 + F_\tau\right]dt, \tag{12}$$

where $C(Z)$ indicates convenience yield. Under the no-arbitrage principle of efficient market, the investment return of above portfolio equals the market return, which means:

$$\frac{1}{2}F_{PP}\sigma^2 P^2 + F_P(rP - C) - F_\tau = 0. \tag{13}$$

The boundary condition is:

$$F(P, G, 0) = P. \tag{14}$$

With Eq. (10), Eq. (11), and Eq. (8) we have:

$$dF = F_P[P(\mu - r) + C]dt + F_P P\sigma dz. \tag{15}$$

Deepwater oil and gas exploration involves special risks and tremendous investments, and the economic value is mainly affected by oil price P, geology technology G, accumulative investment I, and time t. Taking V for the value of the petroleum asset, and with Eq. (7), we get:

$$V \equiv V(G, P, I, t) = V(Z, I, t). \tag{16}$$

With Ito's lemma Eq. (16) is changed into:

$$dV = V_Z dZ + V_I dI + V_t dt + \frac{1}{2}V_{ZZ}(dZ)^2. \tag{17}$$

Let q be per unit investment, λ the average income tax rate, γ the rate of success in exploration, thus the after-tax cash flow of the exploration project will be:

$$\gamma V - q - \lambda V. \tag{18}$$

In order to get the partial differential equation of project value (V), we build another investment portfolio: buy one unit of oil asset and sell the same unit of oil futures, then the investment return will be:

$$dV + [\gamma V - q - \lambda V]dt - (V_P/F_P)dF$$
$$= \frac{1}{2}\sigma^2 Z^2 V_{ZZ} - qV_I + V_t + (rP - C)V_P + [\gamma V - q - \lambda V]. \tag{19}$$

In light of the no-arbitrage principle, the portfolio return is equal to the market return r_V. With Eq. (11) we have:

$$\frac{1}{2}\sigma^2 Z^2 V_{ZZ} - qV_I + V_t + (rP - C)V_Z + q$$
$$- (r + \lambda - \gamma)V = 0. \tag{20}$$

Taking deepwater oil and gas exploration projects as perpetual real options, then the operational period t is infinite. When an investor is operating one project, he also keeps seeking for other potential exploration blocks to ensure continuous cash inflow, which means t in the equation is not a variable ($V_t = 0$), and the value of real options is only related to its price and the geology technology uncertainties.

The value of a deepwater project is defined as $V(Z, I)$ under perpetual operation, then the maximum project value with optimal output level satisfies the following requirement:

$$\frac{1}{2}V_{ZZ}Z^2\sigma_Z^2 + (rZ - C)V_Z + qV_I - q - (r + \lambda + \gamma)V = 0. \tag{21}$$

The boundary conditions are:

$$V(0, I) = 0$$
$$V_Z(0, I) = 0 \tag{22}$$
$$\lim_{Z \to \infty} V_{ZZ}(Z, I) = 0.$$

Taking Eq. (21) together with Eq. (22) to establish the multi-factor real options model under conditions of uncertainty, and it is generally difficult to obtain analytical solutions for this model. A numerical simulation method is adopted instead for this model.

4 Variable simulation and parameter analysis

4.1 Variable simulation

In Sect. 2 the oil price, geological factor and technological factor are described by using stochastic differential equation (SDE), and the SDE is solved to get the key parameters of the equation. Also, the three factors are integrated together to build up the integration model. Figure 2 shows the simulation results of the factors based on the SDE and the Gauss fitting result.

We collect 1500 WTI (West Texas Intermediate) weekly oil price data and fit these data for comparison analysis, as shown in Fig. 3.

Figures 2 and 3 indicate that the simulated movements and actual oil price movements share the same characteristics of stochastic process if excluding unpredictable sudden jumps caused by political or economic emergencies. Therefore, the above simulation model and the given values acquired successfully to reflect the characteristics of

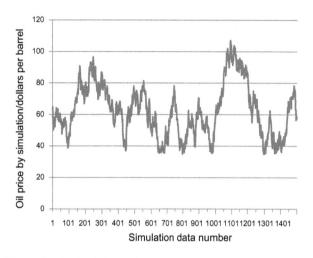 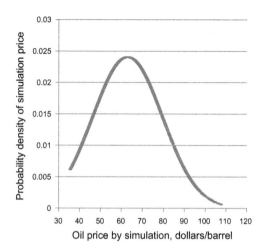

Fig. 2 The stochastic simulation and Gauss fitting of oil price

real oil price movements, and the basic form of the model conforms to a Wiener process, thus it can be used in financial models for oil and gas asset evaluation.

Simulation results of the geology technology factor *G* and the integrated three factors are shown in Figs. 4 and 5.

According to Figs. 2, 3, 4 and 5, the integration made in this work meets the hypothesis of Ito's lemma, and can be the basis of the parametric estimate in the multi-factor real options model. In addition, the fitting results of integrated and single factors are different from each other, which also prove that the study for multi-factor real options model makes sense.

4.2 Parameter analysis

Wang and Li (2010) analyzed the parameters of the real options model and demonstrated the significant influence

of the parameters to the valuation result. In order to make the multi-factor real options model applicable, accurate and understandable, study of the parameters value must be first conducted. Deepwater oil and gas exploration and development projects involve interests of many parties, which are stipulated in complicated contract clauses. The study of convenience yield is computed based on these clauses, which is analyzed in the research by Liu et al. (2012).

4.2.1 Investment rate and its influence on project value

In deepwater oil and gas exploration projects, investors are confronted with many uncertain factors, which on the other hand provide them with great flexibility in project management: they can expand or hold down the investment volume according to updated geological and technological conditions, and increase or reduce their

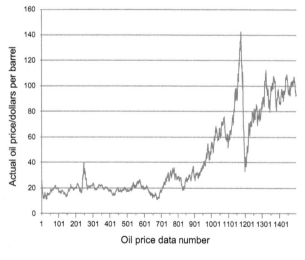 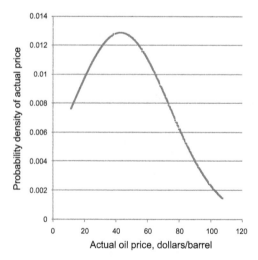

Fig. 3 The plot and Gauss fitting of WTI oil price

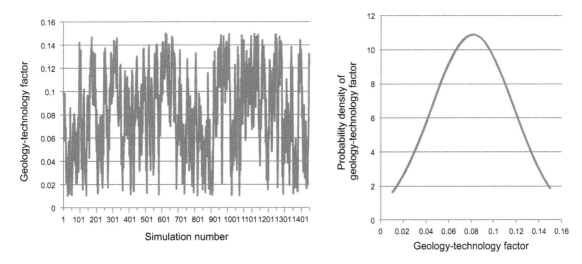

Fig. 4 The stochastic simulation and Gauss fitting of geology technology factor

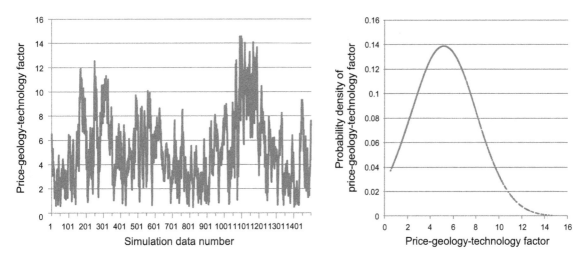

Fig. 5 The stochastic simulation and Gauss fitting of oil price-geology technology factor

production in consistence with market dynamics. Whereas there is one thing in common for most projects: some of the investment is irreversible despite succeeding investment policies as the initial investment is sunk or at least partially sunk. The investment in exploration is totally irreversible no matter if it succeeded or failed in finding recoverable resources.

The investment rate in exploration indicates the capital expenditure in search of recoverable oil and gas resources, and it is a crucial factor for total investment returns in the real options evaluation model. When oil companies increase their investment in exploration, they will acquire information about recoverable reserves with higher volume and better accuracy, and they will also have more confidence to realize greater economical value

of the project. Thus it is reasonable to assume that the increase in exploration investment adds value to the project. To be more simplified, we suppose the exploration investment and project value are in positive linear correlation.

4.2.2 Success rate in exploration

Success rate in exploration changes with many factors such as location, reservoir conditions, exploration engineering and equipment, and oil companies have different success rates in their exploration works in different areas and different blocks. It is fairly difficult to calculate or predict with limited geology and engineering information, while an alternative is to use empirical values from projects with similar location, water depth and other parameters.

4.2.3 Convenience yield rate

Convenience yield rate is a sensitive parameter in a real options evaluation model. Gibson and Schwartz (1990) suggested that convenience yield was the value-added cash flow naturally derived from products, and it belonged to the holders of such products rather than holders of derivative contracts; Convenience yield depended on the volume of inventory: products of less inventory and higher spot price could achieve higher convenience yield, and vice versa. Paddock et al. (1988) concluded that investment return of developed reserves consisted of two parts: operating profits from production sales and capital gains from intrinsic value growth of remaining reserves. Thus let convenience yield be:

$$C_t = \frac{\omega[\Pi_t - V_{bt}]}{V_{bt}}, \tag{23}$$

where ω is the decline rate of production in percentage; Π_t is after-tax profit of oil sales; V_{bt} is oil value of developed reserves per barrel.

4.2.4 Market investment return

We can adopt the return of investment used in discounted cash flow analysis and adjust it according to the characteristics of assets in the same region.

Table 1 Parameter estimation

Market price of developed reserves V_b, \$/bbl	15.38
After-tax earnings Π, \$/bbl	19.01
Total cost D, \$/bbl	13.04
Exploration cost E, \$/bbl	2.96
Influence of investment changes to project value V_I	1.9
Success rate in exploration γ	0.2
Production decline rate ω, %	6.25
Investment rate q, %	23.00
Market investment return r, %	10.00
Convenience yield C, %	1.50

5 Evaluation and decision analysis

An overseas deepwater asset located in West Africa is an example for this multi-factor real options model. This project is under a production-sharing contract (PSC). At the initial stage of investment, the oil price was at a relatively low level but moving upward, so the exploration scheme was conservatively designed for modest oil production since only pre-exploration had been conducted, and detailed exploration data were not available. At the initial decision point, the discounted cash flow analysis did not show too much promise even though adjacent oil blocks showed promising economic returns. So the decision making of this project does not completely refer to DCF analysis results.

Parameters in the real options model are calculated (see Table 1) according to the contract and the analysis in Sect. 4.

Assume that oil price is at USD65\$ with volatility of 0.18 and factor G equals 0.1 with volatility of 0.25, then the value of this project with geology technology uncertainties amounts to USD13.53\$ per barrel. In a more pessimistic scenario with higher geology technology risks, assuming factor G to be 0.07, the evaluation result from the real options model suggests a project value of USD6.48\$ per barrel.

While a single-factor real options model is often applied to evaluate the flexibility value with uncertainties in oil price, this study adopts both multi-factor real options model and net present value method to better demonstrate the uncertainties. The comparison results are shown in Table 2.

The results in Table 2 show that the multi-factor real options model can better reflect the flexibility value. Besides, projects in adjacent blocks are endowed with excellent geological conditions. They have exhibited great potential and gained much higher economic returns than their initial evaluation results, which provide evidence for the effectiveness of multi-factor real options model to some extent.

On the other hand, the results under higher geology technology risks scenario by a multi-factor real options model are much more conservative than those given by the single-factor model, thereby it confirms that multi-factor model is more capable in reflecting the impact of

Table 2 Results of three evaluation methods

Evaluation method	NPV	Single-factor real options model	Multi-factor real options model with high G rate	Multi-factor real options model with low G rate
Result, USD/barrel	5.76	7.38	13.53	6.48
Deviation from NPV result	–	28.13 %	135 %	12.5 %

engineering factors on projects' flexibility value while single-factor model only focuses on the impact of oil prices but ignores the impact of geology and technology.

6 Conclusions

The management of flexibility value is not only embodied in the feasibility to elevate the economic value of reserves by adjusting production to oil prices. It's also shown in the flexibility to design exploration schemes according to the uncertainty of geological information and technology, especially for deepwater oil and gas exploration projects. We analyzed several influential factors for project value with reference to the characteristics of real options in deepwater projects. We established a multi-factor real options model under uncertain conditions for project evaluation, and employed the idea of multi-uncertainty factors integration to make the model practical.

The model has been successfully applied to a real deepwater project. The evaluation result shows that the multi-factor real options model could be more accurate than the single-factor model. This multi-factor model gives investors more reliable theoretical supports to make reasonable decisions. Our sample project has been operated for more than five years, and the real practice has also showed that the estimated value with multi-factor real options model is a better approximation to reality. So the multi-factor real options model could be a good reference approach for investment decisions about deepwater oil and gas projects.

Acknowledgments The authors would like to thank the experts for their helpful discussion and suggestions on the 18th Annual International Conference on Real Options. This paper is supported from the National Science and Technology Major Project under Grant No. 2011ZX05030.

References

Brennan MJ, Schwartz ES. Evaluating natural resource investments. J Bus. 1985;58(2):135–57.

Clapp R, Stibolt R. Useful measures of exploration performance. J Pet Technol. 1991;10:1252–7.

Cortazar G, Schwartz ES, Casassus J. Optimal exploration investments under price and geological-technical uncertainty: a real options model. R&D Manag. 2001;31:181–9.

Dias MA. Valuation of exploration and production assets: an overview of real options models. J Pet Sci Eng. 2004;44:93–114.

Dixit AK, Pindyck RS. Investment under uncertainty. Princeton: Princeton University Press; 1994.

Fan Y, Zhu L. A real options based model and its application to China's overseas oil investment decisions. Energy Econ. 2010;32:627–37.

Gibson R, Schwartz ES. Stochastic convenience yield and the pricing of oil contingent claims. J Financ. 1990;45(3):959–76.

Itō K. Stochastic process. Beijing: Postal and Telecom Press; 2010 (**in Chinese**).

Itō K. Probability theory. Beijing: Postal and Telecom Press; 2011 (**in Chinese**).

Liu M, Wang Z, Zhao L, et al. Production sharing contract: an analysis based on an oil price stochastic process. Pet Sci. 2012;9(3):408–15.

Myers SC. Determinants of corporate borrowing. J Financ Econ. 1977;5(2):147–75.

Schmit TM, Luo J, Conrad JM. Estimating the influence of U.S. ethanol policy on plant investment decisions: a real options analysis with two stochastic variables. Energy Econ. 2011;33:1194–205.

Paddock JL, Seigel DR, Smith JL. Option valuation of claims on real assets: the case of offshore petroleum. Q J Econ. 1988;103(3):479–508.

Wang Z, Li L. Valuation of the flexibility in decision-making for revamping installations: a case from fertilizer plants. Pet Sci. 2010;7(3):428–34.

14

Experimental investigation of the effects of various parameters on viscosity reduction of heavy crude by oil–water emulsion

Talal Al-Wahaibi · Yahya Al-Wahaibi ·
Abdul-Aziz R. Al-Hashmi · Farouq S. Mjalli ·
Safiya Al-Hatmi

Abstract The effects of water content, shear rate, temperature, and solid particle concentration on viscosity reduction (VR) caused by forming stable emulsions were investigated using Omani heavy crude oil. The viscosity of the crude oil was initially measured with respect to shear rates at different temperatures from 20 to 70 °C. The crude oil exhibited a shear thinning behavior at all the temperatures. The strongest shear thinning was observed at 20 °C. A non-ionic water soluble surfactant (Triton X-100) was used to form and stabilize crude oil emulsions. The emulsification process has significantly reduced the crude oil viscosity. The degree of VR was found to increase with an increase in water content and reach its maximum value at 50 % water content. The phase inversion from oil-in-water emulsion to water-in-oil emulsion occurred at 30 % water content. The results indicated that the VR was inversely proportional to temperature and concentration of silica nanoparticles. For water-in-oil emulsions, VR increased with shear rate and eventually reached a plateau at a shear rate of around 350 s^{-1}. This was attributed to the thinning behavior of the continuous phase. The VR of oil-in-water emulsions remained almost constant as the shear rate increased due to the Newtonian behavior of water, the continuous phase.

Keywords Viscosity reduction · Phase inversion ·
Non-newtonian fluid · Oil-in-water emulsions ·
Heavy crude oil

T. Al-Wahaibi (✉) · Y. Al-Wahaibi · A.-A. R. Al-Hashmi ·
F. S. Mjalli · S. Al-Hatmi
Petroleum & Chemical Engineering Department, College of
Engineering, Sultan Qaboos University,
P.O. Box 33, Al-Khod 123, Oman
e-mail: alwahaib@squ.edu.om

Edited by Xiu-Qin Zhu

1 Introduction

The flow of oil and water in pipelines is a challenging subject that is rich in physics and practical applications. The early work of Russell and Charles (1959) shows that the introduction of water into crude oil pipelines under certain conditions can reduce pressure drop and improve the transportation of the oil. This work has attracted the interest of researchers as well as the industry. The concurrent flow of oil and water is encountered in many industries such as the oil and chemical industries. Many researchers have reported the flow of both light oil and heavy oil with water in pipelines (Nädler and Mewes 1997; Angeli and Hewitt 2000; Lovick and Angeli 2004; Al-Wahaibi and Angeli 2011; Al-Wahaibi et al. 2012; Yusuf et al. 2012a, b). However, most of these studies have mainly focused on investigating the flow patterns, pressure drop, and hold-up of the flow.

Heavy crude and extra-heavy crude oils are becoming more important with the increasing demand for world energy, the large amount of heavy oil reserves, and the decline of conventional oils. Heavy oils have to be transported by pipelines from the well head to refineries or ports. However, transporting the heavy oil as a single phase in pipelines is very expensive because of the huge pumping costs. In some cases, it is impossible to transport these crude oils due to their low mobility and flowability and pipeline blockages from wax and/or asphaltene deposition. Different methods have been proposed to facilitate transportation of heavy crude oils. These methods include heating the crude (Yaghi and Al-Bemani 2002; Saniere et al. 2004), dilution with light oil (Yaghi and Al-Bemani 2002; Iona 1978), formation of stable oil-in-water emulsions (Yaghi and Al-Bemani 2002; Lappin and Saur 1989; Gregoli et al. 2006), or imposing core annular flow (Joseph

Experimental investigation of the effects of various parameters on viscosity reduction of heavy crude...

187

et al. 1997). Yaghi and Al-Bemani (2002) have experimentally concluded that transporting heavy crude by heating or dilution is an expensive option. They found that it is cheaper to transport it as oil-in-water emulsion with an optimum oil content of around 70 %. Langevin et al. (2004) have reported that oil-in-water emulsions reduce the viscosity of heavy crude oils and bitumens and may provide an alternative to the use of diluents or heat. Also, hydrocarbon diluents or lighter crudes may not be available or limited, while water is readily available for emulsification. Joseph et al. (1997) have reported that transporting heavy oil as a core annular flow pattern has some operational problems especially if a start-up is required. Unlike core annular flow, the report by Simon and Poynter (1970) has shown that restarting a pipeline after an emergency shutdown and re-emulsification of oil does not pose major problems. The formation of oil-in-water emulsions can cause a reduction of oil viscosity by more than 2 orders of magnitude (Yaghi and Al-Bemani 2002). In addition, Simon and Poynter (1970) reported that since water is the continuous phase, crudes have no contact with the pipe wall and this can reduce the pipe corrosion especially for crude having high sulfur content.

Emulsions are formed traditionally by homogenization (shaking, stirring or some other kinds of intensive dynamic and/or static mixing processes). The energy input for emulsification can be significantly reduced using surfactants, which can reduce the interfacial tension (IFT). Surfactants can also stabilize the emulsions formed. Stability of these emulsions can be further enhanced using high molecular weight polymers and/or nanoparticles to induce higher rigidity and viscosity of the interfacial films. In general, non-ionic surfactants represent a good choice because they are relatively cheap and do not produce any undesirable organic residues that affect oil properties (Rivas et al. 1998).

The subject of heavy crude oil transportation using the emulsification technique has recently attracted researchers to investigate the effect of various parameters such as surfactant concentration, speed and mixing rate, brine pH, and oil content (Ashrafizadeh and Kamran 2010; Hasan et al. 2010; Abdurahman et al. 2012).

In this study, extensive experiments were conducted at various water contents, shear rates, temperatures, and concentrations of nanoparticles to investigate their effects on the viscosity reduction (VR) of heavy crude oil via stable emulsions. Stabilized emulsions were produced using Triton X-100 as a surfactant. Triton X-100 was selected as it has been widely used by different investigators and shows good stability (Ashrafizadeh and Kamran 2010; Hasan et al. 2010; Abdurahman et al. 2012).

2 Experimental set-up

2.1 Materials

Heavy crude oil, surfactant (Triton X-100), and solid nanoparticles (Aerosil 200) are the primary materials used in this study. The crude oil was obtained from the Omani heavy oil field. It has a density of 938 kg/m^3 at 20 °C. Triton X-100 with chemical formulation of $C_{33}H_{60}O_{10}$ was used as an emulsifying agent and an emulsion stabilizer. Triton X-100 is a non-ionic water-soluble surfactant obtained from Sigma-Aldrich Canada Ltd. Aerosil 200 was used to investigate the influence of solid nanoparticles on the VR of the emulsion and to enhance emulsion stability. Aerosil 200 is hydrophilic fumed silica with a specific surface area of 200 m^2/g. It is highly dispersed with spherical-shaped particles of an average diameter of 12 nm and a bulk density of approximately 30 g/L.

2.2 Emulsion preparation

Triton X-100 aqueous phase surfactant solution was prepared with tap water and used throughout the experiments. Hydrochloric acid and sodium hydroxide were used when necessary to keep the aqueous phase at pH 7. A known amount of the Triton X-100 aqueous surfactant solution was then transferred to a beaker. The Triton X-100 aqueous solution and crude oil were mixed using a homogenizer (Tokushu Kika Kogyo Ca. LTD) at a speed of 1,000 rpm for 30 min to prepare the emulsions. Stability analysis was conducted to find the surfactant concentration necessary for preparing a stable oil–water emulsion. After the minimum surfactant concentration was determined, the effects of temperature, shear rates, and solid nanoparticles on both oil-in-water emulsion and water-in-oil emulsion were investigated using this minimum surfactant concentration throughout the experiments.

2.3 Stability of emulsion

The stability of the emulsion was determined by measuring the amount of water separated from the emulsion after 24 h. This was done by transferring the prepared emulsions into graduated bottles and allowing them to rest for 24 h to promote phase separation.

2.4 Viscosity measurements

The viscosity measurements were carried out using a Haake Viscometer VT500. This device is used to determine the viscosity of oil and oil–water emulsions at different shear rates and temperatures.

3 Results and discussion

3.1 Crude oil viscosity

It is important to measure the viscosity of crude oil prior to preparing the emulsion as it will provide the basis to determine the degree of VR. In this study, the crude oil viscosity was measured in the range of shear rate from 27 to 2,700 s^{-1} at 20, 30, 40, 50, 60, and 70 °C. As shown in Fig. 1, the crude oil exhibited a shear thinning behavior where the apparent viscosity was found to decrease with shear rate for all the investigated temperatures. Ghannam and Esmail (2006) reported that the molecular chains in heavy crude oils would be disentangled, stretched, and reoriented with the driving force at high shear rates and this was the reason for the decrease in viscosity with shear rate. The viscosity results also revealed that the strongest shear thinning was observed at 20 °C where the viscosity of the crude decreased from 1,920 to 1,190 mPa s at shear rates of 27 and 349 s^{-1}, respectively. Paso et al. (2009) attributed this behavior to the interaction between the heavy components of the crude oil.

The temperature was found to have a strong effect on crude oil viscosity. At a certain shear rate, the apparent viscosity decreased sharply with temperature. The degree of this decrease was higher at lower shear rates than that at higher values (see Fig. 1). Figure 2 shows the apparent crude oil viscosity profile as a function of temperature at a shear rate of 27 s^{-1}. The general trend of the viscosity profile can be fitted by an Arrhenius equation (Eq. 1)

$$\mu = Ae^{(E_\mu/RT)}, \tag{1}$$

where μ is the apparent viscosity, A is the pre-exponential constant, E_μ is the activation energy, R is the ideal gas constant, and T is the temperature in Kelvin.

The activation energy and the pre-exponential constant were calculated as 37.6 kJ/mol and 3.3×10^{-4}, respectively. The apparent viscosity values were found to fit the Arrhenius model reasonably well in most cases.

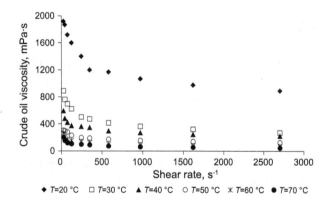

Fig. 1 Viscosity behavior of heavy crude oil at different temperatures

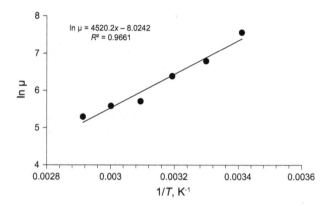

Fig. 2 Apparent viscosity of the heavy crude as a function of $(1/T)$ with the Arrhenius model

3.2 Emulsion stability

The aim of this study is to investigate the influence of temperature, shear rate, and nanoparticles on the VR of stable oil–water emulsion. Therefore, it is important to find the minimum surfactant concentration required to prepare a stable oil–water emulsion. This concentration will be used throughout the study to investigate the aforementioned parameters on oil–water emulsion. The emulsion stability can be quantified using the following equation (see Abdurahman et al. 2012):

$$\text{Emulsion stability (\%)} = 100 \times \left(1 - \frac{\text{Amount of separated water}}{\text{Initial water content}}\right) \tag{2}$$

The effect of Triton X-100 concentration on the stability of the emulsion was examined at three oil–water contents: 40 % oil + 60 % water, 60 % oil + 40 % water, and 80 % oil + 20 % water. The emulsion was prepared under a constant stirring speed of 1,000 rpm for 30 min. Figure 3 shows the stability of the emulsions at 1, 2, 3, and 4 wt% Triton X-100 concentrations. It was clear that emulsion stability increased with increasing surfactant concentrations. This was because with increasing the surfactant concentration, the IFT between oil and water decreased. Thus, the probability for the drops to break increased, which results in more stable emulsion (Sakka 2002). No water separation was observed after 6 days for all the investigated samples when 3 wt% surfactant concentrations were used in the emulsion. This concentration (3 wt%) was considered as the minimum surfactant concentration necessary to stabilize the emulsion for different oil–water contents.

3.3 Viscosity reduction (VR)

The effectiveness of the emulsification process in reducing the viscosity of the crude oil can be quantified by

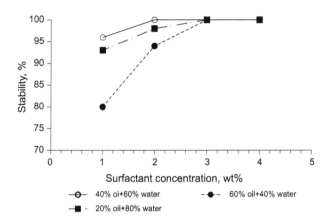

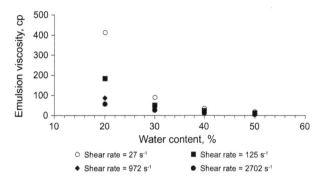

Fig. 3 Effect of surfactant concentration on the emulsion stability of different oil–water contents (30 °C, 1,000 rpm mixing speed, 30 min mixing time)

Fig. 4 Effect of water content on the emulsion apparent viscosity of Omani heavy crude oil (30 °C, 3 wt% surfactant concentration, 1,000 rpm mixing speed, 30 min mixing time)

calculating the viscosity difference caused by emulsification as described by the following equation:

$$\%VR = 100 \times \left(\frac{\mu_o - \mu_{o/w}}{\mu_o} \right), \tag{3}$$

where μ_o is the measured crude oil viscosity and $\mu_{o/w}$ is the emulsion viscosity measured at the same conditions of shear rate and temperature.

In this study, the emulsion viscosities were measured over a wide range of shear rates and temperatures. All the emulsions were prepared using 3 wt% Triton X-100 concentrations while stirring at 1,000 rpm for 30 min

3.3.1 Emulsion and phase inversion

Figure 4 shows the apparent viscosity of the crude oil emulsion at different water contents. In this study, the water content in the emulsion was varied from 50 to 20 wt% in 10 % increments. At a shear rate of 27 s^{-1} and 30 °C, the viscosity of the crude oil was 890 mP s before emulsification with water. Adding 20 wt% water content, the viscosity of the oil emulsion decreased to around 410 mP s at 27 s^{-1} and 30 °C, a VR of around 53 %. The viscosity of the emulsion decreased further to around 90 mPa s, a VR of around 90 %, using 30 wt% water content. This sudden increase in the VR was due to phase inversion of the emulsion from water dispersed in oil (water-in-oil emulsion) to oil dispersed in water (oil-in-water emulsion). This flow behavior was in accordance with which phase is the continuous phase in the emulsion. It is well known that the emulsion viscosity is a strong function of the continuous phase and a weak function of the dispersed phase.

The effect of water content on VR for different shear rates at 30 °C is presented in Fig. 5. The VR was found to increase by increasing the water content and reached its

maximum value at 50 wt% water. At 20 wt% water content (i.e., water-in-oil emulsion), the VR could be enhanced by the increase in shear rate, whereas the VR of the oil-in-water emulsions (i.e., water content higher than 30 %) was almost independent of shear rate.

From a practical point of view and to achieve effective and economic transportation of crude oil, the emulsion viscosity should be as low as possible and the water content should be as low as possible. The experimental results revealed that the apparent viscosity of the emulsion decreased with water content. Thus, the amount of VR would increase and this can lead to lower energy consumption in pumping. In this study, the maximum oil content in the emulsion reached 70 %, above this value, inversion to water in oil emulsion would occur. However, due to the large pressure gradient which might occur at the phase inversion, it is important to operate well below the phase inversion point. Therefore, transportation of oil and water in pipes should be far from the phase inversion point (Ashrafizadeh and Kamran 2010; Hasan et al. 2010; Abdurahman et al. 2012).

3.3.2 Effect of shear rate

The effect of shear rate on the percentage VR at 20, 30, 40, and 50 wt% input water fraction is presented in Fig. 6. It is clear that the VR for the water-in-oil emulsion (i.e., 20 wt% water) increased gradually with increasing shear rates and eventually reached a plateau where a further increase in shear rate had no effect on VR. For the oil-in-water emulsion, the gradual increase of VR with shear rate was not large. The VR reached a plateau for the whole range of shear rates. This is because the flow behavior of the emulsion is a strong function of the continuous phase. For the oil-in-water emulsion, since water was the continuous phase, the decrease in viscosity with shear rate was not significant due to the Newtonian behavior of the water phase. On the other hand, the emulsion viscosity was

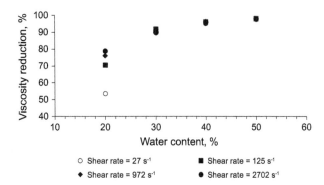

Fig. 5 Effect of water content at different shear rates on the viscosity reduction of Omani heavy crude oil emulsion (30 °C, 3 wt% surfactant concentration, 1,000 rpm mixing speed, 30 min mixing time)

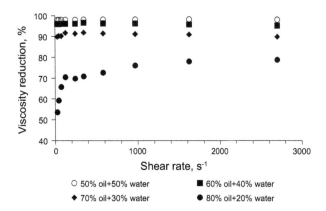

Fig. 6 Effect of shear rates at different water input fractions on the viscosity reduction of Omani heavy crude oil emulsion (30 °C, 3 wt% surfactant concentration, 1,000 rpm mixing speed, 30 min mixing time)

observed to exhibit the shear thinning behavior of the continuous phase at 20 wt% water content (i.e., similar to that of oil as shown in Fig. 1).

3.3.3 Effect of temperature

The influence of temperature on the effectiveness of the emulsion prepared in this study was investigated by measuring the apparent viscosity at 20, 30, 40, 50, 60, and 70 °C. A comparison of the obtained results is presented in Figs. 7 and 8 for oil-in-water emulsion (60 % oil and 40 % water) and water-in-oil emulsion (80 % oil and 20 % water), respectively. Both figures reveal a decrease in VR as temperature increased. This was attributed to the decrease in crude oil viscosity with temperature (see Fig. 1). The reduction in viscosity was more pronounced at lower temperatures. For the oil-in-water emulsion of 60 % oil and 40 % water, it is observed that the VR varied from 83 % at 70 °C to around 99 % at 20 °C. It is also noticed that the VR changed very little with shear rates at all the

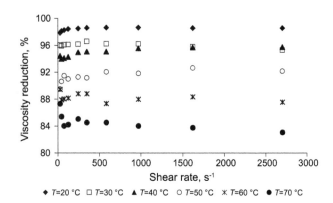

Fig. 7 Effect of shear rates at different temperatures on the viscosity reduction of stable oil-in-water emulsion (60 % oil + 40 % water, 30 °C, 3 wt% surfactant concentration, 1,000 rpm mixing speed, 30 min mixing time)

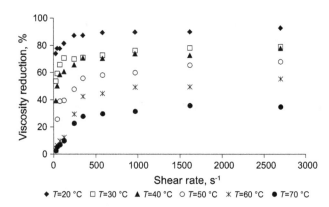

Fig. 8 Effect of shear rates at different temperatures on the viscosity reduction of stable water-in-oil emulsion (80 % oil + 20 % water, 30 °C, 3 wt% surfactant concentration, 1,000 rpm mixing speed, 30 min mixing time)

investigated temperatures (see Fig. 7). On the other hand, for the water-in-oil emulsion of 80 % oil and 20 % water, the VR varied in a wide range from 3 % at 70 °C to around 90 % at 20 °C. This was also attributed to the viscosity of the continuous phase. The VR was also found to increase with shear rates until it reached a plateau at shear rates greater than 400 s^{-1} where a further increase in shear rates had no effect on VR (see Fig. 8).

3.3.4 Effect of solid concentration

Stable oil-in-water emulsion and water-in-oil emulsion might contain fine particles while flowing in pipelines. Therefore, it is important to understand the influence of solid particles on the VR of these emulsions. This effect was investigated using three different solid concentrations; 0.05, 0.1, and 0.2 wt%. The VR at stable oil-in-water emulsion (60 % oil and 40 % water) and stable water-in-oil emulsion (80 % oil and 20 % water) with respect to shear rates is shown in Figs. 9 and 10. It is clear that the VR

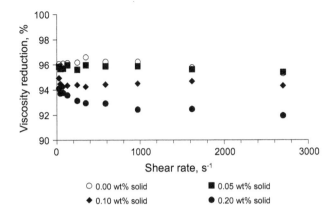

Fig. 9 Effect of shear rates at different solid concentrations on the viscosity reduction of stable oil-in-water emulsion (60 % oil + 40 % water, 30 °C, 3 wt% surfactant concentration, 1,000 rpm mixing speed, 30 min mixing time)

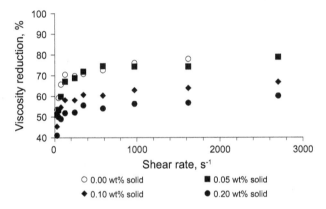

Fig. 10 Effect of shear rates at different solid concentrations on the viscosity reduction of stable water-in-oil emulsion (80 % oil + 20 % water, 30 °C, 3 wt% surfactant concentration, 1,000 rpm mixing speed, 30 min mixing time)

decreased with an increase in solid concentration. The reason might be that the agglomeration of the solid particles increased, hence resulted in the decrease in VR of the emulsion. However, the decrease in VR for the oil-in-water emulsion was insignificant (less than 5 %) as shown in Fig. 9. On the contrary, the decrease in VR was pronounced for the water-in-oil emulsion and could be as high as 25 % (see Fig. 10). This is also attributed to the continuous phase of the emulsion, since the flow behavior of the emulsion is a strong function of the continuous phase (see discussions in Sect. 3.3).

4 Conclusions

The VR induced by emulsifying Omani heavy crude oil using a non-ionic water soluble surfactant (Triton X-100) was experimentally investigated. The effects of water

content, shear rate, temperature, and solid particles concentrations on the performance of the emulsion in reducing the viscosity of the crude were examined. Based on the experimental findings, the following can be concluded:

- The crude oil exhibits a shear thinning behavior where the apparent viscosity is found to decrease with shear rates for all the investigated temperatures. The strongest shear thinning was observed at 20 °C where the viscosity of the crude falls from 1,920 to 1,190 mPa s at shear rates of 27 and 349 s^{-1}, respectively.

- The percentage viscosity reduction was found to increase with water content reaching a maximum value at around 50 wt% input water fraction. The emulsification was able to induce viscosity reduction for all the conditions investigated and regardless of the continuous phase.

- The percentage viscosity reduction was found to decrease with increasing the temperature and the solid concentrations of the emulsion.

- Two behaviors of viscosity reduction were observed with respect to shear rate. For water-in-oil emulsion, the viscosity reduction increased as shear rate increased and eventually reached a plateau at a shear rate of around 350 s^{-1}. For oil-in-water emulsion, viscosity reduction remained almost constant as shear rate increased.

References

Abdurahman NH, Rosli YM, Azhari NH, et al. Pipeline transportation of viscous crudes as concentrated oil-in-water emulsions. J Petrol Sci Eng. 2012;90–91:139–44.

Al-Wahaibi T, Angeli P. Experimental study on interfacial waves in stratified horizontal oil-water flow. Int J Multiph Flow. 2011;37(8):930–40.

Al-Wahaibi T, Yusuf N, Al-Wahaibi Y, et al. Experimental study on the transition between stratified and non-stratified horizontal oil-water flow. Int J Multiph Flow. 2012;38(1):126–35.

Angeli P, Hewitt GF. Flow structure in horizontal oil–water flow. Int J Multiph Flow. 2000;26(7):1117–40.

Ashrafizadeh SN, Kamran M. Emulsification of heavy crude oil in water for pipeline transportation. J Petrol Sci Eng. 2010;71(3–4):205–11.

Ghannam MT, Esmail N. Flow enhancement of medium-viscosity crude oil. J Pet Sci Technol. 2006;24(8):985–99.

Gregoli AA, Hamshar JA, Olah AM, et al. Preparation of stable crude oil transport emulsions. 2006. US Patent no. 4725287.

Hasan SW, Ghannam MT, Esmail N. Heavy crude oil viscosity reduction and rheology for pipeline transportation. Fuel. 2010;89(5):1095–100.

Iona M. Process for producing low-density low sulfur crude oil. 1978. US Patent no. 4092238.

Joseph DD, Bai R, Chen KP, et al. Core annular flows. Annu Rev Fluid Mech. 1997;29:65–90.

Langevin D, Poteau S, Hénaut I, et al. Crude oil emulsion properties and their application to heavy oil transportation. Oil Gas Sci Technol. 2004;59(5):511–21.

Lappin GR, Saur JD. Alpha olefins applications handbook. New York: CRC Press; 1989.

Lovick J, Angeli P. Experimental studies on the dual continuous flow pattern in oil-water flows. Int J Multiph Flow. 2004;30(1): 139–57.

Nädler M, Mewes D. Flow induced emulsification in the flow of two immiscible liquids in horizontal pipes. Int J Multiph Flow. 1997;23(1):55–68.

Paso K, Silset A, Sørland G, et al. Characterization of the formation, flowability, and resolution of Brazilian crude oil emulsions. Energy Fuels. 2009;23(1):471–80.

Rivas H, Gutierrez X, Cardenas AE, et al. Natural surfactant with amines and ethoxylated alcohol. 1998. US Patent 5792223.

Russell TWF, Charles ME. The effect of the less viscous liquid in the laminar flow of two immiscible liquids. Can J Chem Eng. 1959;37(1):18–24.

Sakka S. Sol-gel science and technology topics in fundamental research and applications., Sol–gel prepared ferroelectrics and related materialsNew York: Kluwer Academic Publisher; 2002. p. 33–5.

Saniere A, Hénaut I, Argillier JF. Pipeline transportation of heavy oils, a strategic, economic and technological challenge. Oil Gas Sci Technol. 2004;59(5):455–66.

Simon R, Poynter WG. Pipelining oil/water mixtures. 1970. US Patent 3519006.

Yaghi BM, Al-Bemani A. Heavy crude oil viscosity reduction for pipeline transportation. Energy Sources. 2002;24(2):93–102.

Yusuf N, Al-Wahaibi Y, Al-Wahaibi T, et al. Effect of oil viscosity on the flow structure and pressure gradient in horizontal oil–water flow. Chem Eng Res Des. 2012a;90(8):1019–30.

Yusuf N, Al-Wahaibi T, Al-Wahaibi Y, et al. Experimental study on the effect of drag reducing polymer on flow patterns and drag reduction in a horizontal oil–water flow. Int J Heat Fluid Flow. 2012b;37:74–80.

15

Numerical simulation of the impact of polymer rheology on polymer injectivity using a multilevel local grid refinement method

Hai-Shan Luo[1] · Mojdeh Delshad[1] · Zhi-Tao Li[1] · Amir Shahmoradi[1]

Abstract Polymer injectivity is an important factor for evaluating the project economics of chemical flood, which is highly related to the polymer viscosity. Because the flow rate varies rapidly near injectors and significantly changes the polymer viscosity due to the non-Newtonian rheological behavior, the polymer viscosity near the wellbore is difficult to estimate accurately with the practical gridblock size in reservoir simulation. To reduce the impact of polymer rheology upon chemical EOR simulations, we used an efficient multilevel local grid refinement (LGR) method that provides a higher resolution of the flows in the near-wellbore region. An efficient numerical scheme was proposed to accurately solve the pressure equation and concentration equations on the multilevel grid for both homogeneous and heterogeneous reservoir cases. The block list and connections of the multilevel grid are generated via an efficient and extensible algorithm. Field case simulation results indicate that the proposed LGR is consistent with the analytical injectivity model and achieves the closest results to the full grid refinement, which considerably improves the accuracy of solutions compared with the original grid. In addition, the method was validated by comparing it with the LGR module of CMG_STARS. Besides polymer injectivity calculations, the LGR method is applicable for other problems in need of near-wellbore treatment, such as fractures near wells.

✉ Hai-Shan Luo
haishan.luo@utexas.edu

[1] Center for Petroleum and Geosystems Engineering, The University of Texas at Austin, 200 E Dean Keeton St C0300, Austin, TX 78712, USA

Edited by Yan-Hua Sun

Keywords Polymer rheology · Polymer injectivity · Chemical EOR · Local grid refinement · Non-Newtonian flow

1 Introduction

Polymer flooding has become one of the most widely used enhanced oil recovery (EOR) methods because of its adaptability to a wide range of oil viscosity (Wassmuth et al. 2007), relative simplicity for operations (Mohammadi and Jerauld 2012), and offshore applicability (Morel et al. 2012). For polymer flooding as well as most other chemical flooding processes such as surfactant-polymer flood, alkaline-surfactant-polymer flood, and alkaline-cosolvent-polymer flood, the polymer injectivity is a key index for reservoir management, e.g., deciding the upper limit of the polymer injection rate to optimize the project economics (Seright et al. 2009). Factors affecting polymer injectivity include polymer degradation (Seright et al. 2009; Zaitoun et al. 2012), induced fractures near the injector (Seright et al. 2009; van den Hoek et al. 2012), polymer crosslinking to form gel (Bekbauov et al. 2013; Goudarzi et al. 2013b), and especially polymer rheology (Delshad et al. 2008; Sharma et al. 2011; Kulawardana et al. 2012). A polymer solution is a non-Newtonian fluid whose viscosity is non-linearly related to the flow rate or the in situ shear rate. For example, hydrolyzed polyacrylamide (HPAM) solutions exhibit pseudoplastic behavior at low shear rates and dilatant behavior at high shear rates when flowing through porous media as shown in Fig. 1 (Delshad et al. 2008). In addition, polymer rheology exhibits Newtonian behavior when the flow is at very low or high rates (Stahl and Schulz 1988; Sorbie 1991). This behavior leads to a complex relationship between the pressure drop and

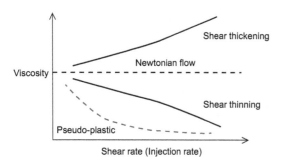

Fig. 1 Rheological relation between viscosity and shear rate for polymer solutions

the local velocity. Consequently, the polymer injectivity is often erroneously calculated from numerical simulations using a gridblock size practical for full field simulations where the flow rate decreases drastically from the wellbore (Sharma et al. 2011; Li and Delshad 2014). It is crucial for a numerical simulator to capture near-wellbore polymer rheology more accurately to improve the estimation of injection rate, shorten the project life, enhance the economics, and prevent or carefully design the injection induced fractures depending on the operators' decisions (Gadde and Sharma 2001; Lee et al. 2011).

The inaccuracy in calculated polymer injectivity mainly results as the flow rate is smeared within a coarse well gridblock. This is especially severe for common reservoir simulations in which the gridblock size is used up to several dozens of feet while the wellbore radius is only about 0.5 ft. In situ shear rates reach as high as 10^4 s^{-1} near the wellbore and decrease sharply to about 1–10 s^{-1} within a well gridblock. To eliminate the grid effects, several empirical or analytical models were proposed based on effective properties of the well blocks. For instance, Sharma et al. (2011) proposed to use an effective well radius to calculate the shear rate and match the polymer injectivity from very fine-grid simulation results; Li and Delshad (2014) proposed an effective viscosity using mathematical integration of in situ viscosity by assuming a radial velocity distribution within the well block. However, these approaches are not rigorous for other near-well effects apart from polymer rheology, e.g., non-zero skin factor, polymer permeability reduction, and injection induced fractures near the wellbore, which are often encountered during injection of polymer solutions. Therefore, in order to have a more accurate polymer injectivity adaptive to most reservoir conditions, it is necessary to refine simulation grids. However, grid refinement for the whole reservoir model leads to excessive computational costs. It is thus important to develop a local grid refinement (LGR) technique (or similar unstructured gridding approaches), such as shown in Fig. 2, so that the grid refinement is only applied to the regions where it is needed.

LGR and similar unstructured gridding approaches have continuously played an important role in reservoir simulations. Successful applications can be found in water flood (Oliveira and Reynolds 2014), miscible gas flood (Suicmez et al. 2011), steam flood (Christensen et al. 2004; Nilsson et al. 2005), etc. LGR methods are classified into cell-based and patch-based approaches (Berger and Oliger 1984), while the former is more frequently used in simulations of flow in porous media. Therefore, in the scope of this paper, we only discuss the cell-based LGR approach. Forsyth and Sammon (1986) developed an LGR algorithm with a rigorous analysis of discretization of flow equations upon the composite grid geometry. However, the accuracy of their numerical scheme is reported to be low because a direct subtraction of pressures of two adjacent blocks is used to calculate the Darcy velocity across the block interface (Rasaei and Sahimi 2009). Nacul et al. (1990) proposed an LGR technique using a domain decomposition method, in which overlapping boundaries are used for the subdomains. Karimi-Fard and Durlofsky (2012) presented an unstructured LGR method, and the well block is fully refined and solved at a fine scale to determine the effective properties that can be used for coarse-grid simulations over the reservoir domain.

In this paper, we propose an LGR method applied to chemical EOR simulations, especially more accurate calculation of polymer rheological viscosity (polymer injectivity) under different reservoir conditions. Meanwhile, for a necessary complement to the scope of LGR approaches, this paper presents details on the numerical schemes to couple the mass conservation equations on the multilevel grid, as well as the indexing to the gridblocks and interfaces. In short, the proposed method includes the following features:

(a) An efficient numerical scheme developed to calculate the velocity and the mass flux across the block interface between different grid levels of the composite grid, which is also applied in the heterogeneous cases.

(b) An algorithm on how to index the gridblock list and gridblock connections under the LGR composite grid presented in detail. The numerical computations under the LGR grid structure can benefit from this data management, which may also be extended to the classical unstructured grid and provide a good basis for the successive simulator development.

This paper is organized as follows: In the next two sections, we will give the mass balance equations and the chemical flood simulation models. The subsequent section presents the details of the proposed efficient LGR algorithm. We will then test several examples simulated with different levels of refinement to demonstrate the

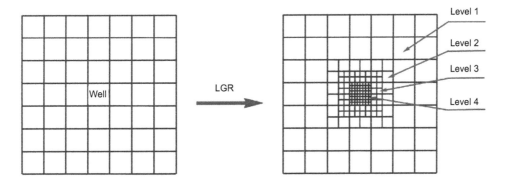

Fig. 2 Schematic of multilevel local grid refinement

improvement in numerical results. The LGR simulations are also compared to those using the analytical injectivity model proposed by Li and Delshad (2014).

2 Mathematical model

In this section, we briefly present the mathematical framework of the University of Texas Chemical Flooding Simulator, UTCHEM (Delshad et al. 1996) and formulations for modeling polymer rheology and injectivity. UTCHEM is a three-dimensional multi-phase multi-component compositional simulator with the capability of modeling geochemical reactions, complex phase behavior, etc. The governing balance equations include (1) the mass conservation equation for each species; (2) the pressure equation obtained by summing up all mass conservation equations for all volume-occupying species; and (3) the energy conservation equation which will not be discussed here.

2.1 Mass conservation equations

We write the mass conservation equation for each component κ,

$$\frac{\partial}{\partial t}(\phi \tilde{C}_\kappa \rho_\kappa) + \nabla \cdot \left[\sum_{l=1}^{n_p} \rho_\kappa (C_{\kappa l} \boldsymbol{u}_l - \tilde{\boldsymbol{D}}_{\kappa l}) \right] = R_\kappa, \quad (1)$$

where ϕ is the porosity, ρ_κ is the density of component κ, $C_{\kappa l}$ is the concentration of component κ in phase l, n_p is the phase number, and $\boldsymbol{u}_l$ is the Darcy flux of phase l which is calculated using Darcy's law:

$$\boldsymbol{u}_l = -\frac{k_{rl}\boldsymbol{k}}{\mu_l} \cdot (\nabla P_l - \gamma_l \nabla h), \quad (2)$$

where $\boldsymbol{k}$ is the intrinsic permeability tensor, k_{rl} is the relative permeability, μ_l is the viscosity, γ_l is the specific weight of phase l, and h represents the vertical depth.

$\tilde{C}_\kappa$ is the overall concentration of component κ in the mobile and stationary phases expressed as

$$\tilde{C}_\kappa = \left(1 - \sum_{\kappa=1}^{n_{cv}} \hat{C}_\kappa \right) \sum_{l=1}^{n_p} S_l C_{\kappa l} + \hat{C}_\kappa \quad \text{for } \kappa = 1, \ldots, n_c,$$

$$(3)$$

where S_l is the saturation of phase l, n_{cv} is the total number of volume-occupying components, and $\hat{C}_\kappa$ is the adsorbed concentration of component κ. In UTCHEM, the liquid phase l includes aqueous ($l = 1$), oleic ($l = 2$), and microemulsion ($l = 3$).

$\tilde{\boldsymbol{D}}_{\kappa l}$ is the dispersive flux which is assumed to have a Fickian form:

$$\tilde{\boldsymbol{D}}_{\kappa l} = \phi S_l \boldsymbol{K}_{\kappa l} \cdot \nabla C_{\kappa l}, \quad (4)$$

where the dispersion tensor $\boldsymbol{K}_{\kappa l}$ is calculated as

$$\boldsymbol{K}_{\kappa lij} = \frac{D_{\kappa l}}{\tau} \delta_{ij} + \frac{\alpha_{Tl}}{\phi S_l} |\boldsymbol{u}_l| \delta_{ij} + \frac{(\alpha_{Ll} - \alpha_{Tl})}{\phi S_l} \frac{u_{li} u_{lj}}{|\boldsymbol{u}_l|}, \quad (5)$$

where $D_{\kappa l}$ is the molecular diffusion, τ is the tortuosity factor of the porous media, α_{Ll} and α_{Tl} are phase l longitudinal and transverse dispersivities, and δ_{ij} is the Kronecker delta function.

R_κ is the source term which is a combination of all rate terms for a particular component κ. It may be expressed as

$$R_\kappa = \phi \sum_{l=1}^{n_p} S_l r_{\kappa l} + (1 - \phi) r_{\kappa s} + Q_\kappa, \quad (6)$$

where $r_{\kappa l}$ and $r_{\kappa s}$ are the reaction rates for component κ in phase l and solid phase s, respectively, and Q_κ is the injection/production rate for component κ per bulk volume.

2.2 Pressure equation

Summing the mass balance equations from Eq. (1) over all the volume-occupying components, substituting Eq. (2)

and using aqueous phase pressure as a reference pressure, we obtain the pressure equation:

$$\mu_{el} = \mu_{max}\left\{1 - \exp\left[-(\lambda_2 \tau \dot{\gamma}_{eff})^{n_2 - 1}\right]\right\}, \tag{13}$$

$$\phi C_t \frac{\partial P_1}{\partial t} + \nabla \cdot \boldsymbol{k} \cdot \left(\sum_{l=1}^{n_p} \lambda_{rlc}\right) \nabla P_1 = \nabla \cdot \left(\sum_{l=1}^{n_p} \vec{\bar{k}} \cdot \lambda_{rlc}(\nabla P_{cl1} - \gamma_l \nabla h)\right) + \sum_{l=1}^{n_{cv}} Q_\kappa \tag{7}$$

where P_{cl1} is the capillary pressure between phase l and phase 1 (the aqueous phase), and λ_{rlc} is the relative mobility expressed as

$$\lambda_{rlc} = \frac{k_{rl}}{\mu_l}\sum_{l=1}^{n_{cv}} \rho_\kappa C_{\kappa l}, \tag{8}$$

and C_t represents the total compressibility which is the volume-weighted sum of the rock matrix (C_r) and component compressibilities (C_κ^0):

$$C_t = C_r + \sum_{l=1}^{n_{cv}} C_\kappa^0 \tilde{C}_\kappa, \tag{9}$$

where $\phi = \phi_R[1 + Cr(P_R - P_{R0})]$, P_R and P_{R0} are rock and reference rock pressures.

2.3 Rheological viscosity of the polymer solution

Non-Newtonian polymer rheology (shear-thinning behavior) is modeled using Meter's equation (Meter and Bird 1964):

$$\mu_{app} = \mu_\infty + \frac{\mu_p^0 - \mu_\infty}{1 + \left(\frac{\dot{\gamma}_{eff}}{\dot{\gamma}_{1/2}}\right)^{P_\alpha - 1}}, \tag{10}$$

where μ_{app} is the apparent viscosity of the polymer solution; μ_∞ is the polymer solution viscosity at infinite shear rate which is assumed to be brine viscosity; $\dot{\gamma}_{1/2}$ is the shear rate at which the apparent viscosity is the average of μ_∞ and μ_p^0; P_α is a fitting parameter. For the synthetic polymer, e.g., HPAM, polymer solutions show shear-thinning behavior at intermediate shear rates and shear-thickening (dilatant) behavior at high rates. To remediate the deficiency of Meter's equation, Delshad et al. (2008) developed a comprehensive polymer viscosity model which covers the whole shear-rate regime. The apparent viscosity consists of two parts:

$$\mu_{app} = \mu_{sh} + \mu_{el}, \tag{11}$$

where the shear-thinning model uses the Carreau model (Carreau 1968):

$$\mu_{sh} = \mu_\infty + \left(\mu_p^0 - \mu_\infty\right)\left[1 + (\lambda_1 \dot{\gamma}_{eff})^2\right]^{(n_1 - 1)/2}, \tag{12}$$

and the shear-thickening model is

where a_1, a_2, and τ are all fitting model parameters obtained by matching experimental data; μ_{max} is given as

$$\mu_{max} = \mu_b\left(AP_{11} + AP_{22}\ln C_p\right)C_{SEP}^{S_p}, \tag{14}$$

where $C_{SEP}^{S_p}$ is the polymer viscosity dependence on salinity and hardness; AP_{11} and AP_{22} are fitting parameters. When AP_{11} and AP_{22} are zero, the comprehensive polymer viscosity model reduces to the Carreau model.

The effective shear rate ($\dot{\gamma}_{eff}$) correlates viscosity measured in a viscometer to an apparent in situ viscosity in porous media and is defined using a capillary bundle model (Cannella et al. 1998) as

$$\dot{\gamma}_{eff} = C\left(\frac{3n + 1}{4n}\right)^{\frac{n}{n-1}} \frac{4|\boldsymbol{u}_w|}{\sqrt{8\bar{k}k_{rw}\phi S_w}}, \tag{15}$$

where n is the slope of the linear portion of bulk polymer viscosity vs. shear rate plotted on a log–log scale (bulk power-law index); $\boldsymbol{u}_w$ is the Darcy flux of the aqueous polymer solution; $\bar{k}$ is the average permeability; k_{rw} is the aqueous phase relative permeability; S_w is the aqueous phase saturation; ϕ is the porosity; C is a shear correction factor used to explain the deviation of the porous medium from an ideal capillary bundle model (Wreath et al. 1990; Sorbie 1991) and should be a function of permeability, porosity, and polymer molecule properties.

2.4 Analytical polymer injectivity model

According to Peaceman's well model (Peaceman 1983), the relationship between the injection rate Q_{inj} and the pressure difference between injector and well block ($P_{inj} - P_{wb}$) can be expressed by

$$Q_{inj} = I(P_{inj} - P_{wb}), \tag{16}$$

where I is the well injectivity:

$$I = \frac{2\pi h\sqrt{k_x k_y}}{\ln\left(\frac{r_o}{r_w}\right) + s}\sum_{l=1}^{n_p} \frac{k_{rl,wb}}{\mu_{l,wb}}, \tag{17}$$

where h represents the thickness of the well block; r_o represents the Peaceman equivalent radius; r_w is the well radius; s is the skin factor; and $k_{rl,wb}$ and $\mu_{rl,wb}$ are the relative permeability and viscosity of phase l of well block, respectively.

In traditional simulation models, the polymer solution viscosity ($\mu_{w,wb}$) of the well block is directly calculated from Eqs. (10) or (11), using the averaged shear rate of the block. Thus, the shear rate is smeared and consequently gives significant error in well injectivity depending on the flow rate and the size of the gridblocks.

To overcome this limitation, Li and Delshad (2014) proposed a rigorous analytical injectivity model to calculate the equivalent apparent viscosity of polymer solution based on the assumption that after conversion of coordinates to account for the effects of non-square grids and anisotropic permeability, radial flow dominates the near-wellbore region, i.e.,

$$u(\bar{r}) = \frac{Q_{\text{inj}}}{2\pi h \bar{r}}, \tag{18}$$

where $\bar{r}$ is the distance from the wellbore after conversion of coordinates.

It can then be derived that the equivalent apparent viscosity of the polymer solution has the following expression:

$$\bar{\mu}_{w,wb} = \frac{\int_{r_w}^{\bar{r}_0} \mu_{\text{app}}(r) \frac{dr}{r}}{\ln\left(\frac{\bar{r}_0}{r_w}\right)}, \tag{19}$$

in which $\mu_{\text{app}}(r)$ adopts the form of Eqs. (10) or (11) using the shear rate calculated from the local velocity expressed by Eq. (18). For the detailed derivation, one can refer to Li and Delshad (2014).

3 UTCHEM flowchart

UTCHEM uses the finite volume method (FVM) and the implicit pressure explicit concentration (IMPEC) approach. The flowchart of the simulator is shown in Fig. 3.

In each time step, the simulator first solves the pressure equation (Eq. 7) implicitly and then solves concentration equations for each component (Eq. 1) explicitly using a third-order scheme with a flux limiter. After that, phase behavior calculations will be performed if a surfactant is present. In the last step, properties are updated by taking into account water reactions and polymer adsorption, as well as other chemical and physical changes. All the newly updated variables and properties will be provided for the initial values of the next time step. This continues until it reaches the final time.

4 Local grid refinement algorithm

The current form of the UTCHEM simulator is developed based on a structured grid, and the use of LGR will transform the grid from structured to unstructured as the connections between blocks are no longer regular. This makes it necessary to change the original data structure and computational model for solving the pressure equation and concentration equations.

To adapt the original computational structure to LGR and to maintain a good memory management, we designed a new flowchart for UTCHEM in Fig. 4. Compared to the original flowchart shown in Fig. 3, this new algorithm automatically generates the block list and connections according to the well location and refinement levels after the initialization step. An LGR module is also used to replace the original modules for solving the pressure equation and concentration equations. The other parts remain unchanged because those calculations are block based and not relevant to the grid structure. Features of the LGR algorithm will be presented in the following two subsections.

4.1 Block list and connections

Computations with an unstructured grid and LGR are normally based on a block list which gives the indices of

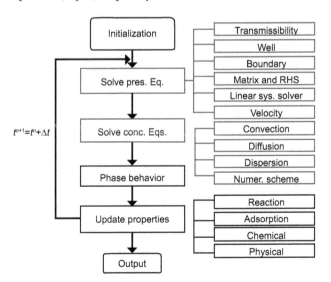

Fig. 3 Flowchart of UTCHEM

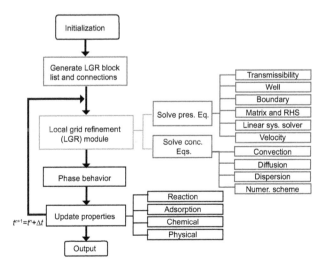

Fig. 4 Flowchart of UTCHEM using the LGR module

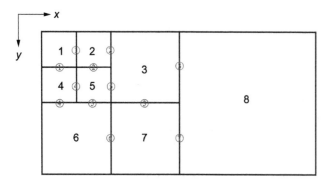

Fig. 5 An example of a block list and connections

gridblocks or cell numbering and connections which give the indices of block interfaces linking to a pair of adjacent blocks. Considering LGR has a special grid topology composed of rectangular blocks at different levels, we developed a fast algorithm to generate the block list and connections as illustrated in Fig. 5 with a 2D example case. The domain is originally covered by two coarse blocks, and then it is refined to 8 blocks. The numbering of the block list is advanced by each coarse block. For each coarse block, the numbering starts first along the x-direction and then the y-direction.

Different from the common unstructured grid, the connections in our LGR algorithm are divided into two types: x-direction connections (marked in red in Fig. 5) and y-direction connections (marked in blue in Fig. 5). A summary of the block list and connections is given in Table 1.

The indexing of a block list and connections facilitates the search for neighboring blocks and the assignment of properties evaluated at the block interfaces, such as transmissibility, velocity, and mass flux using the list of connections.

4.2 Coupling of governing equations

As the IMPEC scheme is used, the pressure equation and concentration equations are solved separately during computations.

4.2.1 Coupling of the pressure equation

The pressure equation needs to be solved implicitly and the calculation of velocities across the block interfaces of the composite grid is a common issue. Let us take the block connection in Fig. 6 as an example. The lengths of the coarse block are Δx and Δy and the lengths of the fine blocks are half. For the sake of simplicity to describe our approach, we assume in Fig. 6 isotropic permeabilities without a gravity effect and define λ as the total fluid mobility, i.e., $\lambda = k_{\text{abs}} \sum_{l=1}^{n_p} \frac{k_{rl}^{\text{ups}}}{\mu_l}$, where k_{rl}^{ups} is the relative permeability of phase l defined on the block interface with an upstream scheme. The upstream scheme to obtain k_{rl}^{ups} is the same as that to obtain the upstream concentration, C_κ^{ups}, which we will explain in the next subsection.

To calculate the fluxes across the block interfaces, such as $u_{(m)}$ and $u_{(n)}$, an early approach (Forsyth and Sammon 1986) used the pressures at the block centers to obtain the pressure difference in Darcy's law. However, it was pointed out that it generated high truncations (Rasaei and Sahimi 2009). Gerritsen and Lambers (2008) proposed in their anisotropic grid adaptivity method to use bilinear interpolation to obtain pressures of the auxiliary points (such as $P_{(i1)}$ and $P_{(i2)}$ in Fig. 6) for calculating the interfacial velocity using Darcy's law. This method proves to be second-order accurate when solving the pressure equation for homogeneous cases. However, the accuracy of bilinear interpolation is insufficient for heterogeneous cases because the discontinuity of the pressure gradient across the block interface is not taken into account. Actually, handling heterogeneity is an important factor to weigh up the reliability of the numerical scheme. As far as we know, there has not been a rigorous numerical scheme in the scope of the cell-centered finite volume method for accurately coupling the pressure equations with the LGR composite grid.

In Appendix 1, we derive a simple but efficient numerical scheme to couple pressure equations for the blocks with different grid levels. The expression of the velocities across the interface is as follows:

$$\begin{cases} u_{(m)} = -T_{(m)} \left[\left(\frac{P_{(j)} + P_{(k)}}{2} - P_{(i)} \right) \frac{u_{(mn)}^0}{u_{(mn)}^0 + u_{(jk)}^0} + \left(\frac{\lambda_{(j)} P_{(j)} + \lambda_{(k)} P_{(k)}}{(\lambda_{(j)} + \lambda_{(k)})} - P_{(i)} \right) \frac{u_{(jk)}^0}{u_{(mn)}^0 + u_{(jk)}^0} \right], \\ u_{(n)} = -T_{(n)} \left[\left(\frac{P_{(j)} + P_{(k)}}{2} - P_{(i)} \right) \frac{u_{(mn)}^0}{u_{(mn)}^0 + u_{(jk)}^0} + \left(\frac{\lambda_{(j)} P_{(j)} + \lambda_{(k)} P_{(k)}}{(\lambda_{(j)} + \lambda_{(k)})} - P_{(i)} \right) \frac{u_{(jk)}^0}{u_{(mn)}^0 + u_{(jk)}^0} \right], \end{cases} \tag{20}$$

Table 1 Mutual indexing of block list and connections

Block No.	Connections (x-direction)	Connections (y-direction)	Connection No. (x-direction)	Block pair	Connection No. (y-direction)	Block pair
1	1	1	1	1, 2	1	1, 4
2	1, 2	2	2	2, 3	2	2, 5
3	2, 5, 3	3	3	3, 8	3	3, 7
4	4	1, 4	4	4, 5	4	4, 6
5	4, 5	2, 5	5	5, 3	5	5, 6
6	6	4, 5	6	6, 7		
7	6, 7	3	7	7, 8		
8	3, 7	–				

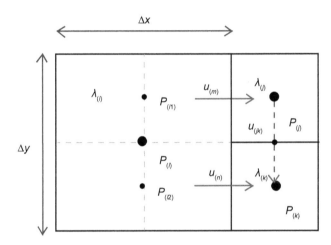

Fig. 6 Schematic block connection and the position of pressure and velocity

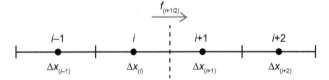

Fig. 7 Schematic of a third-order upstream scheme (Leonard's scheme) for a structured grid

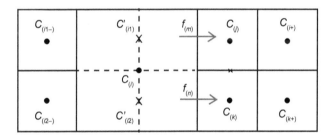

Fig. 8 Schematic of Leonard's scheme for an LGR case

where the meanings of $T_{(m)}$, $T_{(n)}$, $u_{(mn)}^0$, and $u_{(jk)}^0$ are given in Appendix 1.

This numerical scheme has the following advantages:

- It has a simple form as it does not require any additional information from other blocks except for the current three connected blocks.
- It is easy to use as it does not need any interpolation/extrapolation.
- It is based on the continuity of mass flux across the interfaces and it is rigorously self-consistent under the homogeneous condition or the condition that fine-block permeabilities are identical. The latter condition is often met for most LGR applications when the permeabilities of the refined blocks are directly from the coarse block permeability.

4.2.2 Coupling of mass conservation equations

To guarantee the numerical stability, upstream schemes are mainly used to solve mass conservation equations. In the UTCHEM simulator, there are several options for the upstream schemes. These are first-order upstream scheme, second-order upstream scheme, and a third-order upstream scheme named Leonard's scheme (Saad 1989; Liu et al. 1994). Because higher order upstream schemes are more accurate to integrate concentration equations, we only discuss about how to couple concentration equations using the Leonard scheme in this paper. Under the structured grid in Fig. 7, the Leonard scheme to calculate the mass flux

$$
f_{\left(i+\frac{1}{2}\right)} = \begin{cases} u_{\left(i+\frac{1}{2}\right)}\left[C_{(i)} - \dfrac{\Delta x_{(i)}\left(C_{(i-1)} - C_{(i)}\right)}{3\left(\Delta x_{(i)} + \Delta x_{(i-1)}\right)} - \dfrac{2\Delta x_{(i)}\left(C_{(i)} - C_{(i+1)}\right)}{3\left(\Delta x_{(i)} + \Delta x_{(i+1)}\right)}\right] & \text{if } u_{\left(i+\frac{1}{2}\right)} > 0 \\[4mm] u_{\left(i+\frac{1}{2}\right)}\left[C_{(i+1)} - \dfrac{\Delta x_{(i+1)}\left(C_{(i)} - C_{(i+1)}\right)}{3\left(\Delta x_{(i)} + \Delta x_{(i+1)}\right)} - \dfrac{2\Delta x_{(i+1)}\left(C_{(i+1)} - C_{(i+2)}\right)}{3\left(\Delta x_{(i+2)} + \Delta x_{(i+1)}\right)}\right] & \text{if } u_{\left(i+\frac{1}{2}\right)} < 0, \end{cases}
\tag{21}
$$

across the interface at $i + \frac{1}{2}$ is expressed by where C represents the component concentration.

For the LGR grid, we take the block combination in Fig. 8 as one example. In this case, because the block center points are not in the same line, we utilize bilinear interpolation to obtain the concentration values C' at the auxiliary points, e.g., $i1$ and $i2$. After that, we extend Leonard's scheme to this case:

Adsorption and permeability reduction are also considered. The reservoir and well descriptions are given in Table 2. The basic grid used for simulation is $15 \times 15 \times 1$, and the grid with a 4-level refinement is shown in Fig. 9. It shows that the well block is refined to 8×8 finest blocks and several transitional blocks connect the original coarse blocks and finest blocks.

$$
f_{(m)} = \begin{cases} u_{(m)}\left[C'_{(i1)} - \dfrac{\Delta x_{(i)}\left(C_{(i1-)} - C'_{(i1)}\right)}{3\left(\Delta x_{(i)} + \Delta x_{(i-)}\right)} - \dfrac{2\Delta x_{(i)}\left(C'_{(i1)} - C_{(j)}\right)}{3\left(\Delta x_{(i)} + \Delta x_{(j)}\right)}\right] & \text{if } u_{(m)} > 0 \\[4mm] u_{(m)}\left[C_{(j)} - \dfrac{\Delta x_{(j)}\left(C'_{(i1)} - C_{(j)}\right)}{3\left(\Delta x_{(i)} + \Delta x_{(j)}\right)} - \dfrac{2\Delta x_{(j)}\left(C_{(j)} - C_{(j+)}\right)}{3\left(\Delta x_{(j+)} + \Delta x_{(j)}\right)}\right] & \text{if } u_{(m)} < 0, \end{cases}
\tag{22}
$$

$$
f_{(n)} = \begin{cases} u_{(n)}\left[C'_{(i2)} - \dfrac{\Delta x_{(i)}\left(C_{(i1-)} - C'_{(i2)}\right)}{3\left(\Delta x_{(i)} + \Delta x_{(i-)}\right)} - \dfrac{2\Delta x_{(i)}\left(C'_{(i2)} - C_{(k)}\right)}{3\left(\Delta x_{(i)} + \Delta x_{(k)}\right)}\right] & \text{if } u_{(n)} > 0 \\[4mm] u_{(n)}\left[C_{(k)} - \dfrac{\Delta x_{(k)}\left(C'_{(i2)} - C_{(k)}\right)}{3\left(\Delta x_{(i)} + \Delta x_{(k)}\right)} - \dfrac{2\Delta x_{(j)}\left(C_{(k)} - C_{(k+)}\right)}{3\left(\Delta x_{(k+)} + \Delta x_{(k)}\right)}\right] & \text{if } u_{(n)} < 0, \end{cases}
\tag{23}
$$

where $f_{(m)}$ and $f_{(n)}$ are mass fluxes across the interfaces m and n.

5 Case study

To validate the LGR method proposed in this paper, we tested four simulation examples. These examples show comparisons of simulation results using the LGR method with those using the analytical polymer well model, and full grid refinement (FGR) where the whole model has the smallest grid size of the LGR.

5.1 Case 1: Polymer flooding in a 2D homogeneous reservoir

We start with a base case for polymer flooding. The polymer solution is assumed to be shear thinning.

In this case, the injection rate is constant, so the injection pressure varies with different polymer viscosities and thus well injectivities. Figure 10 shows a comparison of injection pressures using different grids or well models. For water flooding periods, there are no obvious differences in injection pressures among different simulations. However, when shifted to polymer flooding, it is observed that the injection pressures are remarkably differentiated using different grids or well models because of polymer rheology. It also shows that using the original grid leads to the highest injection pressure. The reason is that the averaged viscosity within the well block area is artificially amplified due to a lower smeared flux rate caused by the coarse block size, which leads to an over-prediction of injection pressure that triggers the pressure limit, for example, in the case that the facility's injection pressure limit is 6000 psi. By contrast with grid refinement around the well block, the injection pressure gradually decreases which results in a much "safer" injection pressure. Also, the variation of the

Table 2 Reservoir and well descriptions (Case 1)

Model description	Values
Reservoir size	450 ft × 450 ft × 10 ft
No. of gridblocks	15 × 15 × 1
Simulation time, day	365
Number of components	3
Permeability in the x or y directions, mD	300
Initial water saturation	0.35
Polymer rheology exponent P_α	1.8
Shear rate at half zero-rate viscosity γ_{hf}, s^{-1}	10
Wells	1 injector; 1 producer
Injection rate, ft^3/day	500
Producer bottomhole pressure (BHP), psi	1000
Water injection	0–150 and 270–365 days
Polymer injection	150–270 days (0.3 wt%)

injection pressure shrinks with an increase in the level of grid refinement, showing a convergent trend. Because simulation results using 3-level LGR and 4-level LGR are relatively close and further refinement may lead to excessive computational times, we regard the simulation result of 4-level LGR as the reference result to evaluate other simulations. Of course, it should be more precise to use the fully refined grid as the reference. Nevertheless, Fig. 10 shows that 4-level FGR gives a very similar injection pressure to the 4-level LGR. We also use the analytical injectivity model (Li and Delshad 2014) and we observe that the simulated injection pressure is between the results of 3-level LGR and 4-level LGR. This result is more accurate than the case without grid refinement and shows consistency with LGR results.

To further demonstrate the accuracy and computational efficiency of the LGR method, we compare simulation results with CMG_STARS (2012). In the above case, rheology parameters, P_α and $\gamma_{1/2}$, used in the polymer rheology equation (Eq. 10), are set as 1.8 and 10 s^{-1}, respectively. These parameters lead to a relatively sharp shear-thinning curve. CMG_STARS uses a different polymer rheology equation, which is a power-law equation:

$$\mu_{app} = \begin{cases} \mu_p^0 & \text{if} \quad u_w < u_{lower} \\ \mu_p^0 \left[\dfrac{u_w}{u_{lower}}\right]^{n_{thin}-1} & \text{if} \quad u_{lower} < u_w < u_{upper} \\ \mu_\infty & \text{if} \quad u_w > u_{upper}, \end{cases} \quad (24)$$

where n_{thin} is the power-law exponent, and u_{lower} is defined by the point on the power-law curve when μ_{app} is equal to μ_p^0. To be close to the UTCHEM polymer equation (Eq. 10) for Case 1 using the CMG_STARS equation, we found out

that n_{thin} must be small and it causes numerical stability issues which are also indicated in the manual of CMG_STARS (2012). Therefore, to achieve a relatively similar polymer rheology curves for both simulators, we use $P_\alpha = 1.5$ and $\gamma_{1/2} = 3.8$ s^{-1} for UTCHEM and $n_{thin} = 0.5$ and $u_{lower} = 0.02$ ft/day for CMG_STARS. Figure 11 shows a comparison of the results between UTCHEM and CMG_STARS using the original grid, 4-level LGR, and 4-level FGR, respectively. It is found that the injection pressure curves for the original grid match very well between UTCHEM and CMG_STARS. In

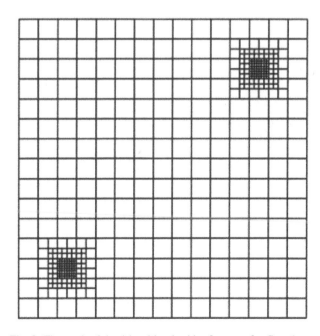

Fig. 9 The mesh of the 4-level local grid refinement for Case 1

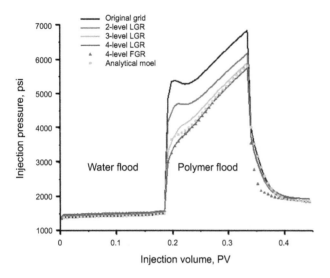

Fig. 10 Comparison of injection pressure using different grids or well models for Case 1

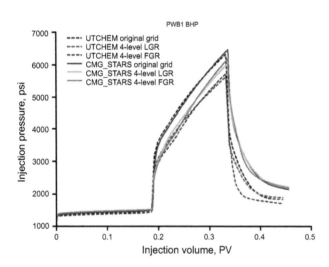

Fig. 11 Comparison of injection pressure between UTCHEM and CMG_STARS using different grids for Case 1

Table 3 CPU times for UTCHEM and CMG_STARS using different grids for Case 1

Grid	Original grid (225 gridblocks)	4-level LGR (471 gridblocks)	4-level FGR (14,400 gridblocks)
CPU time			
UTCHEM	1 min 26 s	3 min 40 s	141 min 39 s
CMG_STARS	4 min 14 s	9 min 12 s	318 min 45 s

addition, the 4-level LGR simulation results of UTCHEM and CMG_STARS are also close, with only a minor difference. This is acceptable because CMG_STARS and UTCHEM use different polymer concentration-dependent

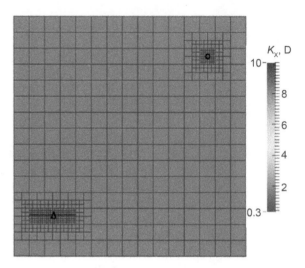

Fig. 12 Permeability field of a fractured reservoir using an LGR grid

viscosity models and shear-thinning models as mentioned in Goudarzi et al. (2013a). Again, for FGR results, we observe both UTCHEM and CMG_STARS match well with LGR results.

In Table 3, we compare the CPU times taken by UTCHEM and CMG_STARS using different grids. For both simulators, we use the same maximum time step (0.01 day) and the same numerical scheme (IMPES) on the same computer for the sake of consistency. We can see that CMG_STARS takes about 3 times that of UTCHEM for the original coarse grid, 2.5 times for the 4-level LGR, and 2.2 times for the 4-level FGR. An increase in CPU times is also in the same order of the increase in gridblock numbers. LGR shows very good computational efficiency compared to FGR. Actually, CMG_STARS can take larger time steps because it can use an adaptive implicit scheme. Therefore, the purpose of the comparison is not to tell which simulator is better in performance but to obtain a sense of the scaling of the CPU times using LGR and FGR. In fact, the advantage of using UTCHEM for modeling polymer flood is that it has more comprehensive polymer models than CMG_STARS, such as more options of rheology models and stricter concentration-dependent and salinity-dependent polymer, and near-well-corrected viscosity models.

5.2 Case 2: Polymer flooding in a 2D reservoir with a fracture near the injector

This case is aimed to analyze the behavior of the LGR in the presence of a planar fracture near the injector, which may often be encountered during polymer injection projects (Manichand et al. 2013; Clemens et al. 2013). We assume that the reservoir has the same condition as case 1 except that there is a fracture initiated from the well block of the injector. We show in Fig. 12 a 4-level LGR grid with

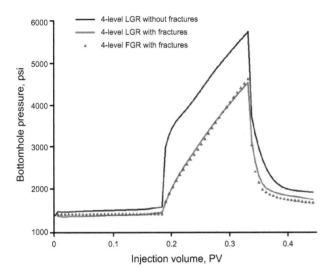

Fig. 13 Comparison of injector bottomhole pressure (BHP) using different grids or well models for Case 2

Table 4 Reservoir and well descriptions (Case 3)

Model description	Values
No. of gridblocks	$17 \times 21 \times 25$
No. of components	6
Total injection volume for simulation, PV	0.32
Polymer injection volume, PV	0–0.16 (0.2 wt%)
Water injection volume, PV	0.16–0.32
BHP, psi	Injectors 4500; Producers 700

permeability field (the fracture permeability is assumed to be 10 Darcy). In this case, it is obviously not proper to use the coarse grid as well as the analytical injectivity model, which could not describe the flow in fractures. Therefore, it is obligatory to refine the gridblocks.

Figure 13 shows simulation results under three conditions: LGR without fractures, LGR with fractures, and FGR with fractures. The LGR with fractures leads to a significantly smaller injection pressure compared to the LGR without fractures, which shows the importance of accounting for fractures near the injector. We also show that the pressure curve of FGR is close to that of LGR, which proves the agreement of the results between using the two types of grids.

5.3 Case 3: Polymer flooding in a 3D heterogeneous reservoir

We study a polymer flooding field case. The polymer solution is assumed to be shear thinning. Adsorption and permeability reduction are also considered. The reservoir and well descriptions are given in Table 4. The permeability field and well locations are shown in Fig. 14. The relevant grid with 4-level LGR is in Fig. 15. Some wells are deviated so that the LGR is expanded in the x–y plane.

In this case, the injection pressure is fixed so that the injection rate varies with well injectivity. Figure 16 shows the simulated overall injection rate using the original grid, 4-level LGR, and the analytical injectivity model. It is observed that we achieve a higher injection rate with grid refinement compared to the original grid, which is consistent with the polymer rheology. This is significant since

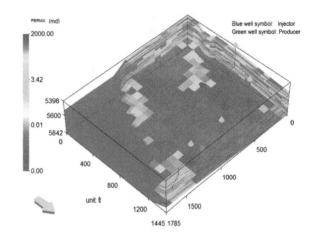

Fig. 14 Permeability distribution and well locations for Case 3

we need to accurately calculate how high a polymer viscosity can be injected and the predicted bottomhole pressure (BHP) for cases that the operators do not plan to inject polymer above the fracture gradient. In this case, we lack the results of FGR because of the excessive simulation time. It is observed that the analytical model slightly overestimates the overall injection rate compared to 4-level LGR. The analytical model is not accurate for this case because the well is inclined.

5.4 Case 4: A pilot of alkaline co-solvent polymer (ACP) flood

The reservoir is a sandstone reservoir at a depth of approximately 1000 ft which has undergone water flooding for several years. The average oil saturation before ACP flood is approximately 44.3 %. The pilot area includes 6 inverted 7-spot well patterns. The polymer solution is assumed to be shear thinning. The reservoir and well descriptions are given in Table 5. The permeability field and well locations are shown in Fig. 17. The relevant grid

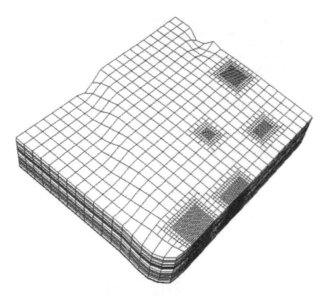

Fig. 15 Well locations and 4-level local grid refinement

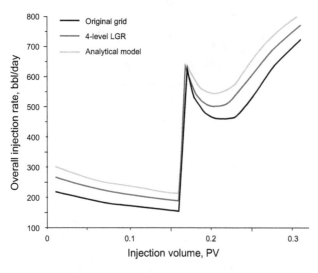

Fig. 16 Comparison of overall injection rate using different grids or well models for Case 3

Table 5 Reservoir and well descriptions (Case 4)

Model description	Values
Reservoir dimension	5512 ft × 4856 ft × 98 ft
No. of gridblocks	42 × 37 × 5
No. of components	12
Total simulation time, day	7300
Optimum salinity, meq/mL	0.26
Wells	6 injectors; 22 producers
ACP injection	0–3650 days
	1.5 wt% co-solvent
	0.275 wt% polymer
Polymer injection	3650–7300 days
	0.225 wt% polymer

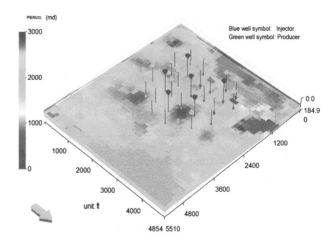

Fig. 17 Permeability distributions and well locations for Case 4

indicated by Table 6, one needs to consider the significant improvement in accuracy to balance the cost of computational time. The analytical injectivity model gives a similar trend but a different profile of the pressure compared to using the original grid and the LGR grids, because there is a non-zero skin factor for the well while the analytical injectivity model originates assuming that skin is equal to 0. This indicates that the analytical polymer injectivity model is not always useful for field cases.

6 Summary and conclusions

We have used an efficient LGR algorithm to improve the accuracy of numerically estimating the near-wellbore solutions when dealing with complex rheology of polymer or emulsion solutions. We present an algorithm to generate the block list and connections and propose an efficient numerical scheme to couple the pressure and mass conservation equations using the LGR composite grid and with consideration of heterogeneous reservoir properties.

with a 3-level LGR is shown in Fig. 18. Injectors are operated at constant injection rates.

Figure 19 shows the BHP of Injector ECN-105i using different grids and well models (we note that other injectors have similar pressure profiles). For the original 42 × 37 × 5 coarse grid, the BHP is highest. When the grid is locally refined near the injector, the BHP decreases significantly, and the pressure change is more gradual, showing a significant improvement for estimating the BHP of the injector using the LGR. The 3-level LGR gives a much smaller BHP compared to the 2-level LGR. The results from the 2-level LGR and 3-level LGR are in very good agreement with the relevant FGR results, respectively, while taking much less CPU time (Table 6). Even though more CPU time is needed using the LGR as

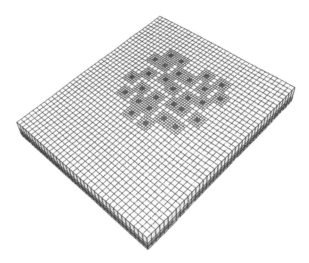

Fig. 18 Schematic of the 3-level local grid refinement for Case 4

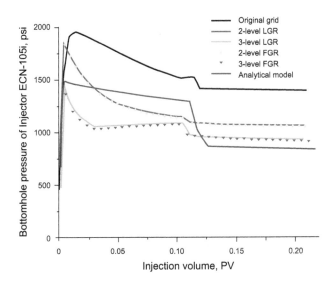

Fig. 19 Comparison of BHP for injector ECN-105i using different grids and well models for Case 4

Several numerical examples are carried out, focusing on polymer flooding, reservoir with fractures near injection wells, and ACP flooding. Simulation results reveal that the LGR is able to obtain more accurate polymer injectivity compared to using the coarse grid and the analytical injectivity model. The LGR can deal with more complex and realistic reservoir conditions such as fractures and skin. CPU time is significantly reduced using LGR compared to FGR. This offers a reliable and efficient solution to handle the general concern of reservoir simulations for the shear-dependent polymer rheology in chemical flooding projects.

Acknowledgments The authors would like to acknowledge the sponsors of the Chemical EOR Industrial Affiliates Project at The University of Texas at Austin.

Appendix 1: A novel numerical scheme to solve the pressure equation on a multilevel grid

To simplify the complex flow in the composite blocks as shown in Fig. 6, we decompose the flow pattern into two flow patterns (for two-dimensional case) with main flow directions along x- and y-directions, which are shown in Fig. 20a, b. For each pattern, a pressure drop is assumed along the main flow direction while making the lateral sides impermeable. Then, we split the flow domain into two parallel parts, such as part A and part B in the x-direction main flow pattern and part C and part D in the y-direction main flow pattern. According to the continuity of mass flux across the interfaces m and n, we draw the pressure curves through the two parts of each pattern shown in Fig. 20c, d by neglecting the cross flows between the two parts. Then, we investigate the expressions of $u_{(m)}$ and $u_{(n)}$ for each pattern based on the pressure curves.

Let us see the curves of Fig. 20c, d. For the flow pattern with a flow direction along x, according to the geometric knowledge, the length of line $i1 - i2$ is equal to the length of line $j–k$ because they are both half the length of line $m–n$. It is easy to infer that line $i1–j$ is equal to and parallel to line $i2–k$. As a result, the line connecting point i and the point at the center of line $j–k$ is equal to and parallel to the previous two lines. Therefore, we can obtain the following relationship:

$$P_{(i1)} - P_{(j)} = P_{(i2)} - P_{(k)} = P_{(i)} - \frac{P_{(j)} + P_{(k)}}{2}. \quad (25)$$

Using Darcy's law, we have the expressions of $u_{(m)}$ and $u_{(n)}$ according to the continuity of mass flux across the interfaces m and n:

$$u_{(m)} = -T_{(m)} \left(\frac{P_{(j)} + P_{(k)}}{2} - P_{(i)} \right)$$
$$u_{(n)} = -T_{(n)} \left(\frac{P_{(j)} + P_{(k)}}{2} - P_{(i)} \right), \quad (26)$$

Table 6 CPU times for Case 4 using original grid, LGR, and FGR

Grid	Original grid (7770 gridblocks)	3-level LGR (13,185 gridblocks)	3-level FGR (124,320 gridblocks)
CPU time, h	0.2	1.9	23

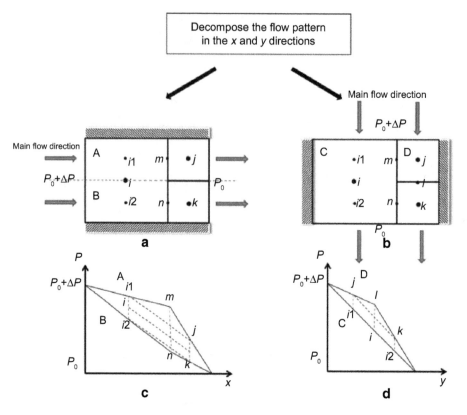

Fig. 20 Decomposition of the flow pattern into two with main flow directions along x- and y-directions; and the approximate pressure curves through the two separated parts (A and B for x-direction pattern, C and D for y-direction pattern) of each pattern

where $T_{(m)}$ and $T_{(n)}$ are transmissibilities on the interfaces m and n defined by

$$T_{(m)} = \frac{4\lambda_{(i)}\lambda_{(j)}}{\left(\lambda_{(i)} + 2\lambda_{(j)}\right)\Delta x}$$
$$T_{(n)} = \frac{4\lambda_{(i)}\lambda_{(k)}}{\left(\lambda_{(i)} + 2\lambda_{(k)}\right)\Delta x}. \tag{27}$$

For the flow pattern with a flow direction along y, according to the geometric relation, the length of line $i1$–j is equal to the length of line $i2$–k because they are both half the length of line i–l. Consequently, the following relationship can be obtained:

$$P_{(i1)} - P_{(j)} = P_{(i2)} - P_{(k)} = P_{(i)} - \frac{\lambda_{(k)}P_{(j)} + \lambda_{(j)}P_{(k)}}{\left(\lambda_{(j)} + \lambda_{(k)}\right)}. \tag{28}$$

Using Darcy's law, we have the expressions of $u_{(m)}$ and $u_{(n)}$ according to the continuity of mass flux across the interfaces m and n:

$$u_{(m)} = -T_{(m)}\left(\frac{\lambda_{(j)}P_{(j)} + \lambda_{(k)}P_{(k)}}{\left(\lambda_{(j)} + \lambda_{(k)}\right)} - P_{(i)}\right)$$
$$u_{(n)} = -T_{(n)}\left(\frac{\lambda_{(j)}P_{(j)} + \lambda_{(k)}P_{(k)}}{\left(\lambda_{(j)} + \lambda_{(k)}\right)} - P_{(i)}\right). \tag{29}$$

A comparison between Eqs. (26) and (29) shows that the expressions of $u_{(m)}$ and $u_{(n)}$ are different under the two flow patterns. Nevertheless, they can achieve agreement under the homogeneous condition or the condition that the total mobilities of the fine blocks are identical, i.e., $\lambda_{(j)} = \lambda_{(k)}$, which indicates that the expressions are rigorously self-consistent under these conditions.

For the heterogeneous condition, we need to choose the appropriate expressions of $u_{(m)}$ and $u_{(n)}$ from Eqs. (26) and (29) by considering the direction of the main flow. The proper velocity across the interface can be a weighting of the x-direction velocity and the y-direction velocity weighted by the magnitude of the local velocities along x and y directions, i.e., $u^0_{(mn)}$ and $u^0_{(jk)}$,

where

$$u^0_{(mn)} = \frac{\left(\left|u^0_{(m)}\right| + \left|u^0_{(n)}\right|\right)}{2}, \tag{30}$$

and $u^0_{(jk)}$ is the flux across the interface of blocks j and k as shown in Fig. 6 which is expressed by

$$u^0_{(jk)} = -\frac{4\lambda_{(j)}\lambda_{(k)}}{\left(\lambda_{(j)} + \lambda_{(k)}\right)\Delta y}\left(P_{(k)} - P_{(j)}\right), \tag{31}$$

where the superscript 0 for the velocities represents the last time step.

Therefore, we design the following numerical scheme to calculate $u_{(m)}$ and $u_{(n)}$:

$$
\begin{cases}
u_{(m)} = -T_{(m)} \left[\left(\dfrac{P_{(j)} + P_{(k)}}{2} - P_{(i)} \right) \dfrac{u^0_{(mn)}}{u^0_{(mn)} + u^0_{(jk)}} + \left(\dfrac{\lambda_{(j)} P_{(j)} + \lambda_{(k)} P_{(k)}}{(\lambda_{(j)} + \lambda_{(k)})} - P_{(i)} \right) \dfrac{u^0_{(jk)}}{u^0_{(mn)} + u^0_{(jk)}} \right] \\[4mm]
u_{(n)} = -T_{(n)} \left[\left(\dfrac{P_{(j)} + P_{(k)}}{2} - P_{(i)} \right) \dfrac{u^0_{(mn)}}{u^0_{(mn)} + u^0_{(jk)}} + \left(\dfrac{\lambda_{(j)} P_{(j)} + \lambda_{(k)} P_{(k)}}{(\lambda_{(j)} + \lambda_{(k)})} - P_{(i)} \right) \dfrac{u^0_{(jk)}}{u^0_{(mn)} + u^0_{(jk)}} \right].
\end{cases}
\tag{32}
$$

We are aware that this numerical scheme is achieved based on the assumption that the cross flows between the separated parts for each flow pattern can be neglected. Actually, the piece-wise pressure curves shown in Fig. 20 may be bent when there is cross flow vertical to the main flow direction. Nevertheless, this numerical scheme has a lot of advantages that will be discussed in the text.

References

Bekbauov BE, Kaltayev A, Wojtanowicz AK, Panfilov M. Numerical modeling of the effects of disproportionate permeability reduction water-shutoff treatments on water coning. J Energy Res Technol. 2013;135(1):011101. doi:10.1115/1.4007913.

Berger MJ, Oliger J. Adaptive mesh refinement for hyperbolic partial differential equations. J Comput Phys. 1984;53(3):484–512. doi:10.1016/0021-9991(84)90073-1.

Cannella WJ, Huh C, Seright RS. Prediction of xanthan rheology in porous media. In: SPE annual technical conference and exhibition, Houston, TX; 1998. doi:10.2118/18089-MS.

Carreau PJ. Rheological equations from molecular network theories. PhD dissertation, University of Wisconsin-Madison. 1968.

Christensen JR, Darche G, Dechelette B, Ma H, Sammon PH. Applications of dynamic gridding to thermal simulations. In: SPE international thermal operations and heavy oil symposium and western regional meeting, Bakersfield; 2004. doi:10.2118/86969-MS.

Clemens T, Deckers M, Kornberger M, Gumpenberger T, Zechner M. Polymer solution injection-near wellbore dynamics and displacement efficiency, pilot test results, Matzen Field, Austria. In: EAGE annual conference and exhibition incorporating SPE Europec, London; 2013. doi:10.2118/164904.

Computing Modeling Group Ltd. User's guide STARS: Advanced Process and Thermal Reservoir Simulator, Calgary, AB; 2012.

Delshad M, Pope GA, Sepehrnoori K. A compositional simulator for modeling surfactant enhanced aquifer remediation, 1. Formulation. J Contam Hydrol. 1996;23(4):303–27. doi:10.1016/0169-7722(95)00106-9.

Delshad M, Kim D, Magbagbeola O, Huh C, Pope G, Tarahhom F. Mechanistic interpretation and utilization of viscoelastic behavior of polymer solutions for improved polymer-flood efficiency. In: SPE symposium on improved oil recovery, Tulsa; 2008. doi:10.2118/113620-MS.

Forsyth PA, Sammon PH. Local mesh refinement and modeling of faults and pinchouts. SPE For Eval. 1986;1(3):275–85. doi:10.2118/13524-PA.

Gadde P, Sharma M. Growing injection well fractures and their impact on waterflood performance. In: SPE annual technical conference and exhibition, New Orleans; 2001. doi:10.2118/71614-MS.

Gerritsen M, Lambers JV. Integration of local-global upscaling and grid adaptivity for simulation of subsurface flow in heterogeneous formations. Comput Geosci. 2008;12(2):193–208. doi:10.1007/s10596-007-9078-2.

Goudarzi A, Delshad M, Sepehrnoori K. A critical assessment of several reservoir simulators for modeling chemical enhanced oil recovery processes. In: SPE reservoir simulation symposium, Woodlands; 2013a. doi:10.2118/163578-MS.

Goudarzi A, Zhang H, Varavei A, Hu Y, Delshad M, Bai B, Sepehrnoori K. Water management in mature oil fields using preformed particle gels. In: SPE Western Regional AAPG Pacific Section meeting, 2013 Joint Technical Conference, Monterey; 2013b. doi:10.2118/165356.

Karimi-Fard M, Durlofsky L. Accurate resolution of near-well effects in upscaled models using flow based unstructured local grid refinement. SPE J. 2012;17(4):1084–95. doi:10.2118/141675-PA.

Kulawardana E, Koh H, Kim DH, Liyanage P, Upamali K, Huh C, Weerasooriya U, Pope G. Rheology and transport of improved EOR polymers under harsh reservoir conditions. In: SPE improved oil recovery symposium, Tulsa; 2012. doi:10.2118/154294-MS.

Lee K, Huh C, Sharma M. Impact of fractures growth on well injectivity and reservoir sweep during waterflood and chemical EOR processes. In: SPE annual technical conference and exhibition, Denver; 2011. doi:10.2118/146778-MS.

Li Z, Delshad M. Development of an analytical injectivity model for non-Newtonian polymer solutions. SPE J. 2014;19(3):381–9. doi:10.2118/163672-MS.

Liu J, Delshad M, Pope GA, Sepehrnoori K. Application of higher-order flux-limited methods in compositional simulation. Transp Porous Media. 1994;16(1):1–29. doi:10.1007/BF01059774.

Manichand RN, Let MS, Kathleen P, Gil L, Quillien B, Seright RS. Effective propagation of HPAM solutions through the Tambaredjo reservoir during a polymer flood. SPE Prod Oper. 2013;28(4):358–68. doi:10.2118/164121-PA.

Meter DM, Bird RB. Tube flow of non-Newtonian polymer solutions: part I. Laminar flow and rheological models. AIChE J. 1964;10(6):878–81. doi:10.1002/aic.690100619.

Mohammadi H, Jerauld G. Mechanistic modeling of the benefit of combining polymer with low salinity water for enhanced oil recovery. In: SPE improved oil recovery symposium, Tulsa; 2012. doi:10.2118/153161-MS.

Morel D, Vert M, Jouenne S, Gauchet R, Bouger Y. First polymer injection in deep offshore field, Angola: recent advances on Dalia/Camelia field case. In: SPE annual technical conference and exhibition, 19–22 September, Florence; 2012. doi:10.2118/135735-PA.

Nacul EC, Lepretre C, Pedrosa OA, Girard P, Aziz K. Efficient use of domain decomposition and local grid refinement in reservoir simulation. In: SPE annual technical conference and exhibition, New Orleans; 1990. doi:10.2118/20740-MS.

Nilsson J, Gerritsen M, Younis R. An adaptive, high-resolution simulation for steam-injection processes. In: SPE western regional meeting, Irvine; 2005. doi:10.2118/93881-MS.

Oliveira DF, Reynolds A. An adaptive hierarchical multiscale algorithm for estimation of optimal well controls. SPE J. 2014;19(5):909–30. doi:10.2118/163645-PA.

Peaceman DW. Interpretation of well-block pressures in numerical reservoir simulation with nonsquare grid blocks and anisotropic permeability. SPE J. 1983;23(3):531–43. doi:10.2118/10528-PA.

Rasaei MR, Sahimi M. Upscaling of the permeability by multiscale wavelet transformations and simulation of multiphase flows in heterogeneous porous media. Comput Geosci. 2009;13(2):187–214. doi:10.1007/s10596-008-9111-0.

Saad N. Field-scale simulation of chemical flooding. Austin: Texas University; 1989.

Seright R, Seheult J, Talashek T. Injectivity characteristics of EOR polymers. SPE Reserv Eval Eng. 2009;12(5):783–92. doi:10.2118/115142-MS.

Sharma A, Delshad M, Huh C, Pope G. A practical method to calculate polymer viscosity accurately in numerical reservoir simulators. In: SPE annual technical conference and exhibition, Denver; 2011. doi:10.2118/147239-MS.

Sorbie KS. Polymer-improved oil recovery. Glasgow: Blackie and Son Ltd.; 1991.

Stahl GA, Schulz DN. Water-soluble polymers for petroleum recovery. New York: Springer; 1988.

Suicmez V, van Batenburg D, Matsuura T, Bosch M, Boersma D. Dynamic local grid refinement for multiple contact miscible gas injection. In: International petroleum technology conference, Bangkok; 2011. doi:10.2523/15017-MS.

van den Hoek PJ, Mahani H, Sorop TG, Brooks AD, Zwaan M, Sen S, Shuaili K, Saadi F. Application of injection fall-off analysis in polymer flooding. In: The 74th EAGE conference & exhibition incorporating SPE Europe, Copenhagen; 2012. doi:10.2118/154376.

Wassmuth F, Green K, Hodgins L, Turta A. Polymer flood technology for heavy oil recovery. In: Canadian international petroleum conference, 12–14 June, Calgary; 2007. doi:10.2118/2007-182.

Wreath D, Pope GA, Sepehrnoori K. Dependence of polymer apparent viscosity on the permeable media and flow conditions. In Situ. 1990;14(3):263–84.

Zaitoun A, Makakou P, Blin N, Al-Maamari RS, Al-Hashmi AAR, Abdel-Goad M. Shear stability of EOR polymers. SPE J. 2012;17(02):335–9. doi:10.2118/141113-MS.

16

Stabilizing and reinforcing effects of different fibers on asphalt mortar performance

Meng-Meng Wu · Rui Li · Yu-Zhen Zhang ·
Liang Fan · Yu-Chao Lv · Jian-Ming Wei

Abstract Physical properties of different fibers (mineral, cellulose, or carbon fiber) and their stabilizing and reinforcing effects on asphalt mortar performance were studied. Scanning electron microscopy was used to study the effect of fiber's microstructure on asphalt mortar's performance. Laboratory tests of mesh-basket draindown and oven heating were designed and performed to evaluate the fibers' asphalt absorption and thermostability. A cone penetration test was used to study the flow resistance of fiber-modified asphalt mortar. Results showed that fiber can form a three-dimensional network structure in asphalt, and this network can be retained at high temperature. This network of fibers favors the formation of a thick coating of mastic without asphalt draining down. Cellulose fiber possessed a greater effect on asphalt absorption and stabilization than did the other fibers (mineral and carbon fiber). A dynamic shear rheometer was used to evaluate their rheological properties and rut resistance. Results indicated that fiber can effectively improve the rut and flow resistance of asphalt mortar. However, the bending beam rheometer results demonstrated that the addition of fiber had negative effects on the creep stiffness and creep rate of asphalt mortar.

Keywords Fiber · Asphalt mortar · SEM · Rheology · Mechanical properties

M.-M. Wu · R. Li · Y.-Z. Zhang (✉) · L. Fan · Y.-C. Lv · J.-M. Wei
State Key Laboratory of Heavy Oil Processing, China University of Petroleum (East China), Qingdao 266555, Shandong, China
e-mail: zhangyuzhen1959@163.com

Edited by Xiu-Qin Zhu

1 Introduction

Asphalt has been widely used for road pavement construction for centuries (Zhang et al. 2009; Wang et al. 2012). Because of its adhesive properties, asphalt has been primarily used as a binder for aggregate particles in asphalt–concrete (AC) mixture, a composite material used for constructing flexible highway pavements. However, AC pavements are subjected to damage from cracking and rutting (permanent deformation) under the effects of repeated vehicle loading and temperature cycling (Xu and Solaimanian 2008). Accordingly, different additives have been used to alter the phase composition and improve the engineering properties of asphalt matrix (Ahmedzade 2013). These additives primarily include organic polymers that have been extensively studied (Hao 2001; Cao and Ji 2011).

Recently, the application of fiber in asphalt modification has attracted many researchers. Experimental results have shown that fibers have better performance than polymers in reducing draindown of AC mixtures, and this is the reason why fibers are widely used in stone matrix asphalt and open-graded friction course (Hassan et al. 2005). In terms of efficiency, fiber–asphalt mixture shows a slight increase in the optimum asphalt binder content compared with the pure asphalt mixture. In this way, adding fibers to asphalt is very similar to the addition of very fine aggregates. Thus, fibers can stabilize asphalt to prevent leakage. This is due to the adsorption of asphalt on fibers (Tapkin et al. 2010; Wu et al. 2014). The reported results indicate that fiber–asphalt mixtures have good moisture resistance, creep compliance, and rutting resistance, low-temperature anti-cracking properties, and durability (Tapkin et al. 2010). The mechanism of how fibers modify asphalt is complicated, yet the impact upon pavement performance is profound (Hassan et al. 2005).

However, studies of the fiber-reinforcing mechanisms are rare, especially those based on physical properties, such as the stability and the microstructure of fiber. This paper aims to study the stabilizing and reinforcing mechanisms of different fiber-modified asphalt mortars. Several laboratory experiments were designed to study the physical properties of fibers and subsequently investigate fiber-reinforcing mechanisms. Softening point, cone penetration, dynamic modulus, and phase angle at high temperatures, as well as the creep stiffness of fiber-modified asphalt mortar at low temperatures were studied.

2 Experimental

2.1 Experimental materials

2.1.1 Asphalt

Straight-run asphalt binder (penetration grade pen-50), provided by the PetroChina Fuel Oil Company Limited (China), was chosen as the base asphalt in the present study. The main physical properties of pen-50 were determined according to the Traffic Industry Standard—Standard Test Methods of Bitumen and Bituminous Mixtures for Highway Engineering (JTG E20-2011, China) and are shown in Table 1, including the thin-film oven test (TFOT) aging properties.

2.1.2 Fibers

Fibers are used to prevent asphalt binder draindown particularly for stone matrix asphalt and porous asphalt during the mixing, transportation, and compaction processes. Three kinds of fibers—mineral, cellulose, and carbon fiber—were used as modifier. The mineral fiber and cellulose fiber were supplied by the Jiangsu TianLong Continuous Basalt Fiber Hi-Tech Co. Ltd. (China), and the

carbon fiber was produced by the Changzhou YueYang Friction Materials Co. Ltd. (China). The physical properties of the three types of fiber used are listed in Table 2.

2.2 Experimental methods

2.2.1 Preparation of fiber–asphalt mortar

The fibers were separately put into a 105 °C oven for 24 h to ensure moisture-free surfaces, and solid asphalt (600 g) stored in a sealed can was preheated at 160 °C for 1 h to liquefy it for mixing. In order to investigate the effects of fibers on asphalt mortar, different fiber contents (i.e., 0, 1, 2, and 3 wt%) in the asphalt mortar were studied. The fiber content of 0 wt% represented pure asphalt mortar. The fiber was weighed and slowly added to the asphalt under stirring at 500 rpm to prevent the fiber from possibly agglomerating. Then, they were mixed at 160 °C for approximately 30 min under stirring to produce homogeneous fiber-modified asphalt mortar. To avoid the adverse effect of excessive heat, the temperature was carefully monitored through thermocouple probe.

2.2.2 Scanning electron microscopy (SEM)

SEM (Hitachi S-4800, Japan) at an accelerating voltage of 5 kV was used in this study for direct observation of fibers. Electron microscopy has been found to be preferable because it can provide a clear view of fibers in their initial state (Williams and Miknis 1998).

2.2.3 Heating treatment in oven

When mixed with AC mixture at high temperature, fibers will coalesce and some can even decompose. Hence, to evaluate the thermostability of fibers, a simple laboratory heating experiment was carried out with an XY oven (Wuxi Petroleum Asphalt Equipment Co. Ltd., China). A

Table 1 Physical properties of pen-50 base asphalt

Properties	JTG E20-2011		Measured values
	Min	Max	
Penetration (25 °C, 100 g, 5 s), 0.1 mm	40	60	48
Softening point, °C	46	–	52.2
Ductility (10 °C, 5 cm/min), cm	10	–	11.8
Ductility (15 °C, 5 cm/min), cm	80	–	>150
Viscosity (60 °C), Pa s	200	–	492
TFOT			
Mass change, %	±0.8		0.26
Retained penetration, %	60	–	71
Ductility (10 °C, 5 cm/min), cm	2	–	5.6

Table 2 Basic physical properties of fibers (provided by manufactures)

Fibers	Carbon fiber	Mineral fiber	Cellulose fiber	Specification
Diameter, μm	12–14	13–16	12–15	ASTM D2130
Length, mm	0.5–1.0	6.0	0.5–2.0	ASTM D204
Tensile strength, MPa	/	2,500–3,500	100–300	ASTM D2256
Density, g/cm^3	1.3–2.0	2.65–3.05	/	ASTM D3800
Melting temperature, °C	700	1,600	/	ASTM D276

D2130: Standard Test Method for Diameter of Wool and Other Animal Fibers by Microprojection in American Society for Testing and Materials (ASTM)

D204: Standard Test Methods for Sewing Threads in ASTM

D3800: Standard Test Method for Density of High-Modulus Fibers in ASTM

D276: Standard Test Methods for Identification of Fibers in Textiles in ASTM

D2256: Standard Test Method for Tensile Properties of Yarns by the Single-Strand Method in ASTM

beaker with 100 g fiber was heated in an oven at 163 °C for 5 h, which was similar to the construction temperature. The color, shape, and volume variations of fiber were recorded continually.

2.2.4 Mesh-basket draindown experiment

A mesh-basket draindown experiment was designed to evaluate adsorption and stabilization by fibers in asphalt binder. It is, described as follows: Ten percent of fiber (by mass) of asphalt binder (about 0.3 wt% fiber of AC mixture including aggregates) was utilized according to engineering practice. Then, a mixing sample (40 g) was uniformly placed into the designed steel mesh-basket with a sieve size of 0.25 mm, and maintained at 25 °C for 2 h. Afterward, the basket was heated in an environmental chamber under higher temperatures, and some asphalt binder would melt, flow, and drop out due to the heating effect. Two temperatures, 130 and 140 °C, were used for all the fibers, while an additional temperature of 170 °C was applied for the cellulose fiber, since no obvious asphalt drops could be measured at lower temperatures. The sample was weighed at 30-min intervals, to determine the weight loss of asphalt binder. Lower weight loss and flowing of asphalt indicated the fiber's greater capacity to adsorb and stabilize asphalt.

2.2.5 Softening point

The softening point of asphalt mortar was obtained by means of a WSY-025D asphalt softening point tester equipped with a steel ball of 3.5 g and steel ring of 20 mm in diameter (Wuxi Petroleum Asphalt Equipment Co. Ltd., China). The sample was cooled in a water bath at 5 °C for 15 min. Then, it was heated at a heating rate of 5 °C/min, and the softening point was taken at the temperature at which the mortar sample became soft enough to allow the

ball to fall to a distance of 25.4 mm. The softening point is also known as the ring-and-ball softening temperature ($T_{R\&B}$).

2.2.6 Cone penetration test

The cone penetration of asphalt mortar was obtained using a WSY-026 penetration tester (Wuxi Petroleum Asphalt Equipment Co. Ltd., China). The cone penetration experiment was conducted to measure fiber-modified asphalt mortar resistance to flow and shear. Cone penetration is expressed in units of 0.1 mm, the penetration depth of a standard cone under a 200 g load into the asphalt mortar sample after a 5-s loading time at a certain temperature. The prepared sample was kept at room temperature for 40 min until it was cooled and solidified, and then put in a water bath at 25 °C for more than 1 h. Afterward, the sample was taken out from the water bath, and an iron cone was put on the sample surface. The cone would gradually penetrate into the asphalt mortar until it was stable without further sinking, and then, the sink depth was measured and recorded. It should be noted that the cone would not completely penetrate into the sample mixture.

2.2.7 Rheological measurements

The dynamic rheological properties of the fiber-modified asphalt mortar were measured using a dynamic shear rheometer (DSR, AR-500, Carri-Med Ltd., UK) over a wide range of temperatures (Tan et al. 2010). For tests at 64 °C and higher, two 25-mm-diameter parallel plates with a 1-mm gap were selected to conduct the DSR measurement, as shown in Fig. 1. A sinusoidal strain was applied, and the strain was kept low enough so that all the tests were performed within the linear viscoelastic range. The actual strain and torque were measured for calculating various

Fig. 1 Schematic
representation of DSR
measurement

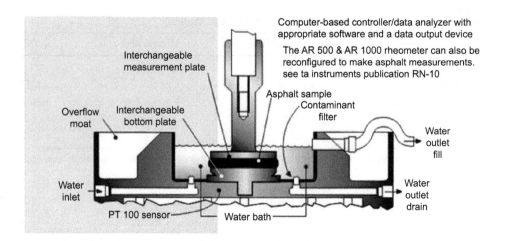

viscoelastic parameters, including complex modulus (G^*) and phase angle (δ). G^* is defined as the ratio of maximum shear stress to maximum strain, and it provides a measurement of the total resistance to deformation, which is the "sum" of the elastic part and viscous part of the mortar. The δ is the time lag between the applied stress and strain responses during a test, and it is a measure of the viscoelastic balance of the material behavior. In addition, the rutting factor $G^*/\sin\delta$ can be obtained to characterize the anti-deformation ability of the asphalt mortars according to the Strategic Highway Research Program (SHRP) specifications (USA) (Tan et al. 2010).

The low-temperature creep test was carried out to determine the low-temperature performance of the asphalt mortar using a bending beam rheometer (BBR, Cannon Instrument Company, Japan). In this test, the BBR sample beams ($125 \times 12.5 \times 6.25$ mm) were cooled in a methanol bath for 60 min at a constant temperature of -6, -12, -18, and $-24\ °C$, respectively. Then, the sample beam was placed on two stainless steel supports and loaded with 100 g. The deflection was monitored with time and used for calculation of the stiffness as a function of time. The creep stiffness (S) and the creep rate (m) of the binders were determined at a loading time of 60 s. The two parameters (creep stiffness and creep rate) were used as the low-temperature characteristics of the asphalt mortars (Hao et al. 2000).

3 Results and discussion

3.1 Microstructure of fiber

SEM images of the three fiber types are shown in Fig. 2. The cellulose fiber consists of ribbons, porous with a relatively flat cross section. Some cellulose filaments have been torn apart, resulting in an increase of surface area. The

1.8 m^2/g surface area of cellulose fiber was larger by >10-fold than that of the mineral and carbon fibers, which were only 0.13 and 0.05 m^2/g, respectively. These surface characteristics can explain the efficiency of cellulose fiber in binding greater amount of asphalt. The microstructures of mineral and carbon fibers were different from that of cellulose fiber since the cross sections of mineral and carbon fibers were quite round, with a smooth surface texture and a small surface area. In addition, the mineral fiber was more rigid than the flexible carbon fiber, and tended to be aligned with entanglement.

3.2 Thermostability

Thermal test results showed that the cellulose fiber was the most susceptible to coagulation, and it shrank noticeably at high temperature, while the other fibers showed no apparent change. It was noteworthy that the color of cellulose fiber changed from light gray to yellow during heating in the oven, and the carbon fiber became darker. However, the color of mineral fiber did not undergo any noticeable change. These results suggested that the cellulose fiber had the lowest thermostability, and the mineral fiber had the highest thermostability. This indicated that the mixing temperature was important for a specific fiber when used in asphalt mixture, or an appropriate fiber should be selected for a specific mixing temperature of AC material. It should be noted that in this study only a qualitative observation was performed, and more advanced test methods like differential scanning calorimetry (DSC) could be used to evaluate the thermostability quantitatively in future research.

3.3 Adsorption and adhesion of asphalt

Asphalt consists of asphaltenes, resins, aromatic hydrocarbons, and saturates (Morozov et al. 2004; Fritschy and

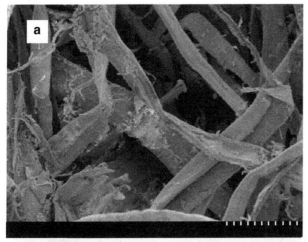

Fig. 2 SEM images of different fibers: **a** cellulose fiber; **b** mineral fiber; **c** carbon fiber

Papirer 1979). The fiber's adsorption of asphalt components can change the rheological behavior of the asphalt binder and the optimal asphalt content for the mixture design, and it will play a key role in the formation of the interface bonding between fiber and asphalt (Chen and Lin 2005). The mesh-basket draindown experiment results are presented in Table 3. It can be seen that cellulose fiber had the lowest asphalt drop and separation, that is to say, the highest asphalt adsorption and stabilization, followed by mineral fiber, and carbon fiber. This result should be attributed to their different specific surface areas and lengths. Compared with the cellulose fiber, the mineral fiber had a much lower adsorption and stabilization effect on asphalt, although it also had high specific surface area, mainly due to its smooth surface texture with a lubricant effect and low soaking effect. Although the carbon fiber had a relatively small specific surface area, it still had good effect on asphalt stabilization and great interface adhesion, because carbon fiber and asphalt are carbon-based products, so that carbon fiber has good adsorption of asphalt.

Interface strength is one of the primary factors determining fiber's reinforcing effect based on the mixing law of composite materials (Wo 2000). Fibers are uniformly dispersed in the asphalt, and their large surface areas could form an interface with asphalt. The acidic resin component in asphalt is a surface-active substance, and its adsorption onto the fiber's surface, and the physical and the chemical bond effects made some asphalt components be distributed as a monolayer on the fiber surface, forming a strongly binding "fiber–asphalt" interface layer. The adhesive property of the "fiber–asphalt" was stronger than that of the discrete asphalt outside of the interface layer, leading to an improvement of the asphalt mortar. The interfacial properties of fiber-modified asphalt depend on the molecular distribution, the chemical properties of the fiber, and the molecular structure and chemical composition of asphalt. The main role of the fiber–asphalt interface is to connect the two phases, and to pass and buffer the stress between them. The fiber–asphalt interface plays a key role in the physical and mechanical properties of fiber-modified asphalt mastic material. Damage to AC material can occur at the interface, such as the pullout of fiber from the asphalt binder, and shear slide at the aggregate surface. Fiber's adsorption of asphalt can increase asphalt's viscosity and

Table 3 Experimental results of mesh-basket draindown test (asphalt separation, %)

Temperature, °C	Fiber type	Asphalt separation, %			
		30 s	60 s	90 s	120 s
130	Carbon fiber	3.5	7.4	8.7	9.5
	Mineral fiber	1.7	4.6	5.7	6.3
	Cellulose fiber	0	0	0	0
140	Carbon fiber	11.1	12.6	13.1	14.0
	Mineral fiber	7.3	8.0	8.2	8.8
	Cellulose fiber	0	0	0	0
170	Cellulose fiber	0	0.1	0.2	0.8

improve interface adhesion between asphalt and fiber or aggregate. Moreover, the increased fiber–asphalt interface adhesion improves the anti-cracking strength of AC mixture (Putman and Amirkhanian 2004).

3.4 Softening point

The ring-and-ball softening point ($T_{R\&B}$) is another important performance criterion for asphalt mortars. The $T_{R\&B}$ increased rapidly with the addition of fibers as shown in Fig. 3. It was also found that the softening point eventually increased in the following order: carbon fiber < cellulose fiber < mineral fiber. As the softening point is used to evaluate asphalt's resistance to flow and deformation, the present results indicated that the addition of fiber to asphalt mortar increases resistance to flow and deformation. Mineral fiber-modified asphalt mortar showed the strongest resistance, followed by cellulose fiber-modified asphalt mortar, and carbon fiber-modified asphalt mortar showed the weakest resistance. The increase in $T_{R\&B}$ was related to fiber length and content (Chen and Lin 2005). The maximum temperature of road pavements is approximately 60 °C in summer; hence, the fiber content needs to be at least 1.0 wt% by mortar weight for cellulose and mineral fibers, and 2 % for carbon fibers, to ease the effect of hot summer temperatures.

Mineral fiber-modified asphalt mortar has a higher softening point than the other two fiber-modified asphalt mortars because of its fiber entanglement. In the cone penetration test, similar results were obtained for $T_{R\&B}$, as shown in Fig. 4. It was found that cellulose and carbon fibers had the same fiber length, but resulted in completely different softening points and cone penetrations (cone sink depth). The reason may be that carbon fibers cannot become entangled with each other.

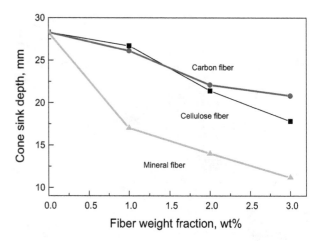

Fig. 4 Cone penetration of fiber-modified asphalt mortar

3.5 Cone penetration test

Based on the force balance, the shear stress of fiber-modified asphalt mortar at the direction tangential to the cone surface can be determined as follows:

$$\tau = \left[G\cos^2(\alpha/2) \right] / \left[\pi h^2 \tan(\alpha/2) \right]$$

where G is the cone weight (1.025 kN), h is the sink depth (m), and α is the cone angle (30°). For each fiber, experiments were repeated twice. Here the cone penetration and the calculated shear stress according to the force balance were used to evaluate fiber-modified asphalt mortar resistance to flow and deformation, and they are presented in Table 4. It can be seen that all the fiber-modified asphalt mortars exhibited a significant decrease in cone sink depth and increase in shear stress. The cone penetration of fiber-modified asphalt mortar decreased in the order (or increased in the order for shear stress) as follows: cellulose fiber, carbon fiber, and mineral fiber.

These experimental results are explained as follows: fibers are dispersed in the asphalt, and the asphalt is adsorbed over the fiber's surface as a monolayer, forming a strong binding force between the "fiber–asphalt" interfacial layer and three-dimensional spatial network. This network can be retained at high temperature, resulting in the formation of a thick mastic coating without asphalt draining down (Wiljanen 2003). The mineral fiber had the

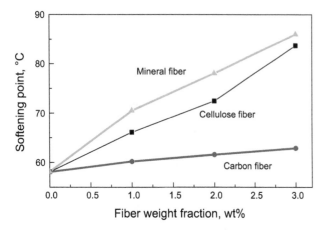

Fig. 3 Softening point of fiber-modified asphalt mortars

Table 4 Cone penetration and shear stress of fiber-modified asphalt mortar

Items	Asphalt mortar	Carbon fiber	Mineral fiber	Cellulose fiber
Fiber fraction, wt%	0	1	1	1
Cone sink depth, mm	2.83	2.61	1.70	2.67
Shear stress, MPa	141	166	393	159

greatest effect on reducing the cone sink depth and improving the shear stress among these three fibers, due to its great networking structure. Moreover, fibers can absorb the light components in asphalt and increase the viscosity of asphalt mortar; therefore, fiber increases the shear strength of asphalt mortar. In addition, the fiber has high tensile strength which can hold asphalt and resist flowing and crack propagation more effectively (Li 1992).

3.6 Viscoelastic behavior

The viscoelastic behavior of the samples was characterized by complex modulus (G^*) and phase angle (δ) of asphalt binder measured with DSR. G^* is a measurement of the overall resistance of a material to deformation, while δ is a measurement of the viscoelastic characteristics (Tan et al. 2010). Figure 5 shows the curves of G^* and δ versus temperature for fiber–asphalt mortars, respectively. G^* decreased, whereas δ increased with the increasing temperature, indicating that the mortars lose their elastic property and behave more like a liquid at high temperature. In addition, δ was always higher than 45°, indicating that the loss modulus ($G'' = G^*\sin\delta$) was higher than the storage modulus ($G' = G^*\cos\delta$) within the range of testing temperatures.

According to the SHRP specification, the rutting factor, $G^*/\sin\delta$, was used to measure the contribution of asphalt binder to rutting performance (Tan et al. 2010). Higher $G^*/\sin\delta$ indicated higher rutting resistance of asphalt–fiber mixture. Therefore, much greater attention was given to this parameter than other parameters. Figure 6 presents the rutting factors of three fiber-modified mortars at different testing temperatures. Regardless of the testing temperatures, the $G^*/\sin\delta$ was improved with the addition of fiber, indicating the fiber-modified asphalt mortars had improved permanent deformation resistance at high temperature. The cellulose fiber-modified asphalt mortar had the highest $G^*/\sin\delta$, which was attributed to its highest absorption of light

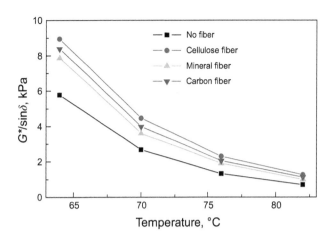

Fig. 6 Rutting factors of different fiber-modified asphalt mortars

components in asphalt to improve the stiffness. The mineral fiber-modified asphalt mortar had the lowest $G^*/\sin\delta$, which was attributed to its smooth surface texture, although it had a relatively larger specific surface area. It can also be seen that the $G^*/\sin\delta$ of the fiber-modified asphalt mortars decreased with the increasing temperature. As higher $G^*/\sin\delta$ represented better high-temperature performance, the rutting resistance of asphalt mortar decreased when temperature rose. In addition, $G^*/\sin\delta$ eventually increased in the following order of pure asphalt mortar < mineral fiber-modified asphalt mortar < carbon fiber-modified asphalt mortar < cellulose fiber-modified asphalt mortar. The results of the present study showed that the addition of fiber increases $G^*/\sin\delta$ for the asphalt mortar, that is to say, the temperature sensitivity decreases.

The low-temperature creep stiffness (S) and creep rate (m) for fiber-modified asphalt mortars were determined using the BBR at four different temperatures, and the result is shown in Fig. 7. It can be seen that the stiffness (S) of asphalt mortar increased with the addition of fibers, indicating that the addition of fibers increased the cracking potential of the asphalt mortars, thereby having a negative effect on the cracking resistance. For m-value, although there were some variations, the whole trend was that m-value was reduced with the addition of fibers. It is well known that the higher the creep rate (m), the quicker the stress could be released, and hence the better the cracking resistance of the asphalt mortars. So, increasing the temperature would improve the cracking resistance of asphalt mortar, and this was in accordance with the low-temperature creep stiffness analysis. However, it has been reported that asphalt mixtures containing fiber have good low-temperature properties in terms of low-temperature flexural tests (Tapkin et al. 2010). Hence, more research needs to be pursued to investigate the effect of fiber on the low-temperature performance of asphalt mixtures.

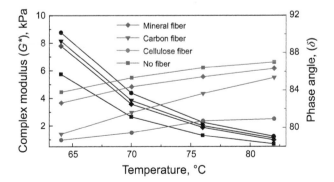

Fig. 5 The complex modulus (G^*) and phase angle (δ) of different fiber-modified asphalt mortars

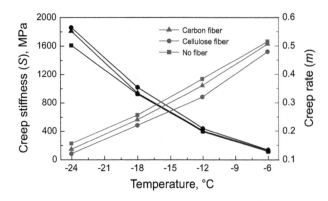

Fig. 7 The creep stiffness (S) and creep rate (m) of fiber-modified asphalt mortars

4 Conclusions

A comprehensive laboratory investigation was carried out to study the reinforcement effects of three kinds of fibers on asphalt mortars and their mechanisms, and the following results were obtained.

Fibers are dispersed in the asphalt, and some asphalt components are adsorbed over the fiber's surface as a monolayer, forming a strong binding force "fiber–asphalt" interface layer and three-dimensional spatial networks. Fibers can reinforce asphalt mortar through a spatial network and the reinforcing effects also depended on the fiber's structures and properties including shape, size and tensile strength. Due to its larger specific surface area, cellulose fiber showed more asphalt absorption than the other two fibers. The cone penetration experiment showed that all fiber–asphalt mortars present pronouncedly decreased sink depth while showing the increasing shear stress. The DSR tests showed that fiber-modified asphalt mortars have larger complex shear modulus (G^*) and rutting parameters ($G^*/\sin\delta$), and have smaller phase angle (δ) compared with the pure asphalt mortar used in this study. Hence it can be noted that fibers considerably improved asphalt mortar rutting resistance and resistance to flow. However, fibers had a negative effect on the creep stiffness and creep rate of the asphalt mortar. Our future efforts will be to perform mechanical testing focusing on fatigue, static and repeated creeps, and wheel tracking tests to further characterize the asphalt mixtures made by fiber-modified asphalt mortars discussed in this paper.

Acknowledgments The work is supported by the National Natural Science Foundation of China (51008307), the Fundamental Research Funds for the Central Universities (09CX04039A), and the Graduate Student Innovation Project of China University of Petroleum (East China) (12CX06055A).

References

Ahmedzade P. The investigation and comparison effects of SBS and SBS with new reactive terpolymer on the rheological properties of bitumen. Constr Build Mater. 2013;38:285–91.

Cao DW, Ji J. Evaluation of the long-term properties of sasobit modified asphalt. Int J Pavement Res Technol. 2011;4(6): 384–91.

Chen JS, Lin KY. Mechanism and behavior of bitumen strength reinforcement using fibers. J Mater Sci. 2005;40:87–95.

Fritschy G, Papirer E. Dynamic-mechanical properties of a bitumen-silica composite. Rheol Acta. 1979;18(6):749–55.

Hao PW. Study on compatibility between SBS modifier and asphalt. Pet Process Petrochem. 2001;32(3):54–6.

Hao PW, Zhang DL, Hu XN. Evaluation method for low temperature anti-cracking performance of asphalt mixture. J Xi'an Highway Univ. 2000;20(3):1–5.

Hassan HF, Oraimi SA, Taha R. Evaluation of open-graded friction course mixtures containing cellulose fibers and styrene butadiene rubber polymer. J Mater Civil Eng. 2005;17(4):415–22.

Li VC. Postcrack scaling relations for fiber reinforced cementitious composites. J Mater Civil Eng. 1992;4(1):41–57.

Morozov V, Starov D, Shakhova N, et al. Production of paving asphalts from high-wax crude oils. Chem Tech Fuels Oil. 2004;40(6):382–8.

Putman BJ, Amirkhanian SN. Utilization of waste fibers in stone matrix asphalt mixtures. Resour Conserv Recycl. 2004;42(3): 265–74.

Tan YQ, Li XL, Zhou XY. Interactions of Granite and Asphalt Based on the Rheological Characteristics. J Mater Civil Eng. 2010;22(8):820–5.

Tapkin S, Çevik A, Uşar Ü. Prediction of Marshall test results for polypropylene modified dense bituminous mixtures using neural networks. Expert Syst Appl. 2010;37:4660–70.

Wang HN, You ZP, Mills-Beale J, et al. Laboratory evaluation on high temperature viscosity and low temperature stiffness of asphalt binder with high percent scrap tire rubber. Constr Build Mater. 2012;26:583–90.

Wiljanen BR. The pavement performance and life-cycle cost impacts of carbon fiber modified hot mix asphalt, MS Thesis, Michigan Technological University. 2003.

Williams TM, Miknis FP. Use of environmental SEM to study asphalt–water interactions. J Mater Civil Eng. 1998;10(2):121–4.

Wo D. Comprehensive composite materials. Beijing: Chemistry Industry Press; 2000.

Wu MM, Li R, Zhang YZ, et al. Reinforcement effect of fiber and deoiled asphalt on high viscosity rubber/SBS modified asphalt mortar. Pet Sci. 2014;11:454–9.

Xu Q, Solaimanian M. Measurement and evaluation of asphalt thermal expansion and contraction. J Test Evaluat. 2008;36:1–6.

Zhang BC, Xi M, Zhang DW, et al. The effect of styrene-butadiene-rubber/montmorillonite modification on the characteristics and properties of asphalt. Constr Build Mater. 2009;23:3112–7.

17

Injection of biosurfactant and chemical surfactant following hot water injection to enhance heavy oil recovery

Yahya Al-Wahaibi[1] · Hamoud Al-Hadrami[1] · Saif Al-Bahry[2] · Abdulkadir Elshafie[2] · Ali Al-Bemani[1] · Sanket Joshi[2]

Abstract This study investigates the potential of enhancing oil recovery from a Middle East heavy oil field via hot water injection followed by injection of a chemical surfactant and/or a biosurfactant produced by a *Bacillus subtilis* strain which was isolated from oil-contaminated soil. The results reveal that the biosurfactant and the chemical surfactant reduced the residual oil saturation after a hot water flood. Moreover, it was found that the performance of the biosurfactant increased by mixing it with the chemical surfactant. It is expected that the structure of the biosurfactant used in this study was changed when mixed with the chemical surfactant as a probable synergetic effect of biosurfactant-chemical surfactants was observed on enhancing oil recovery, when used as a mixture, rather than alone. This work proved that it is more feasible to inject the biosurfactant as a blend with the chemical surfactant, at the tertiary recovery stage. This might be attributed to the fact that in the secondary mode, improvement of the macroscopic sweep efficiency is important, whereas in the tertiary recovery mode, the microscopic sweep efficiency matters mainly and it is improved by the biosurfactant-chemical surfactant mixture. Also as evidenced by this study, the biosurfactant worked better than the chemical surfactant in reducing the residual heavy oil saturation after a hot water flood.

✉ Yahya Al-Wahaibi
ymn@squ.edu.om

[1] Petroleum and Chemical Engineering Department, Sultan Qaboos University, 112 Muscat, Oman

[2] Department of Biology, College of Science, Sultan Qaboos University, 112 Muscat, Oman

Edited by Yan-Hua Sun

Keywords Hot water injection · Biosurfactant · Chemical surfactant · Enhanced oil recovery

1 Introduction

In the oil industry, biosurfactants are used for enhancing oil recovery, bioremediation, dispersion, and transfer of crude oils (Gautam and Tyagi 2005; Lee et al. 2007). These biosurfactants are complex molecules comprising different structures which include lipopeptides, phospholipids, glycolipids (such as rhamnolipids, trehalose lipids, and sophorolipids), fatty acids, and neutral lipids (Gautam and Tyagi 2005).

A *Bacillus subtilis* strain C9 from the Korean Collection for Type Cultures (KCTC) was found to produce biosurfactants that lowered the surface tension of water from 72 to 28.5 mN/m and proved to be stable under various ranges of salinity and pH. There are other *B. subtilis* strains that produced lipopeptide biosurfactants, similar to surfactins or lichenysins, such as *B. subtilis* strain C-1, *B. subtilis* strain PTCC 1696 (Ghojavand et al. 2008), five different *Bacillus* strains (Joshi and Desai 2013), and the surfactin ATCC 6633 produced by a *B. subtilis* strains which is one of the most powerful biosurfactants that reduces the surface tension of water from 72 to 27.9 mN/m (Noudeh et al. 2005; Gautam and Tyagi 2005).

The extraction of crude biosurfactant from the grown microbial broth depends on its ionic charge, water solubility, and location (intracellular, extracellular, or cell bound). There are many recovery methods available such as acetone precipitation, solvent extraction, acid precipitation, and crystallization (Gautam and Tyagi 2005). The most widely used technique in batch mode process is extraction with chloroform–methanol, dichloromethane-methanol, butanol, or acetic acid, which are relatively expensive solvent-based methods. In contrast, the acid precipitation method is comparatively inexpensive and reported for extraction of lipopeptide biosurfactant like surfactin, where lipopeptide biosurfactants that are not soluble under highly acidic conditions (pH 2.0–4.0) are precipitated (Makkar and Cameotra, 1997).

There are various experiments at laboratory scale using sand-pack columns or corefloods and field trials that have proved the effectiveness of using biosurfactants for microbial enhanced oil recovery (MEOR). Lichenysin produced by *Bacillus licheniformis* strain JF-2 showed residual oil recovery from cores up to 40 %. Similarly, four different strains of *Pseudomonas* showed over 50 % recovery of crude oil at 70 °C in saturated sand-pack experiments. Although chemically synthesized surfactants have long been used in the petroleum industry, they are commonly environmentally toxic and not biodegradable. Biosurfactants have the benefit of being biodegradable and relatively inexpensive. However, there are some limitations which reduce the attractiveness of using biosurfactants widely in petroleum field applications. These limitations include the quantity and quality of the biosurfactants compared to chemical surfactants in addition to scale up complications in producing large amounts of biosurfactants for field applications as biosurfactants are generally produced in small amounts even at an industrial level. However, the discovery of new biosurfactants and development of new fermentation and recovery processes may allow more biosurfactants to be used for MEOR (Torres et al. 2011; Joshi and Desai 2013). This paper reports the ability of the biosurfactant produced by *B. subtilis* strain W19 to enhance oil recovery by interaction in porous media using original rock and fluid samples from an Omani oil field in coreflood experiments. In addition, the possibility of enhancing the performance of the biosurfactant for oil recovery by mixing it with commercially available chemical surfactants that are used in the Omani oil fields is investigated. Different mixture solutions are prepared at ratios of 25:70, 50:50, and 75:25 of the biosurfactant to the chemical surfactant, respectively. The mixing is done to better prove the applicability of biosurfactant for enhancing oil recovery by increasing its performance by adding chemical surfactants.

Surfactant loss due to adsorption is a major limitation during a surfactant flood for enhancing oil recovery since it causes surfactant retention which affects the economical feasibility of this process. Excessive surfactant retention results in adverse phase behavior properties, which cause the mobilized oil to be trapped again (Daoshan et al. 2004). This study includes adsorption analysis to quantify the amount of biosurfactant adsorbed in milligrams per gram of solid or crushed rocks. This was done to assess the applicability of using this biosurfactant for enhancing oil recovery and comparing its adsorption tendency to that of the commercially available chemical surfactants.

Al-Sulaimani et al. (2010, 2011a, b, 2012) and Al-Bahry et al. (2013a, b) reported that the biosurfactant had potential for enhancing oil recovery since it yielded a total production of 23 % of residual oil. In this study, possibility of enhancing the oil recovery from a Middle East heavy oil field by biosurfactant following hot water injection was investigated. Additionally, the biosurfactant performance was compared with the performance of a commercially available chemical surfactant. Previous studies reported that biosurfactants could potentially be used in conjunction with synthetic surfactants to provide more cost-effective enhanced oil recovery and subsurface remediation (Daoshan et al. 2004). The economic efficiency of biosurfactants depends on the use of low cost raw materials, such as molasses or cheese whey, which account for 10 %–30 % of the overall cost (Joshi et al. 2008). Portwood (1995) reviewed hundreds of projects and concluded that the cost of MEOR process, including biosurfactants, ranges from $ 0.25 to $ 0.50 per barrel of oil produced and does not go up as oil production increases. A more recent study reported that the price of biosurfactants ranges between US$ 2 and 3 per kg (Hazra et al. 2011). It was reported that the reduction in interfacial tension (IFT) by the surfactants has to be ultra low, where the IFT values should be in the range of 10^3 mN/m, to enhance oil recovery by increasing the capillary number (Aoudia et al. 2006; Curbelo et al. 2007; Zhu et al. 2009; Iglauer et al. 2010; Lu et al. 2014a, c, d). Although the minimum IFT value obtained by the biosurfactant in this study is not ultra low, other recovery mechanisms are expected to take place.

Recently, wettability alteration has been proposed as one of the mechanisms of MEOR where several studies reported the relation between IFT reduction and alteration of wetting conditions following microbial treatment (Sayyouh et al. 1995; Zekri et al. 2003; Kowalewski et al. 2006; Zargari et al. 2010). Al-Sulaimani et al. (2012) concluded that the ability of the biosurfactant used in this study to alter the wettability of rocks and surfaces is one of the mechanisms for enhancing oil recovery.

2 Materials and methods

2.1 Biosurfactant production and extraction

The procedure for bacterial growth and biosurfactant production is described in previous studies (Al-Sulaimani et al. 2010, 2011a, b). Briefly, the *Bacillus subtilis* strain W19 was grown in a minimal media (Table 1) containing 2 % (w/v) glucose and incubated for 16 h at 40 °C and at

Table 1 Composition of the production minimal medium

Composition	Concentration, g/L
Glucose	20.0
NH_4NO_3	4.002
KH_2PO_4	4.083
Na_2HPO_4	7.119
$MgSO_4$	0.197
$CaCl_2$	0.00077
$FeSO_4 \cdot 7H_2O$	0.0011
$MnSO_4 \cdot 4H_2O$	0.00067
Na_2-EDTA	0.00148

160 rpm. The bacterial cells were separated from the broth by centrifuging at 10,000 rpm for 20 min at 20 °C in a high speed centrifuge (Beckman, USA, JLA 16.250 rotor).

For biosurfactant extraction, the cell-free broth was concentrated by a precipitation method (Youssef et al. 2007a, b). The precipitated biosurfactant was collected by centrifuging at 10,000 rpm, and finally, biosurfactant powder was obtained by spray drying following a standardized protocol at 160–100 °C using a Mini spray dryer (Buchi, Switzerland), as previously reported by Al-Sulaimani et al. (2010; 2011a) .

2.2 Rock and fluid samples

Core plugs from a Middle East heavy oil field were used in coreflood experiments (Table 2). They are heterogeneous and consolidated (i.e., they do not produce fines). On average, the core plugs are 5.17 cm long with a diameter of 3.75 cm. In order to understand better their mineralogy, XRD analyses were conducted on core plugs No. 4 and 7 and their mineral compositions are listed in Table 3. The salinity of formation water was between 7 % and 9 % and its chemical composition is shown in Table 4. Formation

water was filter sterilized, prior to use, by a Millipore Filtration Unit with a membrane pore size of 0.45 µm. Original crude oil from the Middle East heavy oil field was used to saturate core samples. The characteristics of the crude oil are given in Table 5. The chemical surfactant used in this work is ethoxylated sulfonate, S-8B, kindly provided by a local oil company (active concentration of ∼23.9 %).

2.3 Coreflood experiments

Eleven core plugs obtained from the Middle East heavy oil field were used in coreflood experiments. Formation brine and crude oil used in all experiments were obtained from the same field (characteristics of crude oil are shown in Table 5). Initially, the core was cleaned using the Soxhlet extraction method where chloroform and methanol were solvents used as an azeotropic mixture in the proportion of 75:25. These solvents are constantly evaporated and condensed. The condensed solvent passed through the core sample removing all the oil and any other soluble material from the core before returning back for another cycle. This process was repeated until a clear color solvent was observed.

After cleaning, the core was dried at 65 °C for 24 h before use. The core was evacuated and then saturated with filtered formation brine for 24 h in a vacuum desiccator and the pore volume (PV) was determined using the dry and wet weights of the core. The core was then flooded with oil at 6 cm³/h until no more water was produced to establish residual water saturation. The oil initially in place (OIIP) was determined which was indicated by the volume of water displaced. After that, the core was subjected to hot waterflood at 6 cm³/h until no further oil was produced. The residual oil saturation to hot water was then calculated by measuring the amount of oil produced from the hot waterflood. Then, the chemical surfactant or the cell-free

Table 2 Properties of core plugs used in this study

Core No.	Length, cm	Diameter, cm	Porosity, %	Pore volume, cm³	Liquid permeability, mD	Initial wettability
1	4.85	3.75	23.0	12.30	144	Oil wet
2	5.25	3.75	22.8	13.20	162	Oil wet
3	4.85	3.60	30.1	14.84	173	Oil wet
4	4.85	3.60	26.8	13.24	149	Oil wet
5	5.47	3.80	23.0	14.27	177	Oil wet
6	5.19	3.82	22.9	13.64	152	Oil wet
7	5.35	3.78	23.0	13.81	159	Oil wet
8	5.40	3.65	23.0	13.00	151	Oil wet
9	4.40	3.90	23.0	12.10	163	Oil wet
10	5.10	3.70	23.0	12.60	158	Oil wet
11	5.20	3.90	23.0	14.30	149	Oil wet

Table 3 Mineral composition of core plugs determined by X-ray diffraction

Core No.	Quartz, %	Albite, %	Orthoclase, %	Calcite, %	Muscovite, %	Clinochlore, %
4	68.7	16.1	12.2	0.3	1.2	1.4
7	67.4	16.4	11.2	0.5	1.0	3.5

Table 4 Composition of formation water

Component	Concentration, kg/m^3
Sodium	25.08
Calcium	3.76
Magnesium	0.878
Iron	0.045
Chloride	47.72
Sulfate	0.247
Carbonate	0
Bicarbonate	0.079

Table 5 Middle East heavy oil characteristics

Characteristics	Values
Density at 15 °C, kg/L	0.98
Specific gravity @ 60/60 °F	0.98
API gravity @ 60 °F, °API	13.5
Pour point, °F	62
Flash point, °F	≥ 240
Kinematic viscosity @ 140 °F, cST	2500
Total salts, ppm	80,000

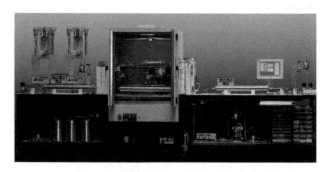

Fig. 1 Core flow set up at the Sultan Qaboos University

3) Also housed inside the oven are Hastelloy tubing coils for the temperature equilibrium while injecting fluids. The coil can hold up to 600 mL of fluid. All the fittings inside the oven are acid-resistant Hastelloy fittings.

4) Two twin Isco pumps which have a working pressure of 7500 psi and a flow rate ranging from 0.001 to 50 mL/min. The pumps are calibrated beforehand and found to be producing the expected rate to within 0.1 %.

5) Two high-pressure accumulators.

6) Back pressure regulator.

7) Data-logging system.

Apart from the above-mentioned major parts, the system is equipped with high precision pressure and differential pressure transducers. Besides, there is a high-pressure nitrogen gas compressor (4500 psi) used to pressurize the system to the reservoir condition.

3 Results and discussion

Worldwide petroleum companies are struggling to develop new economical technologies to recover heavy oil from maturing on-shore and off-shore oil fields. Among different technologies currently used, miscible gas like CO_2 injection, steam injection, and use of chemical surfactants are quite successful. There are certain issues related to availability and cost-effectiveness for gas injection or steam injection; thus, chemical surfactants are preferred for EOR operations. Usage of chemical surfactants also has its pros and cons: it is effective in enhancing the oil recovery but is not so environmentally friendly and comparatively costly. Biosurfactants can be an environmentally friendly and an

supernatant (biosurfactant broth) was injected as a tertiary recovery stage and extra oil recovery was determined. In another set of experiments, the biosurfactant and chemical surfactant mixture solutions were injected at different ratios of 25:75, 50:50, and 75:25 of biosurfactant to chemical surfactant, respectively, all at a final concentration of 0.25 % (w/v). The effluent was collected at regular time intervals in 12- or 20-mL containers and the volumes of effluent were measured.

All corefloods were conducted at 90 °C to mimic the average reservoir temperature of the field of interest. Flow experiments were performed at the coreflood rig housed at Sultan Qaboos University (Fig. 1). The coreflood rig is composed of following components:

1) High-pressure Quizix pumps (up to 10,000 psi working pressure). These pumps are housed inside the oven. They can be used for permeability measurements.

2) A specially designed core holder is placed inside the oven to carry out tests at the reservoir temperature. Working pressure of the core holder is 10,000 psi and it can sustain up to 250 °C.

equally effective alternative to its chemical counterpart. We used chemical surfactant and biosurfactant individually or as a mixture, for their potential in enhancing heavy oil recovery from core plugs taken from the Middle East heavy oil field.

3.1 Core, fluids, chemical surfactant, and biosurfactant properties

Core plugs used contained mainly quartz (38 % − 67 %), and remainder was other components (Table 3). The oil used was very heavy crude with 13.5° API and 2500 cST viscosity. The chemical surfactant was ethoxylated sulfonate and the biosurfactant was a lipopeptide, produced in our laboratory. The chemical surfactant (CS) and biosurfactant (BS) reduced brine/oil IFT values to 3.24 and 3.97 mN/m, respectively, from 36 mN/m. When CS and BS were mixed at different proportions, the brine/oil IFT values were reduced to 3.2 mN/m (CS:BS; 75:25), 3.11 mN/m (CS:BS; 50:50), and 4 mN/m (CS:BS; 25:75), respectively. Thus, we observed a slight reduction in IFT with a 50:50 CS + BS mixture, compared to individual surfactants. The biosurfactant used in this study also showed the ability to change the wettability of sandstone rock surfaces, thus altering it from oil-wet to water-wet (Al-Sulaimani et al. 2012).

3.2 Coreflood experiments using chemical surfactant and biosurfactant

Coreflood experiments were carried out to recover heavy oil, as initially flooded by hot water (as the secondary mode) followed by either chemical surfactant or biosurfactant individually and as a mixture (the tertiary mode). Tables 6, 7 and 8 summarize the initial water and oil saturations, residual oil saturations after the injection of hot water, chemical surfactant, biosurfactant, and mixtures of both surfactants, where it can be observed that the biosurfactant injection recovered more oil compared to chemical surfactant only injection (Tables 6 and 7), which was around 1.4 % − 18.5 % over the residual oil saturation (S_{or}), whereas the mixture of the biosurfactant and the

chemical surfactant at different ratios gave the highest recovery of 27 % − 34 % over S_{or} (Table 8).

Enhancement of oil recovery from Berea sandstone cores treated with cell-free metabolites from a surfactant-producing strain, *Bacillus sp.* JF-2, was reported by Thomas et al. (1993). Joshi et al. (2015) reported additional 37.1 % of heavy oil from Berea sandstone cores at 80 °C was achieved using a lipopeptide-type of biosurfactant. Previous studies reported that biosurfactants could potentially be used in conjunction with synthetic surfactants to provide more cost-effective enhanced oil recovery and subsurface remediation (Youssef et al. 2007a, b). They reported that the activity of biosurfactants depends on their structural components where the 3-hydroxy fatty acid composition of lipopeptides is very important for the biosurfactant activity. Youssef et al. (2007a, b) manipulated the biosurfactant activity by changing the fatty acid composition, knowing the relationship to hydrophobicity/hydrophilicity, of the mixtures with different biosurfactants and synthetic surfactants and achieved an ultra-low IFT. So, it was hypothesized that the activity of the biosurfactant used in this study was enhanced when mixed with the chemical surfactant. Probably due to chemical interactions between the surface charges of the two surfactants and the synergetic effect, the enhancement in oil recovery was greater when the two surfactants were used as a mixture, rather than alone. Lu et al. (2014c) reported that for oils with a high alkane carbon number, surfactants with very large hydrophobes are needed to obtain ultra-low IFT and to reduce the residual oil saturation to nearly zero. They reported new classes of large-hydrophobe surfactants developed for chemical EOR, where both the sulfates and carboxylates were tailored to specific reservoir conditions and oils by adjusting the number of ethylene oxide (EO) or propylene oxide (PO) groups in the surfactant.

Figures 2, 3 and 4 show cumulative oil recoveries of the chemical surfactant flooding, biosurfactant flooding, and mixtures of both following hot water injection. Figure 5 shows the best of the 3 flooding types following hot water injection. All the experiments were carried out at 0.25 % (w/v) concentration of chemical surfactant or biosurfactant. The results revealed that 1.4 % − 11 % of residual oil was produced by the pure chemical surfactant injection (Fig. 2),

Table 6 Residual oil saturations and % of heavy oil recovery enhancement following hot water injection and chemical surfactant injection

Core No.	Initial water saturation, %	Initial oil saturation, %	Residual oil saturation after hot water injection, %	Residual oil saturation after chemical surfactant injection, %	% of heavy oil recovery enhancement
1	11.0	89.0	91.5	89.0	2.5
3	10.5	89.5	83.0	75.0	8.0
5	12.0	88.0	95.8	94.4	1.4
7	10.0	90.0	57.0	46.0	11.0

Table 7 Residual oil saturations and % of heavy oil recovery enhancement following hot water injection and biosurfactant injection

Core No.	Initial water saturation, %	Initial oil saturation, %	Residual oil saturation after hot water injection, %	Residual oil saturation after biosurfactant injection, %	% of heavy oil recovery enhancement
2	13.0	87.0	83.7	76.9	6.8
4	12.5	87.5	76.0	67.0	9.0
6	10.0	90.0	52.0	33.5	18.5
8	11.0	89.0	62.0	51.0	11.0

Table 8 Residual oil saturations and % of heavy oil recovery enhancement following hot water injection and chemical/biosurfactant mixture injection

Core No.	Initial water saturation, %	Initial oil saturation, %	Residual oil saturation after hot water injection, %	Residual oil saturation after chemical surfactant/biosurfactant mixture injection, %	Biosurfactant to chemical surfactant ratio	% of heavy oil recovery enhancement
9	13.0	87.0	67.0	36.0	75:25	31
10	12.5	87.5	50.0	16.0	50:50	34
11	10.0	90.0	71.0	44.0	25:75	27

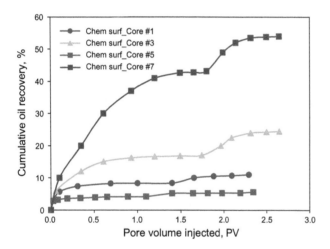

Fig. 2 Cumulative oil recovery via hot water injection (injection was continued until no more oil was recovered) followed by injecting chemical surfactant

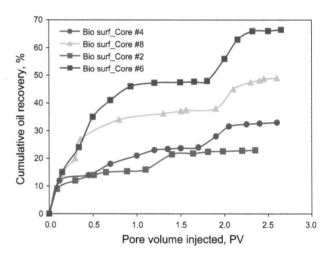

Fig. 3 Cumulative oil recovery via hot water injection (injection was continued until no more oil was recovered) followed by injecting biosurfactant

while the production increased to 6.8 % − 18.5 % of residual oil when the biosurfactant was injected (Fig. 3). However, it was interesting to note that the performance was improved when mixing the biosurfactant with the chemical surfactant at all ratios tested compared to injecting pure solutions. Recovery up to 34 % of residual oil was produced when mixing both surfactants in a ratio of 50:50, while the mixture of 75 % biosurfactant and 25 % chemical surfactant yielded an increased production of 31 %. The least production by the mixed surfactants was in a ratio of 25:75 of the biosurfactant to the chemical surfactant where the recovery was estimated to be 27 % (Fig. 4). However, it is still higher than the production obtained by injecting the biosurfactant or chemical surfactant solutions alone (Tables 6, 7 and 8). Nguyen et al.

(2008) investigated the efficiency of a mixture of rhamnolipid biosurfactant and synthetic surfactant for improving the interfacial activity of the surfactant system against several light non-aqueous-phase liquids (LNAPLs). They reported that the rhamnolipid biosurfactant was quite hydrophilic relative to the hydrocarbons tested and that mixing it with more hydrophobic synthetic surfactants enhanced the interfacial activity of the rhamnolipid against those hydrocarbons. Torres et al. (2011) reported the performance of three biosurfactants (of bacterial and vegetal origin) in comparison to different synthetic surfactants (cationic, anionic, non-ionic, and zwitterionic) for potential use in EOR applications. They also reported that biosurfactants have potential for EOR and analyzing the surface properties (ST and IFT) of pure surfactants and mixtures,

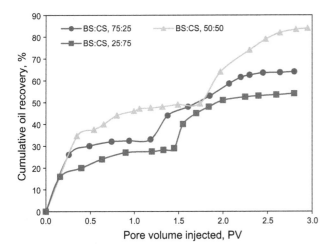

Fig. 4 Cumulative oil recovery via hot water injection (injection was continued until no more oil was recovered) followed by injecting mixtures of chemical surfactant and biosurfactant. Cores # 9, 10, and 11 were used in these tests

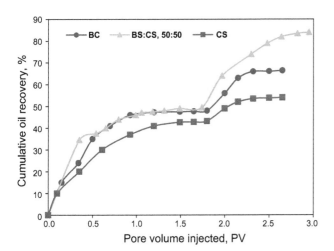

Fig. 5 Comparison of maximum cumulative oil recovery via hot water injection (injection was continued until no more oil was recovered) followed by chemical surfactant, biosurfactant, or mixtures of both. Cores # 6, 7, and 10 were used in these tests

together with other tests will give important information regarding the behavior of surfactants under oil-wet conditions. These assessments will lead to the selection of the right surfactant(s) and mixtures for different oil field applications.

Youssef et al. (2007a, b) reported that biosurfactant and synthetic surfactant mixtures could be formulated to provide appropriate hydrophobic/hydrophilic conditions necessary to reduce the IFT against NAPLs, and that such mixtures produced synergism that made them more effective than individual surfactants alone. They reported that mixtures of lipopeptide biosurfactants with the hydrophobic synthetic surfactant were able to produce low IFT against hexane and decane as compared to an individual

surfactant alone. When we mixed the biosurfactant and the chemical surfactant in mode a ratio of 50:50, a slight reduction in IFT was observed, as compared to individual surfactants. This might explain part of the increase in residual oil recovery. Other recovery mechanisms are expected by the nature of biosurfactant, such as wettability alteration. Al-Sulaimani et al. (2012) conducted experiments which proved the ability of the biosurfactant used in this study to change the wettability of sandstone rock surfaces. The influence of biosurfactants on wettability was studied by contact angle measurements, atomic force microscopy (AFM) technique on few-layer graphene (FLG) surfaces, and Amott wettability tests. It was reported that the biosurfactant altered the wettability of sandstone rocks from oil-wet to more water-wet. Thus, it was concluded that the wettability alteration by the biosurfactant is one of the major mechanisms of microbial enhanced oil recovery. The combined effects of the reduction in IFT and wettability alteration using surfactants have also been discussed in the literature (Anderson 1986; Alveskog et al. 1998; Austad and Standnes 2003; Hirasaki and Zhang 2004; Kowalewski et al. 2006; Zhang and Austad 2006; Lu et al. 2014b). Kowalewski et al. (2006) reported that changes in wetting properties are dependent on the initial wetting conditions where an initially oil-wet system can result in more water-wet conditions and vice versa. Lu et al. (2014b) reported a surfactant formulation (a novel large-hydrophobe alkoxy carboxylate surfactant and an internal olefin sulphonate co-surfactant) developed for carbonate reservoirs under high salinity and temperature, where it reduced the IFT to ultra-low values and also altered the wettability of the rock toward more favorable water-wet conditions, leading to enhanced oil recovery.

3.3 Coreflood experiment using a mixture of chemical and biosurfactant in the secondary or tertiary mode

As revealed from the above results, the maximum reduction in the residual oil saturation was achieved when mixing the chemical surfactant and the biosurfactant in a ratio of 50:50. This surfactant mixture was selected for testing whether starting the injection with the surfactant solution rather than waterflooding is more effective (surfactant injection at the secondary mode). Core plug No. 10 was used in this test. Figure 6 presents a comparison between the secondary (direct chemical surfactant/biosurfactant mixture, without hot water flooding) and tertiary modes (hot water followed by the chemical surfactant/biosurfactant mixture) of surfactant injection. Results show that compared to the tertiary mode, the secondary mode resulted in higher breakthrough recovery by 7 % (Fig. 6). However, the ultimate oil recovery in the secondary mode

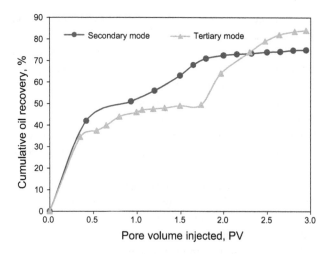

Fig. 6 Comparison of cumulative oil recovery obtained by injecting the 50:50 of biosurfactant to chemical surfactant mixture in the secondary and tertiartry recovery modes. Core # 10 was used in these tests

is 9 % less than that in the tertiary mode. This may be due to the fact that in the secondary mode, the surfactant should improve the macroscopic sweep efficiency besides the microscopic sweep efficiency, whereas in the tertiary mode, the microscopic sweep efficiency is what matters mainly. In other words, the surfactant mixture injected in the secondary mode improved (a) the volumetric sweep of the injection fluid, (b) the displacement efficiency of the injection fluid in the rock volume that is swept, and (c) the capture of the displaced oil at the core sample outlet, whereas in the tertiary mode when the surfactant mixture was injected after hot water injection, it was mainly utilized for improving the displacement efficiency of the injection fluid in the rock volume that is swept. Because of viscous fingering and incomplete areal sweepout (caused by rock pore structure, e.g., dead-end pores filled with oil), the volumetric sweepout of the reservoir volume is always much less than 100 %. Additionally, not all the oil displaced from the swept areas is captured at the core sample outlet. Babadagli et al. (2002; 2005) reported that when the surfactant solution is injected as a secondary recovery fluid, the critical issue is the better penetration of the fluid provided by less emulsion, more water wettability, and less adsorption. On the other hand, when the surfactant is injected as a tertiary recovery fluid, the critical issue is the reduced IFT between oil and water and oil and rock rather than a better penetration causing a better sweep. This is in line with our observations.

Thus, it was concluded that it is not effective and not feasible to inject the surfactant mixture directly at the secondary recovery stage. This is valid when viscous forces dominant the flood in the reservoir rock matrix. If the

fractures dominate the flow, the recovery mechanism will change (Babadagli et al. 2005).

4 Conclusions

(1) Injecting the chemical surfactant and the biosurfactant following hot water injection (the tertiary recovery mode) reduces the heavy oil residual saturation by the maximum of 11 % and 18.5 %, respectively.

(2) Interestingly, the reduction in residual oil saturation after hot water flood increases to 34 % when the chemical surfactant is mixed with the biosurfactant in a ratio of 50:50. This is attributed to the synergetic effect between the two surfactants.

(3) Compared to the tertiary mode, the secondary mode resulted in higher breakthrough recovery but lower ultimate oil recovery. This is maybe due to the fact that in the secondary mode, the surfactant should improve the macroscopic sweep efficiency (volumetric sweep of the injection fluid and capture of the displaced oil at the core sample outlet) rather than the microscopic sweep efficiency (displacement efficiency of the injection fluid in the rock volume that is swept), whereas in the tertiary mode, the microscopic sweep efficiency is what matters mainly.

(4) If viscous forces dominant the flood in the reservoir rock matrix like the cases investigated in this study, it is not rewarding to inject the chemical surfactant/biosurfactant mixture as the secondary recovery stage.

In retrospect, the results presented in this work demonstrate the high potential of injecting a mixture of biosurfactant and chemical surfactant following hot water injection to reduce heavy oil residual saturation. For field scale applications, however, it is imperative to conduct a study to determine the range of conditions at which the proposed technology can be successfully applied.

Acknowledgments Authors acknowledge His Majesty Research Fund (SR/SCI/BIOL/08/01), Sultan Qaboos University, Oman and the Petroleum Development of Oman (CR/SCI/BIOL/07/02) for the research grants. The authors would like also to gratefully thank Mr. Nayyer Afzal, Mr. Adel Al-Rubkhi, and Mr. Ike Sirono for their technical assistance with coreflood experiments.

References

Al-Bahry S, Elshafie A, Al-Wahaibi Y, et al. Microbial consortia in Oman oil fields: a possible use in enhanced oil recovery. J Microbiol Biotechnol. 2013a;23(1):106–17. doi:10.4014/jmb.1204.04021.

Al-Bahry SN, Al-Wahaibi Y, Elshafie A, et al. Biosurfactant production by *Bacillus subtilis* B20 using date molasses and its application in enhanced oil recovery. Int Biodeterior Biodegrad. 2013b;81:141–6. doi:10.1016/j.ibiod.2012.01.006.

Al-Sulaimani H, Al-Wahaibi Y, Al-Bahry S, et al. Experimental investigation of biosurfactants produced by *Bacillus* species and their potential for MEOR in Omani Oil Field. In: Oil and gas West Asia, 11–13 April, Muscat; 2010. doi:10.2118/129228-MS.

Al-Sulaimani H, Al-Wahaibi Y, Al-Bahry S, et al. Optimization and partial characterization of biosurfactant produced by *Bacillus* species and their potential for ex-situ enhanced oil recovery. SPE J. 2011a;16(3):672–82. doi:10.2118/129228-PA.

Al-Sulaimani H, Joshi S, Al-Wahaibi Y, et al. Microbial biotechnology for enhancing oil recovery: current developments and future prospects. Invited Rev Biotechnol Bioinform Bioeng J. 2011b;1(2):147–58.

Al-Sulaimani H, Al-Wahaibi Y, Al-Bahry SN, et al. Residual oil recovery through injection of biosurfactant, chemical surfactant and mixtures of both under reservoir condition: induced wettability and interfacial tension effects. SPE Reserv Eval Eng. 2012;15(2):210–7. doi:10.2118/158022-PA.

Alveskog PL, Holt T, Torsaeter O. The effect of surfactant concentration on the Amott wettability index and residual oil saturation. J Pet Sci Eng. 1998;20:247–52. doi:10.1016/S0920-4105(98)00027-8.

Anderson WG. Wettability literature survey- part 2: wettability measurements. J Pet Technol. 1986;281:1246–62. doi:10.2118/13933-PA.

Aoudia M, Al-Shibli M, Al-Kasimi L, et al. Novel surfactants for ultralow interfacial tension in a wide range of surfactant concentration and temperature. J Surfactants Deterg. 2006;9(3):287–93. doi:10.1007/s11743-006-5009-9.

Austad T, Standnes DC. Spontaneous imbibitions of water into oil-wet carbonates. J Pet Sci Eng. 2003;39:363–7. doi:10.1016/S0920-4105(03)00075-5.

Babadagli T. Dynamics of capillary imbibition when surfactant, polymer and hot water are used as aqeous phase for oil recovery. J Colloids Interface Sci. 2002;246(1):203–13. doi:10.1006/jcis.2001.8015.

Babadagli T, Al-Bemani A, Boukadi F, Al-Maamari R. A laboratory feasibility study of dilute surfactant injection for the Yibal field, Oman. J Pet Sci Eng. 2005;48(1):37–52. doi:10.1016/j.petrol.2005.04.005.

Curbelo FD, Santanna VC, Neto EL, et al. Adsorption of nonionic surfactants in sandstones. Colloids Surf A. 2007;293:1–4. doi:10.1016/j.colsurfa.2006.06.038.

Daoshan L, Shouliang L, Yi L, Demin W. The effect of biosurfactant on the interfacial tension and adsorption loss of surfactant in ASP flooding. Colloids Surf A. 2004;244:53–60. doi:10.1016/j.colsurfa.2004.06.017.

Gautam KK, Tyagi VK. Microbial surfactants: a review. J Oleo Sci. 2005;55(4):155–66. doi:10.5650/jos.55.155.

Ghojavand H, Vahabzadeh F, Roayaei E, Shahrakai A. Production and properties of a biosurfactant obtained from a member of the *Bacillus subtilis* group (PTCC 1696). J Colloid Interface Sci. 2008;324:172–6. doi:10.1016/j.jcis.2008.05.001.

Hazra C, Kundu D, Ghosh P, et al. Screening and identification of *Pseudomonas aeruginosa* AB4 for improved production, characterization and application of a glycolipid biosurfactant using low-cost agro-based raw materials. J Chem Technol Biotechnol. 2011;86:185–98. doi:10.1002/jctb.2480.

Hirasaki G, Zhang DL. Surface chemistry of oil recovery from fractured, oil-wet, carbonate formation. SPE J. 2004;9(2):151–62. doi:10.2118/80988-MS.

Iglauer S, Wu Y, Shuler P, et al. New surfactant classes for enhanced oil recovery and their tertiary oil recovery potential. J Pet Sci Eng. 2010;71:23–9. doi:10.1016/j.petrol.2009.12.009.

Joshi S, Bharucha C, Jha S, et al. Biosurfactant production using molasses and whey under thermophilic conditions. Bioresour Technol. 2008;99(1):195–9. doi:10.1016/j.biortech.2006.12.010.

Joshi SJ, Desai AJ. Bench-scale production of biosurfactants and their potential in ex-situ MEOR application. Soil Sediment Contam. 2013;22(6):701–15. doi:10.1080/15320383.2013.756450.

Joshi SJ, Geetha SJ, Desai AJ. Characterization and application of biosurfactant produced by *Bacillus licheniformis* R2. Appl Biochem Biotechnol. 2015;177:346–61. doi:10.1007/s12010-015-1746-4.

Kowalewski E, Rueslatten I, Steen KH, et al. Microbial improved oil recovery—bacterial induced wettability and interfacial tension effects on oil production. J Pet Sci Eng. 2006;52:275–86. doi:10.1016/j.petrol.2006.03.011.

Lee S, Kim S, Park I, et al. Isolation and structural analysis of bamylocin A, novel lipopeptide from *Bacillus amyloliquefaciens* LP03 having antagonistic and crude oil-emulsifying activity. J Arch Microbiol. 2007;188:307–12. doi:10.1007/s00203-007-0250-9.

Lu J, Britton C, Solairaj S, et al. Novel large-hydrophobe alkoxy carboxylate surfactants for enhanced oil recovery. SPE J. 2014a;19(6):1024–34. doi:10.2118/154261-PA.

Lu J, Goudarzi A, Chen P, et al. Enhanced oil recovery from high-temperature, high-salinity naturally fractured carbonate reservoirs by surfactant flood. J Pet Sci Eng. 2014b;124:122–31. doi:10.1016/j.petrol.2014.10.016.

Lu J, Liyanage PJ, Solairaj S, et al. New surfactant developments for chemical enhanced oil recovery. J Pet Sci Eng. 2014c;120:94–101. doi:10.1016/j.petrol.2014.05.021.

Lu J, Weerasooriya UP, Pope GA. Investigation of gravity-stable surfactant floods. Fuel. 2014d;124:76–84. doi:10.1016/j.fuel.2014.01.082.

Makkar RS, Cameotra SS. Biosurfactant production by a thermophilic *Bacillus subtilis* strain. J Ind Microbiol Biotechnol. 1997;18:37–42. doi:10.1038/sj.jim.2900349.

Nguyen TT, Youssef NH, McInerney MJ, Sabatini DA. Rhamnolipid biosurfactant mixtures for environmental remediation. Water Res. 2008;42:1735–43. doi:10.1016/j.watres.2007.10.038.

Noudeh GD, Housaindokht M, Bazzaz BS. Isolation, characterization and investigation of surface hemolytic activities of a lipopeptide biosurfactant produced by *Bacillus subtilis* ATCC 6633. J Microbiol. 2005;43(3):272–6.

Portwood JT. A commercial microbial enhanced oil recovery technology: evaluation of 322 projects. In: SPE production operations symposium, January, Oklahoma; 1995. doi:10.2118/29518-MS.

Sayyouh MH, Al-Blehed MS. Effect of microorganisms on rock wettability. J Adhes Sci Technol. 1995;9:425–31. doi:10.1163/156856195X00365.

Thomas CP, Bala GA, Duvall ML. Surfactant-based enhanced oil recovery mediated by naturally occurring microorganisms. SPE Reserv Eng. 1993;11:285–91. doi:10.2118/22844-PA.

Torres L, Moctezuma A, Avendaño JR, et al. Comparison of bio-and synthetic surfactants for EOR. J Pet Sci Eng. 2011;76:6–11. doi:10.1016/j.petrol.2010.11.022.

Youssef N, Simpson DR, Duncan KE, et al. In situ biosurfactant production by *Bacillus* strains injected into a limestone petroleum reservoir. Appl Environ Microbiol. 2007a;73:1239–47. doi:10.1128/AEM.02264-06.

Youssef NH, Nguyen T, Sabatini DA, McInerney MJ. Basis for formulating biosurfactant mixtures to achieve ultralow interfacial tension values against hydrocarbons. J Ind Microbiol Biotechnol. 2007b;34(7):497–507. doi:10.1007/s10295-007-0221-9.

Zargari S, Ostvar S, Niazi A, Ayatollahi S. Atomic force microscopy and wettability study of the alteration of mica and sandstone by a biosurfactant-producing bacterium *Bacillus thermodenitrificans*. J Adv Microsc Res. 2010;5:1–6. doi:10.1166/jamr.2010.1036.

Zekri AY, Mamdouh TG, Almehaideb RA. Carbonate rocks wettability changes induced by microbial solution. In: SPE Asia Pacific oil and gas conference and exhibition, 15–17 April, Jakarta; 2003. doi:10.2118/80527-MS.

Zhang P, Austad T. Wettability and oil recovery from carbonates: effects of temperature and potential determining ions. Colloids Surf A. 2006;279:179–87. doi:10.1016/j.colsurfa.2006.01.009.

Zhu Y, Xu G, Gong H, et al. Production of ultra-low interfacial tension between crude oil and mixed brine solution of Triton X-100 and its oligomer Tyloxapol with cetyltrimethylammonium bromide induced by hydrolyzed polyacrylamide. Colloids Surfaces A: Physicochem Eng Aspects. 2009;332:90–7. doi:10.1016/j.colsurfa.2008.09.012.

Permissions

List of Contributors

X.-L. Zhang, L.-Z. Xiao and L. Guo
State Key Laboratory of Petroleum Resources and Prospecting, China University of Petroleum, Beijing 102249, China

Q.-M. Xie
Key Laboratory of Shale Gas Exploration, Ministry of Land and Resources, Chongqing Institute of Geology and Mineral Resources, Chongqing 400042, China

Dun-Long Liu and Dan Tang
College of Software Engineering, Chengdu University of Information and Technology, Chengdu 610225, China

Zi-Yong Zhou
State Key Laboratory of Petroleum Resources and Prospecting, China University of Petroleum, Beijing 102200, China

Qian Wu
Sichuan Institute of Geological Engineering Investigation, Chengdu 610072, China

X.-Q. Pang and W.-Y. Wang
State Key Laboratory of Petroleum Resources and Prospecting, Beijing 102249, China
Basin and Reservoir Research Center, China University of Petroleum, Beijing 102249, China

C.-Z. Jia
Petro China Company Limited, Beijing 100011, China
Research Institute of Petroleum Exploration and Development, Beijing 100083, China

Kang Liu, Guo-Ming Chen, Yuan-Jiang Chang, Ben-Rui Zhu, Xiu-Quan Liu and Bin-Bin Han
Centre for Offshore Engineering and Safety Technology, China University of Petroleum, Qingdao 266580, Shandong, China

Effat Kianpour and Saeid Azizian
Department of Physical Chemistry, Faculty of Chemistry, Bu-Ali Sina University, Hamedan 65167, Iran

Jun-Hong Wu, Jian-Zhong Liu, Yu-Jie Yu, Rui-Kun Wang, Jun-Hu Zhou and Ke-Fa Cen
State Key Laboratory of Clean Energy Utilization, Zhejiang University, Hangzhou 310027, Zhejiang, China

Tian Yang
School of Geosciences, China University of Petroleum, Qingdao 266580, Shandong, China
Laboratory for Marine Mineral Resources, Qingdao National Laboratory for Marine Science and Technology, Qingdao 266071, China
Department of Geoscience, Aarhus University, Høegh-Guldbergs Gade 2, 8000 Aarhus C, Denmark

Ying-Chang Cao, Yan-Zhong Wang and Hui-Na Zhang
School of Geosciences, China University of Petroleum, Qingdao 266580, Shandong, China
Laboratory for Marine Mineral Resources, Qingdao National Laboratory for Marine Science and Technology, Qingdao 266071, China

Ke-Lai Xi
School of Geosciences, China University of Petroleum, Qingdao 266580, Shandong, China
Laboratory for Marine Mineral Resources, Qingdao National Laboratory for Marine Science and Technology, Qingdao 266071, China
Department of Geosciences, University of Oslo, P.O. Box 1047, Blindern, 0316 Oslo, Norway

Henrik Friis
Department of Geoscience, Aarhus University, Høegh-Guldbergs Gade 2, 8000 Aarhus C, Denmark

Beyene Girma Haile
Department of Geosciences, University of Oslo, P.O. Box 1047, Blindern, 0316 Oslo, Norway

Gan Cui, Zi-Li Li, Chao Yang and Meng Wang
College of Pipeline and Civil Engineering, China University of Petroleum (East China), Qingdao 266580, Shandong, China

Saeid Shokri and Mohammad Taghi Sadeghi
Department of Chemical Engineering, Iran University of Science and Technology (IUST), Tehran, Iran

Mahdi Ahmadi Marvast
Process and Equipment Technology Development Division, Research Institute of Petroleum Industry (RIPI), Tehran, Iran

Shankar Narasimhan
Department of Chemical Engineering, IIT Madras, Chennai, India

Bin Gong and Jia-Peng Yu
College of Engineering, Peking University, Beijing 100871, China

Xuan Liu
College of Engineering, Peking University, Beijing 100871, China
Sinopec Petroleum Exploration and Production Research Institute, Beijing 100083, China

Yong-Feng Zhu and Shi-Ti Cui
Petrochina Tarim Oilfield Exploration and Production Research Institute, Korla 841000, Xinjiang, China

Dong-Feng Mao, Ming-Hui Zhang, Yang Yu, Meng-Lan Duan and Jun Zhao
College of Mechanical and Transportation Engineering, China University of Petroleum, Beijing 102249, China

Guang-Zhi Zhang, Xin-Peng Pan, Zhen-Zhen Li, Chang-Lu Sun and Xing-Yao Yin
School of Geosciences, China University of Petroleum (East China), Qingdao 266580, Shandong, China

Xin-Hua Qiu and Qing Xue
School of Business Administration, China University of Petroleum, Beijing 102249, China

Zhen Wang
Academy of Chinese Energy Strategy, China University of Petroleum, Beijing 102249, China

Talal Al-Wahaibi, Yahya Al-Wahaibi, Abdul-Aziz R. Al-Hashmi, Farouq S. Mjalli and Safiya Al-Hatmi
Petroleum and Chemical Engineering Department, College of Engineering, Sultan Qaboos University, P.O. Box 33, Al-Khod 123, Oman

Hai-Shan Luo, Mojdeh Delshad, Zhi-Tao Li and Amir Shahmoradi
Center for Petroleum and Geosystems Engineering, The University of Texas at Austin, 200 E Dean Keeton St C0300, Austin, TX 78712, USA

Meng-Meng Wu, Rui Li, Yu-Zhen Zhang, Liang Fan, Yu-Chao Lv and Jian-Ming Wei
State Key Laboratory of Heavy Oil Processing, China University of Petroleum (East China), Qingdao 266555, Shandong, China

Yahya Al-Wahaibi, Hamoud Al-Hadrami and Ali Al-Bemani
Petroleum and Chemical Engineering Department, Sultan Qaboos University, 112 Muscat, Oman

Sanket Joshi, Saif Al-Bahry and Abdulkadir Elshafie
Department of Biology, College of Science, Sultan Qaboos University, 112 Muscat, Oman

Index

Printed in the USA
CPSIA information can be obtained
at www.ICGtesting.com
JSHW051433221024
72173JS00006B/1455